Comparative Aspects
of Lactation

SYMPOSIA OF THE ZOOLOGICAL SOCIETY OF LONDON
NUMBER 41

Comparative Aspects of Lactation

*(The Proceedings of a Symposium held at
The Zoological Society of London
on 11 and 12 November 1976)*

Edited by

MALCOLM PEAKER

*Department of Physiology, ARC Institute of
Animal Physiology, Babraham, Cambridge, England*

Published for

THE ZOOLOGICAL SOCIETY OF LONDON

BY

ACADEMIC PRESS

1977

ACADEMIC PRESS INC. (LONDON) LTD
24/28 Oval Road, London NW1 7DX

United States Edition published by
ACADEMIC PRESS INC.
111 Fifth Avenue, New York, New York 1003

Library of Congress Catalog Card Number: 74-5683
ISBN: 0-12-613341-7

Printed in Great Britain by
J. W. ARROWSMITH LTD, BRISTOL BS3 2NT

Contributors

BALCH, C. C., *Feeding and Metabolism Department, National Institute for Research in Dairying, Shinfield, Reading, England* (p. 285)

BRAMBELL, M. R., *The Zoological Society of London, Regent's Park, London NW1 4RY, England* (p. 333)

CHADWICK, A., *Department of Pure and Applied Zoology, University of Leeds, Leeds, England* (p. 341)

CLARK, SUSAN, *Department of Biochemistry, University of Leeds, Leeds, England* (p. 43)

CROSS, B. A., *ARC Institute of Animal Physiology, Babraham, Cambridge, England* (p. 193)

DILS, R., *Department of Physiology and Biochemistry, University of Reading, Whiteknights, Reading, England* (p. 43)

DRIFE, J. O., *MRC Unit of Reproductive Biology, 2 Forrest Road, Edinburgh EH1 2QW, Scotland* (p. 211)

FORSYTH, ISABEL A., *Department of Physiology, National Institute for Research in Dairying, Shinfield, Reading, England* (p. 135)

GUNTHER, MAVIS, *77 Ember Lane, Esher, Surrey, England* (p. 277)

HALL, BARBARA M., *Nuffield Institute of Comparative Medicine, The Zoological Society of London, Regent's Park, London NW1 4RY, England* (p. 231)

HANWELL, ANN, *The Orchard, Mallin's Lane, Longcot, Faringdon, Oxon, England* (p. 297)

HAYDEN, T. J., *National Institute for Research in Dairying, Shinfield, Reading, England* (p. 135)

JONES, D. M., *The Zoological Society of London, Regent's Park, London NW1 4RY, England* (p. 333)

JONES, E. A., *Department of Physiology, National Institute for Research in Dairying, Shinfield, Reading, England* (p. 77)

KNUDSEN, J., *Institute of Biochemistry, University of Odense, DK-5230, Odense M, Denmark* (p. 43)

KUHN, N. J., *Department of Biochemistry, University of Birmingham, Birmingham, England* (p. 165)

LASCELLES, A. K., *CSIRO Division of Animal Health, Private Bag No. 1, P.O. Parkville, Victoria 3052, Australia* (p. 241)

MEPHAM, T. B., *Department of Physiology and Environmental Studies, University of Nottingham School of Agriculture, Sutton Bonington, Loughborough, Leics, England* (p. 57)

OXBERRY, JANET M., *Nuffield Institute of Comparative Medicine, The Zoological Society of London, Regent's Park, London NW1 4RY, England* (p. 231)

PEAKER, M., *Department of Physiology, ARC Institute of Animal Physiology, Babraham, Cambridge, England* (pp. 113, 297)

PORTER, J. W. G., *Nutrition Department, National Institute for Research in Dairying, Shinfield, Reading, England* (p. 285)

PUMPHREY, R. S. H., *Department of Bacteriology, Royal Infirmary, Glasgow, Scotland* (p. 261)

SHILLITO WALSER, ELIZABETH E., *Department of Applied Biology, ARC Institute of Animal Physiology, Babraham, Cambridge, England* (p. 313)

SHORT, R. V., *MRC Unit of Reproductive Biology, 2 Forrest Road, Edinburgh EH1 2QW, Scotland* (p. 211)

SMITH, G. H., *Department of Animal Physiology and Nutrition, University of Leeds, Kirkstall Laboratories, Vicarage Terrace, Leeds, England* (p. 95)

TAYLOR, D. J., *Department of Animal Physiology and Nutrition, University of Leeds, Leeds, England* (p. 95)

WOODING, F. B. P., *Cell Biology Group, ARC Institute of Animal Physiology, Babraham, Cambridge, England* (p. 1)

Organizers and Chairmen

ORGANIZERS

M. PEAKER and J. L. LINZELL, on behalf of The Zoological Society
of London

CHAIRMEN OF SESSIONS
A. T. COWIE, *Department of Physiology, National Institute for
Research in Dairying, Shinfield, Reading, England*
D. B. LINDSAY, *Department of Biochemistry, ARC Institute of Animal
Physiology, Babraham, Cambridge, England*
R. D. MARTIN, *Wellcome Institute of Comparative Physiology, The
Zoological Society of London, Regent's Park, London NW1 4RY,
England*
ELSIE M. WIDDOWSON, *Department of Medicine, Level 5, Adden-
brooke's Hospital, Hills Road, Cambridge CB2 2QQ, England*

with concluding remarks by
E. C. AMOROSO, ARC Institute of Animal Physiology, Babraham,
Cambridge, England

James Lincoln Linzell, BSc, PhD, MRCVS, FIBiol., FZS; to whom this Symposium is dedicated.

James Lincoln Linzell—an Appreciation

James Lincoln Linzell, Head of the Department of Physiology at the Agricultural Research Council Institute of Animal Physiology, Babraham, Cambridge, died on 28th December 1975 at the age of 54.

Linzell's work on the fundamental physiology of the mammary gland and milk secretion earned him an international reputation. His great strength was that his approach to a problem was direct and practical, and the answers came from a rare mixture of immense enthusiasm, brain power and hard work. Such an approach meant that he developed an array of techniques to make possible the physiological study of the mammary gland. Although in his early career he did classical physiological experiments on anaesthetized animals to investigate, for example, the innervation of the gland, he soon realized that to study milk secretion effectively most work must be done on conscious, undisturbed animals, and surgical preparations and apparatus were devised accordingly. For example, he developed the first accurate method of measuring mammary blood flow and he gained access to the mammary artery of goats (a vital technique for the investigation of secretory mechanisms) by transplanting the gland to the neck. In addition he developed the first successful perfusion of the isolated mammary gland. These and other techniques enabled him to establish the relationship between nutrient uptake, mammary blood flow and milk secretion, and formed the basis of more recent studies on the mechanism of milk secretion.

It might be imagined that having exploited so effectively the methods he developed his outlook would have become narrow in more recent years. But this was certainly not the case and he would turn his hand to any technique that would enable him to get a better understanding. Moreover, although his main interest was lactation he had a general biological curiosity and an encyclopaedic knowledge of many different fields; he collaborated in studies on reproductive organs, the ruminant liver and avian salt glands. His papers, of which he produced nearly 200 with about 70 collaborators, are clearly and engagingly written in a characteristic style, and demonstrate his like of discovering how organs work and the quantitative relationships between different functions.

Jim Linzell was born in London on 5th February 1921. Although he showed an interest in biology while at school he said that he was persuaded into a veterinary education by a school friend who was about to begin such a course. He is remembered at the Royal Veterinary College as being a keen and conscientious student (he won a prize in anatomy) and he qualified as MRCVS with a BSc in physiology in 1943. A period as house surgeon (large animals) at the College followed and during this time he married his colleague in small animal surgery, Audrey Irene Tyler.

Linzell seemed set for a standard veterinary life by entering practice as an assistant in 1944. However, he found the work frustrating because, as he noted later, he lacked the time to be thorough and he hated doing routine jobs that technicians were perfectly capable of doing. He therefore changed direction by going to the Physiology Department of the University of Edinburgh in 1946 to work under Professor I. de Burgh Daly as an Agricultural Research Council research trainee. He said that he would have liked to study the uterus and placenta but Daly told him to find out everything about the anatomy and physiology of the mammary gland in as many species as he could; this he did for the next 30 years. It was in Edinburgh that he, in parallel with K. C. Richardson working in London, established the presence of myoepithelial cells in the mammary gland and their contractile nature. During this time Daly was appointed Director of the new Institute of Animal Physiology and Linzell was one of the first to be taken onto the staff. He moved to Babraham when laboratory space became ready in 1950 and stayed there for the rest of his life despite a number of offers of university chairs. His PhD was conferred in 1951 and he was elected to the Physiological Society in 1952. He was promoted Principal Scientific Officer in 1954, Senior Principal (Special Merit) in 1963 and Deputy Chief Scientific Officer (Special Merit) in 1974; he became Head of Physiology in 1971. He was a Fellow of the Institute of Biology and a Scientific Fellow of the Zoological Society of London.

His first trip abroad was to the International Physiological Congress in Montreal in 1953. After that he travelled in Europe, North America and Australia to attend meetings, lecture and visit laboratories. Perhaps his most memorable visits were to the USSR in 1961 where he visited laboratories and for which he acquired a working knowledge of the Russian language, his sojourn in Philadelphia in 1965 working on mammary blood flow, and to Switzerland in 1974 to receive the International Prize in Modern Nutrition at a ceremony held in a mediaeval castle. On this occasion cannon were fired in his honour over the valley and the cows in the fields below scattered. In typical Linzell fashion he then lectured the guests on the effects of stress on milk secretion.

Linzell did not seek administrative duties or high office and often failed to understand the reason why others did. His view of scientific organization was that it should be done by working scientists and not by professional administrators who do not understand the way in which research is done. As much as he disliked administration he excelled in the generation of an atmosphere of enthusiasm and friendly co-operation in his department. Similarly his presentation of lectures was dynamic while reflecting his intellectual sincerity.

His reputation sometimes made him a daunting figure for he always spoke his mind and he would have no truck either with those who in his opinion did second-rate work, with those who took short-cuts when he thought they should know better, or with those who failed to give credit to the work of others. In fact he was one of the most approachable of men and he would take endless trouble to discuss work with a beginner or to demonstrate a technique to a scientist from another laboratory, interspersing facts with a string of anecdotes which entertained colleagues and visitors alike. Completely lacking pretentiousness, and abhorring it in others, he was equally at ease with farmers, veterinary surgeons and pure scientists, for having worked on farms in College vacations (and escaped death from the horns of a lactating Ayrshire cow by inches) he was familiar with the terminology used by different groups of people to describe the same phenomenon.

Apart from his work he showed concern for both animal welfare and fauna conservation. He served on the scientific advisory committee of the Universities Federation for Animal Welfare and was a member of the Fauna Preservation Society. In this connection he never failed to assail visiting Japanese and Russian scientists on their governments' attitude to whaling; the blue whale has had no fiercer advocate. To some it may appear contradictory that such a brilliant experimental physiologist should have a high regard for animal life but he believed that the acquisition of knowledge from such experiments ultimately benefits other animals including man.

Jim Linzell was not only a major animal physiologist but also a kind, humane man of absolute integrity.

At his funeral in Cambridge, attended by his widow and three married children, personal tributes were made by Geoffrey Waites, Malcolm Peaker and Barry Cross. A memorial meeting was held on 1st March 1976 in the Physiological Laboratory, University of Cambridge. The chair was taken by Professor R. D. Keynes, FRS and the contributors were Dr B. A. Cross, FRS, who spoke on Linzell's early research on mammary physiology, and Dr M. Peaker who described their collaborative work on the mechanism of milk secretion from 1968 to 1975; a BBC television/Open University film of Jim Linzell with his goats in the laboratory was shown by the producer, Dr Roger Jones.

Publications of J. L. Linzell

1943

1. Diffuse lymphoid leukosis of the skin in a dairy cow. *Vet. Rec.* **55**: 19.

1944

2. Recovery from gas gangrene in the dog. *Vet. Rec.* **56**: 343–344.

1948

3. Diesel oil poisoning in cattle. *Vet. Rec.* **60**: 60 (with E. A. Gibson).
4. Long-continued lactation without pregnancy. *Vet. Rec.* **60**: 511.

1950

5. Vasomotor nerve fibres to the mammary glands of the cat and dog. *Q. Jl exp. Physiol.* **35**: 295–319.

1951

6. Some conditions affecting the blood flow through the perfused mammary gland, with special reference to the action of adrenaline. *Q. Jl exp. Physiol.* **36**: 159–175 (with C. O. Hebb).
7. An extra-uterine foetus in the cat. *Vet. Rec.* **63**: 223–224.

1952

8. The silver staining of myoepithelial cells, particularly in the mammary gland, and their relation to the ejection of milk. *J. Anat.* **86**: 49–57.
9. Pulmonary vasomotor nerve activity. *Q. Jl exp. Physiol.* **37**: 149–161 (with I. de Burgh Daly, H. N. Duke & J. Weatherall).

1953

10. Internal calorimetry in the measurement of blood flow with heated thermocouples. *J. Physiol., Lond.* **121**: 390–402.
11. The blood and nerve supply to the mammary glands of the cat, and other laboratory animals. *Br. Vet. J.* **109**: 427–433.
12. The contractility of the alveoli of the mammary gland. *J. Physiol., Lond.* **123**: 32P.

1954

13. Observations on the perfused living animal (dog) using homologous and heterologous blood. *Q. Jl exp. Physiol.* **39**: 29–54 (with I. de Burgh Daly, P. Eggleton, C. Hebb & O. A. Trowell).

14. Pulmonary vasomotor responses and acid-base balance in perfused eviscerated dog preparations. *J. Physiol., Lond.* **125**: 40P (with I. de Burgh Daly, L. E. Mount & G. M. H. Waites).
15. Pulmonary vasomotor responses and acid-base balance in perfused eviscerated dog preparations. *Q. Jl exp. Physiol.* **39**: 177–183 (with I. de Burgh Daly, L. E. Mount & G. M. H. Waites).
16. Some observations on the use of the perfused lactating mammary gland. *Revue Can. Biol.* **13**: 291–298.

1955

17. Some observations on the contractile tissue of the mammary glands. *J. Physiol., Lond.* **130**: 257–267.
18. Variations in the direction of venous blood-flow in the mammary region of the sheep and goat. *Nature, Lond.* **176**: 37 (with L. E. Mount).

1956

19. Evidence against a parasympathetic innervation of the mammary glands. *J. Physiol., Lond.* **133**: 66–67P.

1957

20. The measurement of udder blood flow in the conscious goat. *J. Physiol., Lond.* **137**: 75–76P.
21. The effects of occluding the carotid and vertebral arteries in sheep and goats. *J. Physiol., Lond.* **138**: 20P (with G. M. H. Waites).

1958

22. The blood flow and milk yield of goats' udders perfused with artificial media. *J. Physiol., Lond.* **143**: 74–75P (with H. E. C. Cargill-Thompson, A. N. Drury, D. C. Hardwick & E. M. Tucker).

1959

23. The innervation of the mammary glands in the sheep and goat with some observations on the lumbo-sacral autonomic nerves. *Q. Jl exp. Physiol.* **44**: 160–176.
24. Physiology of the mammary glands. *Physiol. Rev.* **39**: 534–570.

1960

25. Valvular incompetence in the venous drainage of the udder. *J. Physiol., Lond.* **153**: 481–491.
26. Mammary-gland blood flow and oxygen, glucose and volatile fatty acid uptake in the conscious goat. *J. Physiol., Lond.* **153**: 492–509.
27. The flow and composition of mammary gland lymph. *J. Physiol., Lond.* **153**: 510–521.

28. Some factors affecting milk secretion by the isolated perfused mammary gland. *J. Physiol., Lond.* **154**: 547–571 (with D. C. Hardwick).
29. Transplantation of mammary glands. *Nature, Lond.* **188**: 596–598.
30. Tuberculosis in goats caused by the avian type tubercle bacillus. *Vet. Rec.* **72**: 25–27 (with I. W. Lesslie & E. J. H. Ford).
31. How artificial organs aid research. *New Scient.* **7**: 714–716.
32. The rate of incorporation of ^{32}P into milk and casein by the perfused udder. *Biochem. J.* **75**: 5P (with D. C. Hardwick & S. M. Price).
33. The need of the isolated perfused goat udder for acetate and glucose to produce milk. *Biochem. J.* **78**: 3P (with D. C. Hardwick & S. M. Price).

1961

34. Polypnoea evoked by heating the udder of the goat. *Nature, Lond.* **190**: 173 (with J. Bligh).
35. The effect of glucose and acetate on milk secretion by the perfused goat udder. *Biochem. J.* **80**: 37–45 (with D. C. Hardwick & S. M. Price).
36. Recent advances in the physiology of the udder. *Vet. A.* **3**: 44–53.

1962

37. Review of *The motor apparatus of the mammary gland* by M. G. Zaks. *Q. Jl exp. Physiol.* **47**: 372–374.
38. The contribution of glucose and acetate to milk constituents and carbon dioxide in the isolated perfused udder. *Biochem. J.* **84**: 102P (with D. C. Hardwick).

1963

39. Some effects of denervating and transplanting mammary glands. *Q. Jl exp. Physiol.* **48**: 34–59.
40. Carotid loops. *Am. J. vet. Res.* **24**: 223–224.
41. The incorporation of chylomicra into milk fat by the goat udder. *Biochem. J.* **87**: 4P (with D. C. Hardwick, A. K. Lascelles & T. B. Mepham).
42. A comparison of the transit times between radioactive plasma precursors and labelled milk constituents. *Biochem. J.* **87**: 4–5P (with D. C. Hardwick & T. B. Mepham).
43. The uptake from the blood of triglyceride fatty acids of chylomicra and low-density lipoproteins by the mammary gland of the goat. *Biochem. J.* **87**: 23–24P (with D. S. Robinson, J. M. Barry & W. Bartley).

44. The metabolism of acetate and glucose by the isolated perfused udder. 2. The contribution of acetate and glucose to carbon dioxide and milk constituents. *Biochem. J.* **88**: 213–219 (with D. C. Hardwick & T. B. Mepham).
45. The uptake from the blood of triglyceride fatty acids of chylomicra and low-density lipoproteins by the mammary gland of the goat. *Biochem. J.* **89**: 6–10 (with J. M. Barry, W. Bartley & D. S. Robinson).
46. Granulosa cell tumour of the ovary in a virgin heifer. *J. Endocr.* **27**: 327–382 (with D. R. Shorter & R. V. Short).
47. The oxidation of glucose and acetate by the mammary glands of the goat. *J. Physiol., Lond.* **170**: 65–66P (with E. F. Annison).

1964

48. The accuracy of the indicator absorption method of measuring mammary blood flow by the Fick Principle. *Q. Jl exp. Physiol.* **49**: 219–225 (with F. Rasmussen).
49. The transfer of [^{3}H]stearic acid from chylomicra to milk fat in the goat. *Biochem. J.* **92**: 36–42 (with A. K. Lascelles, D. C. Hardwick & T. B. Mepham).
50. Some observations on general and regional anaesthesia in goats. In *Small animal anaesthesia*:163–175 Graham-Jones, O. (ed.). Oxford: Pergamon.
51. Dehorning goats. *Vet. Rec.* **76**: 853–854.
52. A rapid method of estimating fat in very small quantities of milk. *J. Physiol., Lond.* **175**: 15–17P (with I. R. Fleet).
53. The oxidation and utilization of glucose and acetate by the mammary gland of the goat in relation to their overall metabolism and to milk formation. *J. Physiol., Lond.* **175**: 372–385 (with E. F. Annison).

1965

54. The metabolism of plasma free fatty acids by the lactating mammary gland of the goat. *Biochem. J.* **94**: 21P (with E. F. Annison, S. Fazakerley & B. W. Nichols).
55. Amino acid uptake by the lactating goat mammary gland. *Biochem. J.* **95**: 47P (with T. B. Mepham).
56. The use of goats in research at Babraham, Cambridge. *Surrey Goat Cl. Gaz.* **22**: 5–11.
57. The effects of fasting and glucose infusion on hourly milk yield in the goat. *J. Physiol., Lond.* **179**: 91–92P.
58. Plasma progesterone levels in the goat and mammary uptake during pregnancy. *J. Physiol., Lond.* **180**: 10–11P (with R. B. Heap).
59. Milk fever in goats. *Vet. Rec.* **77**: 767–768.
60. Review of *Comparative endocrinology*, ed. U. S. von Euler & H. Heller. *Br. Vet. J.* **121**: 147.

1966

61. Review of *Comparative biology of reproduction in mammals*, ed. I. W. Rowlands. *Br. Vet. J.* **122**: 459.
62. A quantitative assessment of the contribution of individual plasma amino acids to the synthesis of milk proteins by the goat mammary gland. *Biochem. J.* **101**: 76–82 (with T. B. Mepham).
63. Measurement of udder volume in live goats as an index of mammary growth and function. *J. Dairy Sci.* **49**: 307–311.
64. Measurement of venous flow by continuous thermodilution and its application to measurement of mammary blood flow in the goat. *Circulation Res.* **43**: 745–754.
65. Progesterone production by the ovary and adrenal, and uptake by the mammary gland and uterus in the goat. *J. Endocr.* **35**: xxiv (with R. B. Heap).
66. Infusion and blood sampling techniques for use in minimally-restrained goats. *J. Physiol.* **186**: 79–81P.
67. A histochemical study of the innervation of the mammary glands. *J. Physiol., Lond.* **186**: 82–83P (with C. O. Hebb).
68. Arterial concentration, ovarian secretion and mammary uptake of progesterone in goats during the reproductive cycle. *J. Endocr.* **36**: 389–399 (with R. B. Heap).
69. Composition of zebra milk. *Int. Zoo Yb.* **6**: 262 (with J. M. King).

1967

70. Plasma free fatty acid uptake and release by the goat mammary gland. *Biochem. J.* **102**: 23P (with C. E. West & E. F. Annison).
71. Urea formation by the lactating goat mammary gland. *Nature, Lond.* **214**: 507–508 (with T. B. Mepham).
72. The acetylation of sulphanilamide by mammary tissue of lactating goats. *Biochem. Pharmacol.* **16**: 918–919 (with F. Rasmussen).
73. Mammary metabolism in the lactating sow. *Biochem. J.* **103**: 42P (with T. B. Mepham, C. E. West & E. F. Annison).
74. The effect of very frequent milking and of oxytocin on the yield and composition of milk in fed and fasted goats. *J. Physiol., Lond.* **190**: 333–346.
75. The effect of infusions of glucose, acetate and amino acids on hourly milk yield in fed, fasted and insulin-treated goats. *J. Physiol., Lond.* **190**: 347–357.
76. The oxidation and utilization of palmitate, stearate, oleate and acetate by the mammary gland of the fed goat in relation to their overall metabolism, and the role of plasma phospholipids and neutral lipids in milk-fat synthesis. *Biochem. J.* **102**: 637–647 (with E. F. Annison, S. Fazakerley & B. W. Nichols).

77. The incorporation of acetate, stearate and D(-)-B-hydroxy-bu-tyrate into milk fat by the isolated perfused mammary gland of the goat. *Biochem. J.* **104**: 34–42 (with E. F. Annison, S. Fazaker-ley & R. A. Leng).
78. Ovarian activity in the ewe after autotransplantation of the ovary or uterus to the neck. *J. Physiol., Lond.* **191**: 129–130P (with J. R. Goding, F. A. Harrison & R. B. Heap).
79. Mode of uptake of triglyceride by the goat mammary gland. *Bio-chem. J.* **104**: 59–60P (with C. E. West & E. F. Annison).

1968

80. Ovarian function in the sheep after autotransplantation of the ovary and uterus to the neck. *J. Endocr.* **40**: xiii (with F. A. Harrison & R. B. Heap).
81. The output of spermatozoa and fluid by, and the metabolism of, the isolated perfused testis of the ram. *J. Physiol., Lond.* **195**: ,25–26P (with B. P. Setchell).
82. Mammary and whole animal metabolism of glucose and fatty acids in fasting lactating goats. *J. Physiol., Lond.* **197**: 445–459 (with C. E. West & E. F. Annison).
83. Plasma prolactin levels in the goat: physiological and experimental modification. *J. Endocr.* **40**: iv–v (with G. D. Bryant & F. C. Greenwood).
84. Mammary synthesis of amino acids in the lactating goat. *Biochem. J.* **107**: 18–19P (with T. B. Mepham).
85. The effect of some drugs, hormones and physiological factors on the flow of rete testis fluid in the ram. *J. Reprod. Fert.* **16**: 320–321 (with B. P. Setchell).
86. A comparison of progesterone metabolism in the pregnant sheep and goat: sources of production and an estimation of uptake by some target organs. *J. Endocr.* **41**: 433–438 (with R. B. Heap).
87. The magnitude and mechanisms of the uptake of milk precursors by the mammary gland. *Proc. Nutr. Soc.* **27**: 44–52.
88. Comparison of four methods for measuring mammary blood flow in conscious goats. *Am. J. Physiol.* **214**: 1415–1424 (with M. Reynolds & F. Ramussen).
89. Quantitative measurement of progesterone metabolism in the mammary gland of the goat. *J. Physiol., Lond.* **200**: 38–40P (with R. B. Heap & C. A. Slotin).
90. Quantitative aspects of mammary metabolism in goats. *Int. Congr. Physiol.* **24**: Washington, D.C.

1969

91. Metabolism, sperm and fluid production of the isolated perfused testis of the sheep and goat. *J. Physiol., Lond.* **201**: 129–143 (with B. P. Setchell).

92. Reproductive behaviour in the sheep and pig after transplantation of the ovary. *J. Reprod. Fert.* **20**: 356–357 (with R. M. Binns, F. A. Harrison & R. B. Heap).

93. Discussion opening in *Proceedings of Symposium on machine milking*: 130–132. Hall, H. S. (ed.). Shinfield: National Institute for Research in Dairying.

94. Mammary metabolism in lactating sows: arteriovenous differences of milk precursors and the mammary metabolism of [^{14}C]glucose and [^{14}C]acetate. *Br. J. Nutr.* **23**: 319–332 (with T. B. Mepham, E. F. Annison & C. E. West).

95. Cardiovascular changes during lactation in the rat. *J. Endocr.* **44**: 247–254 (with A. L. Chatwin & B. P. Setchell).

96. Accuracy of the micromethod of estimating milk fat concentration by high-speed centrifugation in capillary tubes. *J. Dairy Sci.* **52**: 1685–1686 (with I. R. Fleet).

97. Milk secretion. *Br. Goat Soc. Yb.* **1969**: 18–20.

1970

98. Synthesis of progesterone by the mammary gland of the goat. *Nature, Lond.* **225**: 385–386 (with C. A. Slotin, R. B. Heap & J. M. Christiansen).

99. Ionic composition of guinea-pig mammary cells. *J. Physiol., Lond.* **207**: 37–38P (with M. Peaker).

100. Avian salt-gland blood flow and the extraction of ions from the plasma. *J. Physiol., Lond.* **207**: 83–84P (with A. Hanwell & M. Peaker).

101. The uptake of glycerol ethers by the lactating goat mammary gland. *Biochem. J.* **117**: 30P (with R. Bickerstaffe, L. J. Morris & E. F. Annison).

102. Cardiac output and organ blood flow in late pregnancy and early lactation in rats. In *Lactogenesis*; 153–156, Reynolds, M. & Folley, S. J. (eds.). Philadelphia: University of Philadelphia Press.

103. Theories of milk secretion: evidence from the electron microscopic examination of milk. *Nature, Lond.* **225**: 762–764 (with F. B. P. Wooding & M. Peaker).

104. Measurement of the quantity of lactose passing into mammary venous plasma and lymphs in goats and a cow. *J. Dairy Res.* **37**: 203–208 (with N. J. Kuhn).

105. Plasma prolactin in goats measured by radioimmunoassay: the effects of teat stimulation, mating behaviour, stress, fasting and of oxytocin, insulin and glucose injections. *Hormones* **1**: 26–35 (with G. D. Bryant & F. C. Greenwood).

106. Salt-gland function in the domestic goose. *J. Physiol., Lond.* **210**: 97–99P (with A. Hanwell & M. Peaker).

107. Innervation of the mammary gland. A histochemical study in the rabbit. *Histochem. J.* **2**: 491–505 (with C. Hebb).
108. Milk secretion (a letter). *Nature, Lond.* **228**: 1007.

1971

109. The isolated perfused liver of the sheep: an assessment of its metabolic, synthetic and excretory functions. *Q. Jl exp. Physiol.* **56**: 53–71 (with B. P. Setchell & D. B. Lindsay).
110. Permeability of mammary ducts in the lactating goat. *J. Physiol., Lond.* **213**: 48–49P (with M. Peaker).
111. Membrane potentials in the mammary gland of the lactating rat. *J. Physiol., Lond.* **213**: 49–50P (with M. H. Evans & M. Peaker).
112. Salt-gland secretion and blood flow in the goose. *J. Physiol., Lond.* **213**: 373–387 (with A. Hanwell & M. Peaker).
113. Cardiovascular responses to salt-loading in conscious domestic geese. *J. Physiol., Lond.* **213**: 389–398 (with A. Hanwell & M. Peaker).
114. Techniques for measuring nutrient uptake by the mammary gland. In *Lactation*: 261–279. Falconer, I. R. (ed.). London: Butterworths.
115. Mammary blood vessels, lymphatics and nerves. In *Lactation*: 41–50. Falconer, I. R. (ed.). London: Butterworths.
116. Location of receptors for salt gland secretion in the goose. *Proc. Int. Union Physiol. Sci.* **9** (*25th International Congress, Munich*): 444 (with A. Hanwell & M. Peaker).
117. The secretion of cortisol and its mammary uptake in the goat. *J. Endocr.* **50·** 493–499 (with J. Y. F. Paterson).
118. The location and nature of the receptors for secretion by the salt glands of the goose. *J. Physiol., Lond.* **216**: 28–29P (with A. Hanwell & M. Peaker).
119. Mechanism of milk secretion. *Physiol. Rev.* **51**: 564–597 (with M. Peaker).
120. Intracellular concentrations of sodium, potassium and chloride in the lactating mammary gland and their relation to the secretory mechanism. *J. Physiol., Lond.* **216**: 683–700 (with M. Peaker).
121. The permeability of mammary ducts. *J. Physiol., Lond.* **216**: 701–716 (with M. Peaker).
122. The effects of oxytocin and milk removal on milk secretion in the goat. *J. Physiol., Lond.* **216**: 717–734 (with M. Peaker).
123. Goats. In *The UFAW Handbook on the care and management of farm animals*: 90–98. UFAW (eds). Edinburgh: Churchill Livingstone.
124. Full-time research. *Incisor* **1971**: 48–50.
125. Milk fat synthesis in cows fed high-roughage and low-roughage diets. *Proc. Nutr. Soc.* **30**: 37A (with R. Bickerstaffe, D. E. Noakes & E. F. Annison).

126. The concentration of total unconjugated oestrogens in the plasma of pregnant goats. *J. Reprod. Fert.* **26**: 401–404 (with J. R. G. Challis).
127. A scientific view on ritual slaughter. *RSPCA Today* **4**: 71.
128. The role of the mammary glands in reproduction. *Res. Reprod.* **3** (6): 2–3.
129. Early detection of mastitis. *Vet. Rec.* **89**: 393–394 (with M. Peaker).

1972

130. Studies on the mode of uptake of blood triglycerides by the mammary gland of the lactating goat. The uptake and incorporation into milk fat and mammary lymph on labelled glycerol, fatty acids and triglycerides. *Biochem. J.* **126**: 477–490 (with C. E. West, R. Bickerstaffe & E. F. Annison).
131. Oxygen consumption by the liver of the conscious sheep. *J. Physiol., Lond.* **222**: 48–49P (with F. A. Harrison & J. Y. F. Paterson).
132. Validation of the thermodilution technique for the estimation of cardiac output in the rat. *Comp. Biochem. Physiol.* **41A**: 647–657 (with A. Hanwell).
133. Estimation of cardiac output by thermodilution in the conscious lactating rat. *Comp. Biochem. Physiol.* **41A**: 659–665 (with A. Hanwell & I. R. Fleet).
134. A simple technique for measuring the rate of milk secretion in the rat. *Comp. Biochem. Physiol.* **43A**: 259–270 (with A. Hanwell).
135. Perfusion of the isolated mammary gland of the goat. *Q. Jl exp. Physiol.* **57**: 139–161 (with I. R. Fleet, T. B. Mepham & M. Peaker).
136. Mammary blood flow and changes in circulation during lactation in goats and cows. In *Handbuch der Tierernährung* **2**: 203–207. Lenkeit, W. & Breiren, K. (eds.). Hamburg: Paul Parey (with F. Rasmussen).
137. Milk yield, energy loss in milk, and mammary gland weight in different species. *Dairy Sci. Abstr.* **34**: 351–360.
138. A modified apparatus for perfusion of the liver. *J. Physiol., Lond.* **226**: 19–20P (with I. R. Fleet, I. G. Jarrett, D. B. Lindsay & B. P. Setchell).
139. Determination of cardiac output and mammary blood flow in the conscious lactating rat. *J. Physiol., Lond.* **226**: 24–25P (with A. Hanwell).
140. Sweat gland function in isolated perfused sheep and goat skin. *J. Physiol., Lond.* **226**: 25–27P (with K. G. Johnson).
141. Nature and location of the receptors for salt-gland secretion in the goose. *J. Physiol., Lond.* **226**: 453–472 (with A. Hanwell & M. Peaker).
142. Day-to-day variations in milk composition as a guide to the detec-

tion of subclinical mastitis. *Br. Vet. J.* **128**: 284–295 (with M. Peaker).

143. The use of an autoanalyser for the rapid analysis of milk constituents affected by subclinical mastitis. *Br. Vet. J.* **128**: 297–300 (with I. R. Fleet & M. Peaker).

144. Elevation of the cardiac output in the rat by prolactin and growth hormone. *J. Endocr.* **53**: lvii–lviii (with A. Hanwell).

145. Mammary and whole-body metabolism of glucose, acetate and palmitate in the lactating horse. *Proc. Nutr. Soc.* **31**: 72A (with E. F. Annison, R. Bickerstaffe & L. B. Jeffcott).

146. Review of *The physiology of lactation* by A. T. Cowie & J. S. Tindal. *Q. Jl exp. Physiol.* **57**: 240–241.

147. Mechanism of secretion of the aqueous phase of milk. *J. Dairy Sci.* **55**: 1316–1322.

1973

148. Sensitivity of the receptors for salt-gland secretion in the domestic duck and goose. *Comp. Biochem. Physiol.* **44A**: 41–46 (with M. Peaker, S. J. Peaker & A. Hanwell).

149. Changes in mammary gland permeability at the onset of lactation in the goat: an effect on tight junctions? *J. Physiol., Lond.* **230**: 13–14P (with M. Peaker).

150. The demands of the udder and adaptation to lactation. In *Production disease in farm animals*: 89–106. Payne, J. M., Hibbitt, K. G. & Sansom, B. F. (eds). London: Baillière, Tindal.

151. Vasopressin and milk secretion: lack of effect of low doses. *J. Endocr.* **57**: 87–95 (with M. Peaker).

152. Innate seasonal oscillations in the rate of milk secretion in goats. *J. Physiol., Lond.* **230**: 225–233.

153. Oestrone metabolism in pregnant and lactating goats. *J. Endocr.* **57**: 451–457 (with J. R. G. Challis).

154. The lack of effect of aldosterone on milk secretion in the goat. *J. Endocr.* **58**: 139–140 (with M. Peaker).

155. Cloudburst: pseudopregnancy in goats. *Br. Goat Soc. J.* **66**: 94.

156. The time course of cardiovascular changes in lactation in the rat. *J. Physiol., Lond.* **233**: 93–109 (with A. Hanwell).

157. The effects of engorgement with milk and of suckling on mammary blood flow in the rat. *J. Physiol., Lond.* **233**: 111–125 (with A. Hanwell).

158. Changes in milk composition preceding oestrus in the goat: comparison with the effects of exogenous oestrogens. *J. Endocr.* **59**: xlvii (with M. Peaker).

1974

159. The metabolism of glucose, acetate, lipids and amino acids in

lactating dairy cows. *J. agric. Sci.* **82**: 71–85 (with R. Bickerstaffe & E. F. Annison).

160. Glucose and fatty acid metabolism in cows producing milk of low fat content. *J. agric. Sci.* **82**: 87–95 (with E. F. Annison & R. Bickerstaffe).

161. Mammary blood flow and methods of identifying and measuring precursors of milk. In *Lactation* **1**: 143–225. Larson, B. L. & Smith, V. R. (eds). New York and London: Academic Press.

162. Hour-to-hour variation in amino acid arteriovenous concentration differences across the lactating goat mammary gland. *J. Dairy Res.* **41**: 95–100 (with T. B. Mepham).

163. Effects of intramammary arterial infusion of essential amino acids in the lactating goat. *J. Dairy Res.* **41**: 101–109 (with T. B. Mepham).

164. Effects of intramammary arterial infusion of non-essential amino acids and glucose in the lactating goat. *J. Dairy Res.* **41**: 111–121 (with T. B. Mepham).

165. The effects of oestrus and exogenous oestrogens on milk secretion in the goat. *J. Endocr.* **61**: 231–240 (with M. Peaker).

166. The use of isotopes in the study of milk secretion. *Proc. Nutr. Soc.* **33**: 17–23.

167. Cortisol secretion rate, glucose entry rate and the mammary uptake of cortisol and glucose during pregnancy and lactation in dairy cows. *J. Endocr.* **62**: 371–383 (with J. Y. F. Paterson).

168. Studies of the efficacy of various methods of organ preservation, perfusion, and transplantation as assessed by the rate of milk secretion of the mammary gland. *Transpl. Proc.* **6**: 241–244.

169. Electrical conductivity of foremilk for detecting subclinical mastitis in cows. *J. agric. Sci.* **83**: 309–325 (with M. Peaker & J. G. Rowell).

170. The development of sucking behaviour in the newborn goat. *Anim. Behav.* **22**: 628–633 (with D. B. Stephens).

171. Improved methods for measuring mammary metabolism in conscious farm animals. *J. Physiol., Lond.* **242**: 1–2P (with I. R. Fleet).

172. Changes in colostrum composition and in the permeability of the mammary epithelium at about the time of parturition in the goat. *J. Physiol., Lond.* **243**: 129–151 (with M. Peaker).

1975

173. Citrate in milk: a harbinger of lactogenesis. *Nature, Lond.* **253**: 464 (with M. Peaker).

174. The distribution and movements of carbon dioxide, carbonic acid and bicarbonate between blood and milk in the goat. *J. Physiol., Lond.* **244**: 771–782 (with M. Peaker).

175. Soluble indicator techniques for tissue blood flow measurement using ^{86}Rb-rubidium chloride, urea, antipyrine (phenazone) derivatives or ^{3}H-water. *Clin. exp. Pharmacol.*, supplement 1: 15–29 (with B. P. Setchell).
176. The physiological background to subclinical mastitis. *Vet. A.* **15**: 42–46.
177. Metabolic clearance rate, production rate and mammary uptake of progesterone in the goat. *J. Endocr.* **64**: 485–502 [with C. A. Bedford (née Slotin) & R. B. Heap].
178. Utilization of amino acids by the isolated perfused sheep liver. *Q. Jl exp. Physiol.* **60**: 141–149 (with D. B. Lindsay, I. G. Jarrett & J. L. Mangan).
179. Effects of prolactin on ion movements across the mammary epithelium of the rabbit. *J. Endocr.* **65**: 26–27P (with J. C. Taylor & M. Peaker).
180. *Salt glands in birds and reptiles.* (Monographs of the Physiological Society, 32.) Cambridge: Cambridge University Press (with M. Peaker).
181. Efficacy of the measurement of the electrical conductivity of milk for the detection of subclinical mastitis in cows: detection of infected cows at a single visit. *Br. Vet. J.* **131**: 447–461 (with M. Peaker).
182. Metabolic clearance rate, production rate and mammary uptake and metabolism of progesterone in cows. *J. Endocr.* **66**: 239–247 (with R. B. Heap & A. Henville).
183. Carnitine secretion into milk of ruminants. *J. Dairy Res.* **42**: 371–380 (with A. M. Snoswell).
184. Secretory activity of goat mammary glands during pregnancy and the onset of lactation. *J. Physiol., Lond.* **251**: 763–773 (with I. R. Fleet, J. A. Goode, M. H. Hamon, M. S. Laurie & M. Peaker).
185. Changes in milk composition during lactation in the guinea-pig, and the effect of prolactin. *J. Endocr.* **67**: 307–308 (with M. Peaker, C. D. R. Jones & J. A. Goode).
186. Methods of measuring the utilization of metabolites absorbed from the alimentary tract. In *Digestion and metabolism in the ruminant*: 306–319. McDonald, I. W. & Warner, A. C. I. (eds). Armidale: University of New England Publishing Unit (with E. F. Annison).
187. The effects of prolactin and oxytocin on milk secretion and on the permeability of the mammary epithelium in the rabbit. *J. Physiol., Lond.* **253**: 547–563 (with J. C. Taylor & M. Peaker).

1976

188. Review of *The blood of sheep. Composition and function.* Blunt, M. H. (ed.). Springer-Verlag: Berlin. *FEBS Lett.* **64**: 239.
189. The secretion of citrate into milk. *J. Physiol., Lond.* **260**: 739–750 (with T. B. Mepham & M. Peaker).

190. Detecting subclinical mastitis. *N. Z. J. Agric.* **132**: 75–76.

1977

191. An analysis of specific stimuli causing the release of prolactin and growth hormone in the goat. *J. Endocr.* **72**: 163–171 (with I. C. Hart).

Preface

Lactation is a complex but vital physiological function. Probably because the mammary glands are not necessary for the survival of adults they have not been studied so intensively as essential organs; nevertheless they are essential for the survival of species, and they are of course a unique mammalian character. But to biochemists, physiologists and endocrinologists an organ that functions only periodically and when doing so synthesizes and secretes fat, protein and carbohydrate in large quantities provides rare opportunities for studying the control of cell function at a fundamental level. It is also becoming increasingly apparent that comparisons between species are yielding important information on the extent to which mechanisms are common to all mammals and on the strategies of lactation employed by different species or groups of animals.

Comparative studies are sometimes criticized as mere extensions of the sort of comparative anatomical investigations that stifled zoology for many decades and which were apparently designed to substantiate evolutionary theories and to plot phylogenetic family trees. I would counter such criticisms by arguing that comparative studies, involving as they do the examination of both the similarities and dissimilarities between animals, permit conclusions to be drawn on general mechanisms and particular adaptations. This is not to say that evolutionary considerations are not important but they now seem to be more of a by-product of the comparative approach than they once were. Furthermore, we can often extrapolate general features to species in which for technical or ethical reasons direct experimentation is not possible, to man himself (or in the present context to woman herself) for example. This wide comparative approach is clearly more meaningful than using just one species as a "model" for another. However, such an approach requires the accumulation of information from, and the testing of hypotheses in, a wide range of species, and it will be evident where, through lack of knowledge, we are restricted to comparisons between very few animals and where general conclusions cannot yet be drawn.

The mammary glands and their secretion impinge directly on such diverse topics as sociology, anthropology, medicine, agriculture and economics. Research in lactation is particularly important as the scientific base of the dairy industry but the high incidence of breast cancer and the role of breast feeding in modern society provide a further impetus to the research effort. Furthermore, with a number of mammalian species in danger of extinction and the preservation of such species by captive breeding becoming ever more important, comparative research in lactation finds another application in wild animal husbandry.

Lactation obviously cannot be divorced from reproduction but the time available did not permit an examination of some of the interactions between these two processes. However, it is well known that marked diversity exists throughout eutherian mammals in the control of reproductive processes; even in closely related species the mechanisms involved in the maintenance of pregnancy and in the induction of parturition may be very different. The evolutionary basis for this diversity has not really been explained satisfactorily but a pertinent question is whether such great diversity exists in lactation and its control. I believe an answer to this question, and to many others, emerges in this symposium.

Dr Linzell and I planned the programme for this symposium shortly before his death. It is therefore particularly fitting that the proceedings should be dedicated to his memory, and it was gratifying that many of his friends, colleagues, collaborators and pupils were able to take part in the meeting.

Finally I would like to thank the Australian High Commission for allowing us to show the CSIRO film, "Comparative Biology of Lactation"; the staff of The Zoological Society of London, and in particular Miss Unity McDonnell for great assistance; Miss W. M. Reynolds and Mrs J. H. Banks of the Institute Library; Academic Press, and the Chairmen and Contributors who have given so freely of their time.

Babraham   MALCOLM PEAKER
September 1977

Contents

Comparative Mammary Fine Structure

F. B. P. WOODING

Comparative Aspects of Milk Fat Synthesis

R. DILS, SUSAN CLARK and J. KNUDSEN

Synthesis and Secretion of Milk Proteins

T. B. MEPHAM

Synthesis and Secretion of Milk Sugars

E. A. JONES

Mammary Energy Metabolism

G. H. SMITH and D. J. TAYLOR

The Aqueous Phase of Milk: Ion and Water Transport

M PEAKER

Comparative Endocrinology of Mammary Growth and Lactation

ISABEL A. FORSYTH and T. J. HAYDEN

Lactogenesis: the Search for Trigger Mechanisms in Different Species

N. J. KUHN

Comparative Physiology of Milk Removal

B. A. CROSS

The Aetiology of Mammary Cancer in Man and Animals

R. V. SHORT and J. O. DRIFE

Comparative Studies on Milk Lipids and Neonatal Brain Development

BARBARA M. HALL and JANET M. OXBERRY

Role of the Mammary Gland and Milk in Immunology

A. K. LASCELLES

A Comparative Study of Plasma Cells in the Mammary Gland in Pregnancy and Lactation

R. S. H. PUMPHREY

Rearing Human Infants: Breast or Bottle

MAVIS GUNTHER

Dairying: Past, Present and Future

C. C. BALCH and J. W. G. PORTER

Physiological Effects of Lactation on the Mother

ANN HANWELL and M. PEAKER

Maternal Behaviour in Mammals

ELIZABETH SHILLITO WALSER

The Management of Young Mammals

M. R. BRAMBELL and D. M. JONES

Comparison of Milk-Like Secretions Found in Non-Mammals

A. CHADWICK

Symp. zool. Soc. Lond. (1977) No. 41, 1–41

Comparative Mammary Fine Structure

F. B. P. WOODING

*ARC Institute of Animal Physiology,
Babraham, Cambridge, England*

SYNOPSIS

The embryology and post-natal development of the mammary gland are very similar in all species studied. The ultrastructure of cells of the gland during pregnancy in any particular species is rather varied, but most of the variation disappears a day or so after parturition with the development of the organelles characteristic of a fully lactating cell. These include, in a large volume of cytoplasm, massive amounts of rough endoplasmic reticulum cisternae, a vastly hypertrophied Golgi body, numerous mitochondria and a considerably infolded basal plasmalemma. Full development of this ultrastructural pattern is closely dependent on the presence of insulin, glucocorticoids and prolactin, and the decrease in progesterone concentration at parturition.

There seems to be no correlation between the ultrastructure, assessed morphometrically, and the composition of the milk, since the glands of animals producing milks of widely different composition have identical ultrastructure.

The mechanism of secretion of the major milk protein casein and possibly other important milk constituents such as lactose and calcium is by way of exocytosis of the Golgi vesicle contents. Lipid droplet release may also require Golgi vesicle exocytosis and a common mechanism is suggested for release of lipid and secretory cell cytoplasmic fragments into milk. It is suggested that immunoglobulins pass across the epithelium by a coated microvesicle system.

The monovalent ionic content of milk can be explained by the observed basolateral distribution of the Na^+–K^+ ATPase enzyme on the plasmalemmas of the secretory cell and duct cell. Involution of the secretory cell by auto- and heterophagocytosis is very similar in all species studied although the degree to which circulating macrophages are involved may vary.

INTRODUCTION

The mammary secretory cell offers an excellent basis for studies on the correlation of structure and function. Its development, mature function and regression are controlled by complex hormonal patterns. It synthesizes and secretes large quantities of protein and lipid as well as controlling the content of the final secretion within very close limits. Results of the search for ultrastructural bases for some of these manifold functions and responses will form the content of this contribution.

EARLY DEVELOPMENT

The mammary gland in mammals originates embryologically from an ectodermal ingrowth which, under the influence of the mesenchyme, produces a branched system of ducts at birth (Raynaud, 1971). In response to the hormonal changes at parturition there may be a transient production of a watery fluid in the neonatal gland—the so-called "witches milk". Tobon & Salazar (1974) have shown that in the human fetus at term there is considerable development of the mammary cell fine structure with production of Golgi vesicles filled with what appears to be a mucus-like secretion. Lipid is also secreted, but the duct system rapidly reverts to a quiescent state in the neonate. The duct system keeps pace with body growth until puberty when there is a considerable specific growth of the duct system until it occupies a major part of the original fat pad volume. At this stage the cells of the gland are simple in structure with a large nucleus and few cytoplasmic organelles (Hollmann, 1974).

CHANGES DURING PREGNANCY

During the initial stages of pregnancy there is a considerable further growth of the duct system, and alveoli are budded off to provide the characteristic lobulo-alveolar structure (Fig. 1).

In the mouse there are two peaks of mitotic activity, in early and mid-pregnancy. The cells lining the alveoli are very simple in fine structure and similar to the duct cells from which they arise. As pregnancy progresses the alveoli increase in size and number, and fat and casein are secreted into the alveolar lumina. The alveolar secretory cells can now be distinguished from those of the ducts by their lack of cytoplasmic fibrils and desmosomes and the presence of numerous lipid droplets. The Golgi apparatus is somewhat larger and mitochondria and endoplasmic reticulum slightly more frequent than found in the mature virgin state (Hollmann, 1974).

The most characteristic feature of the alveolar cell in the second half of pregnancy is the accumulation of lipid droplets in the cytoplasm (Fig. 1c, j). These vary considerably in size but frequently are so numerous that

---

Abbreviations used on figures:
A, mammary alveolus; B, basement membrane; E, (rough) endoplasmic reticulum; G, Golgi body or vesicle; L, lipid droplet; M, myoepithelial cell; MT, mitochondrion; N, nucleus; P, plasmalemma (cell membrane).

FIG. 1. Light micrographs of mammary tissue from various animals at various stages of lactation. Toluidine blue stained 1-μm sections of glutaraldehyde-fixed, araldite-embedded tissue. All ×1000. Note the increase in size of the alveolus (A) with appearance of casein and lipid as pregnancy proceeds in the pig; the great increase in cytoplasmic volume and polarization of organelles which occur at parturition in the pig and cow; and the accumulation of lipid droplets (L) toward the end of pregnancy in pig and sheep.

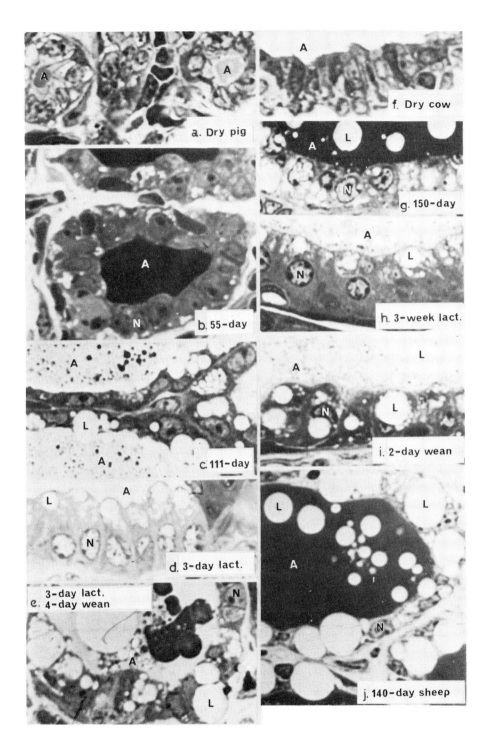

there is little room in the cytoplasm for any great increase in organelles. The lipid may be above or below the nucleus, or both; the cells show no polarization of organelles. The Golgi apparatus varies in position and is usually fairly compact, with only infrequent indication of the presence of casein in the Golgi vacuoles of the cells (Fig. 2). Examination of the alveolar lumen at this stage indicates that an appreciable secretion of both casein miscelles and lipid droplets occurs during pregnancy (Fig. 1c, g). This could be due to a very low level of secretion by the majority of largely undifferentiated cells but is more likely due to a high rate of secretion by a small proportion of more fully developed cells. Mills & Topper (1970) have reported that half-way through pregnancy in mature primiparous mouse mammary glands only about 50% of the alveolar cells are at the same stage of development at any one time. Fully differentiated alveolar cells are present when the majority are only at the low level of differentiation characteristic of mid-pregnancy. The reverse situation, a few undifferentiated alveoli in a gland otherwise fully lactating, has also been reported (Saacke & Heald, 1974); thus considerable caution is necessary in identifying the "typical" cell at any one stage of development of the mammary glands. However it has now been clearly established that the mammary gland cell structural development occurs in a definite sequence in response to hormonal changes. This has been demonstrated most clearly in the mammary gland explant studies of Topper, who showed that insulin is required for division and growth of the alveolar cells; subsequently addition of hydrocortisone stimulates development of the Golgi and rough endoplasmic reticulum membranes, and finally prolactin produces the characteristic polarization of cell organelles with hypertrophy of Golgi vacuoles, containing casein micelles, seen in the fully-developed state at full lactation (Topper & Oka, 1974).

There is some doubt about the stage at which the mammary ductules become competent for full differentiation of a secretory cell. Pre-pubertal and even fetal mammary tissue can be stimulated with hormones to produce "casein-like material", even in the absence of lobulo-alveolar differentiation (Ceriani, 1970). However, the amount of this secretion is very small and the development of full milk production is clearly dependent on the correct sequence of hormonal changes involving oestrogen and progesterone as well as insulin, glucocorticoids and prolactin.

In vivo full lactation has been produced in dry or virgin cows four to 14 days after the administration of oestrogens at high levels but the response is very variable. The milk yield was found to be closely correlated with the degree of development of the fully differentiated alveolar cell structure, which was identical to that found in a normal lactating cow (Croom, Collier, Bauman & Hays, 1976).

The morphological site of action of the hormones involved in mammary-gland development is, at present, uncertain. Insulin can affect *in vitro* unicellular systems even when coupled to sepharose beads and thus must have a primary action on the plasma membrane (Oka &

Topper, 1972). It has been claimed that prolactin can also be localized at the (basal) plasmalemma of the mammary cell by autoradiography but the resolution of the technique was limited (Birkinshaw & Falconer, 1972).

Technical difficulties at present rule out precise localization on the site of action of the steroid hormones, though most can be demonstrated to have specific intracellular binding proteins by biochemical techniques (e.g. Gardner & Wittliff, 1973).

PARTURITION AND LACTATION

Fine Structural and Enzyme Changes Around Parturition

In most species so far examined the development of the typical ultrastructure of the lactating mammary secretory cell begins before parturition and is complete by the third or fourth day post-partum. Development is very slow during pregnancy but varies within the gland and between different species, giving a structurally heterogeneous population of alveolar cells in the last quarter of pregnancy. The most prominent features in the last week before parturition are the accumulation of lipid droplets in the cytoplasm of the alveolar cells (Fig. 1c, g) and, in rats and mice, the occurrence of stasis vacuoles containing large numbers of casein micelles or their breakdown products (Hollmann, 1974; Saacke & Heald, 1974).

In the rat the lipid droplets and stasis vacuoles are abruptly secreted into the alveolar lumen about six hours before parturition. This produces a great increase in alveolar volume and leaves the cells containing little or no secretory product (Chatterton, Harris & Wynn, 1975). The "trigger" for the sudden secretion is unknown. Subsequently (compare Figs 2 and 3) in response to the complex hormonal changes at parturition there is a rapid proliferation of rough endoplasmic reticulum (Fig. 4), an increase in the amount of Golgi membranes and in the number and size of the Golgi vacuoles with their content of casein micelles (Fig. 4). A polarization of the cell organelles occurs with the Golgi apparatus above the nucleus and the majority of the rough endoplasmic reticulum and mitochondria below. The basal plasmalemma of the cells adjacent to the basement membrane usually develops considerable infolding (Fig. 5). These changes at parturition enormously increase the volume of the cytoplasm with respect to the nucleus, synchronize the cells, and produce a structurally homogeneous population of secretory alveolar cells.

Hollmann (1974, for a review) followed these ultrastructural changes using methods of statistically-controlled sampling (morphometry) in the mouse during pregnancy and lactation; some of his results are presented in Fig. 6. The rapid increases in the Golgi apparatus and endoplasmic reticulum at the time of parturition are clearly shown. These morphological changes can be correlated with the increase in the activities of all the enzymes in the cell, including those specifically involved in the synthesis of casein, lactose and milk lipid in rat, mouse, rabbit and guinea-pig (Baldwin & Yang, 1974). In these animals this great increase in the ultrastructural machinery of the cell is accompanied by a wave of mitotic division. The whole process results in a mammary gland capable of its maximum production about three to seven days after parturition. Little or no mitosis occurs in the gland after this time.

Since the activities of all the enzymes increase at the same time it is not possible to establish whether there is a particular enzyme whose increase in activity would be sufficient to trigger milk synthesis. However, in the cow the enzyme activities and levels of available metabolites are sufficient to support full lactation up to two weeks before parturition. Changes in the levels of these key enzymes of milk synthesis at parturition have been reported (Baldwin & Yang, 1974 for a review). The stimulus for the changes would most likely be the rapid fall in progesterone levels at parturition since the changes in corticoids and prolactin seem to be concerned with maintenance of a higher level of enzymes concerned with milk synthesis rather than a triggering effect.

The differences between the cow and the rat, mouse, rabbit and guinea-pig in enzymic capacity for lactogenesis is not reflected in any obvious cytological difference in the alveolar cell immediately pre-partum (Saacke & Heald, 1974). In all animals so far examined such a "typical" cell shows no polarization of organelles, there is very little development of rough endoplasmic reticulum, and few of the Golgi apparatus vesicles are swollen or contain any casein micelles (Fig. 2). The most notable feature is the large number of lipid droplets in the cells. However there is considerable variation from alveolus to alveolus, and, even in a single alveolus, cells with three or four parallel endoplasmic reticulum cisternae and a hypertrophied Golgi apparatus may be found.

This marked variation in the cytological picture largely disappears with the onset of lactation. On the first day after parturition cow mammary cells show a vast increase in cytoplasmic area with arrays of rough endoplasmic reticulum below the nucleus and an extensively hypertrophied Golgi body above (Saacke & Heald, 1974) (Fig. 3). In mammary tissue sampled at the time of maximum milk yield, whatever the animal, the disposition of the organelles involved and the relative amounts of these are remarkably similar (Figs 3, 7–14 and Table I).

There seems to be no correlation between milk composition and the proportions of the various organelles in the secretory cell, as has been suggested by Reid & Chandler (1973). Compare, for example, fat content

---

FIGS 2 and 3. Cow mammary secretory epithelial cells at mid-pregnancy (150 days) Fig. 2, and three weeks after parturition, Fig. 3. Both ×5000. Note the vast increase and polarization of the Golgi body and endoplasmic reticulum in the lactating animal. The cellular fine structure does not change greatly between mid- and late pregnancy.

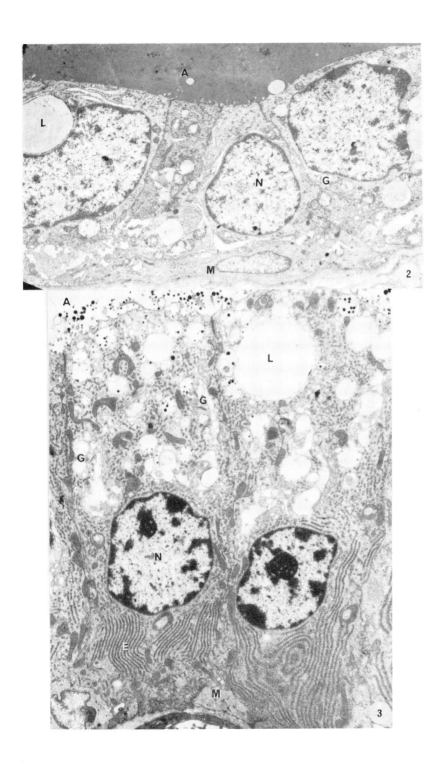

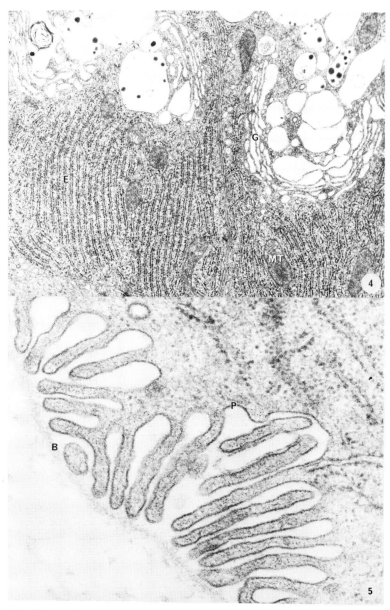

FIGS 4 and 5. Detail of lactating mammary tissue, showing the endoplasmic reticulum and Golgi membranes, and the infolded basal plasmalemma.
Fig. 4, Rat, ×14 600. Fig. 5, Goat, ×55 000.

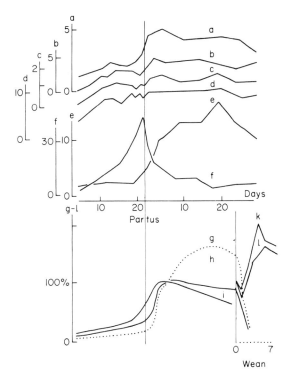

FIG. 6. Changes in ultrastructural variables in the mouse (a–f) (data from Hollmann, 1974) and enzyme and DNA changes in rats (g–1) (data from Baldwin & Yang, 1974; Helminen & Ericsson, 1968) during pregnancy, lactation and involution.
a, Area of rough endoplasmic reticulum surface, μm^2, per unit volume of cytoplasm, μm^3. b, Golgi vesicle volume as % of cytoplasmic volume. c, Area of Golgi membrane surface μm^2, per unit volume of cytoplasm, μm^3. d, Mitochondrial volume as % of cytoplasmic volume. e, Nucleocytoplasmic ratio. f, Lipid droplet volume, as % of cytoplasmic volume (nucleus excluded). g, UDPG pyrophosphorylase. h, Aspartate aminotransferase. i, DNA concentration. g-i, Expressed as a percentage of the levels at day 5 of lactation. k, Acid phosphatase, aryl sulphatase and acid deoxyribonuclease. l, Cathepsin D. k and l, Expressed as a percentage of the levels immediately prior to the start of weaning (day 0).

in the fur seal and horse, or protein in man and the cow. The similarity in fine structure of the lactating alveolar cell extends to the mechanism by which the cell elaborates and secretes the protein and lipid and other constituents of the milk.

Protein Secretion

Autoradiographic studies have established that the primary site of casein synthesis is the ribosomes of the rough endoplasmic reticulum (Saacke & Heald, 1974). Label from amino acid precursors then passes to the Golgi cisternae and vacuoles, and eventually to the casein micelles in the lumen. The Golgi body may modify the protein by the addition of carbohydrate groups and concentrates the result for export from the cell in the Golgi vesicle. In the Golgi cisternae small granules and threadlike material are first seen at the inside of the membrane and this material appears to aggregate to form the casein micelle (Fig. 32). This may be an artifact of fixation or it may represent a real process that occurs *in vivo*, suggesting that the casein micelle has an identical subunit substructure. The progress of labelled protein material into the rat or mouse mammary-gland lumen is about ten times as rapid as that found in pancreas or thyroid by similar methods.

It is generally accepted that the Golgi vesicle fuses with the apical plasma membrane to release its casein micelles and other contents of the lumen (Hollmann, 1974; Saacke & Heald, 1974) (Fig. 7). There is cytochemical (Tan, Young, Goldsmith & Livingston, 1973) and immunohistochemical evidence that the dark granules seen in the apical Golgi vacuoles are casein, and that β-lactoglobulin is present in the same area of the cell (Kihm, 1973).

Lactoperoxidase, an enzyme which commonly occurs in milk, has also recently been shown to be present in the endoplasmic reticulum and the Golgi vacuoles containing casein micelles, and is presumably secreted by the same route as the casein (Anderson, Trantalis & Kang, 1975).

Calcium Transport

The majority of the calcium in milk is associated with the casein micelle, and there is preliminary evidence that the Golgi cisternae and vesicles are the major site of calcium localization in the cell. Electron microscopical histochemical localization with pyroantimonate shows a definite precipitate in cisternae and vacuoles of the Golgi apparatus (Figs 15, 16 and 17). It has now been found that the precipitate definitely contains calcium pyroantimonate by electron microscopic microanalysis (Wooding & Morgan, unpublished results). Baumrucker & Keenan (1975) have demonstrated an ATP-dependent calcium accumulation by Golgi apparatus-enriched subcellular fractions from the bovine mammary gland, so the biochemical evidence would support the localization.

FIGS 7–14. Secretory epithelial cells from various animals in full lactation. All show extensive rough endoplasmic reticulum, a hypertrophied Golgi body above the nucleus with numerous granule (casein) containing Golgi vacuoles and infoldings of the basal plasmalemma. An example of Golgi vesicle exocytosis is inset on Fig. 7.

Fig. 7, Goat ×4700; inset ×24 000. Fig. 8, Fur-seal ×4500. Fig. 9, Horse ×4400. Fig. 10, Guinea-pig ×4400. Fig. 11, Rat ×4800. Fig. 12, Sheep ×5000. Fig. 13, Pig ×4600. Fig. 14, Wallaby ×5000.

Material from which Fig. 8 was taken by courtesy of the British Antarctic Survey. Fixed in formol-calcium for 12 months prior to my receiving it.

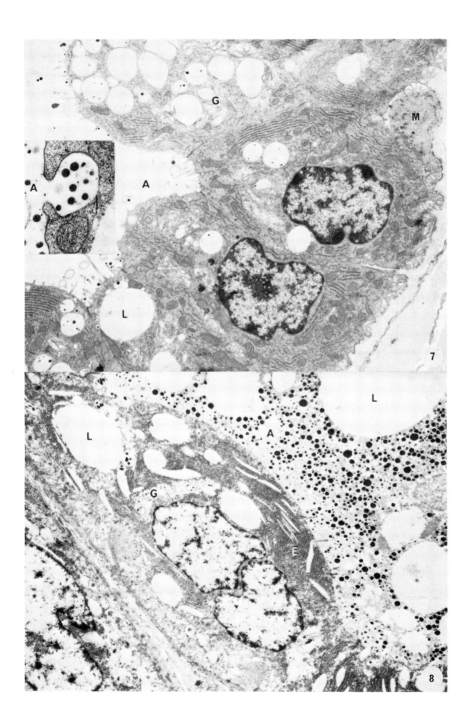

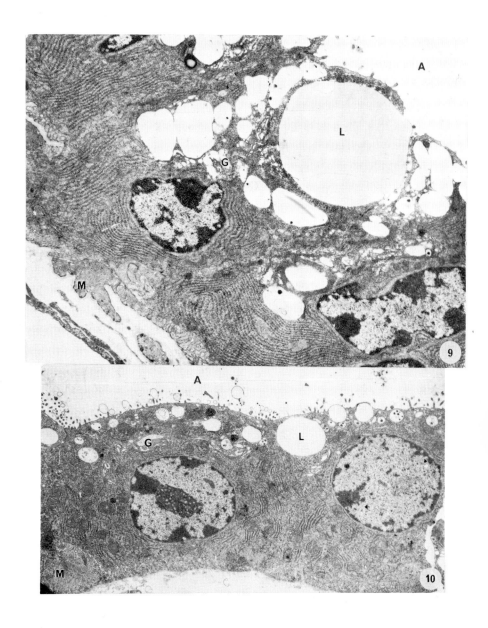

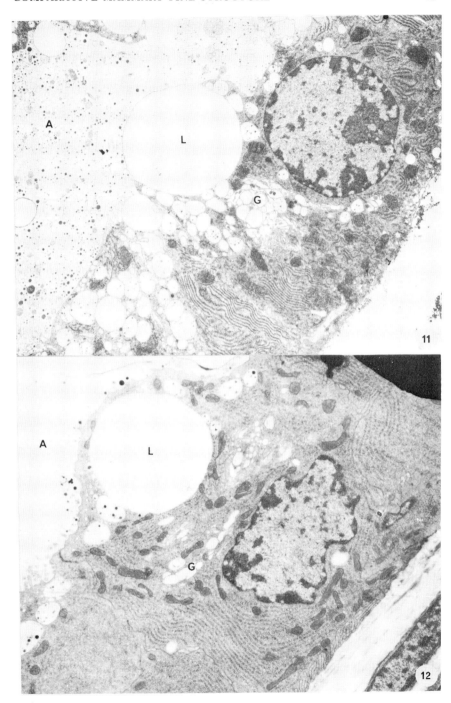

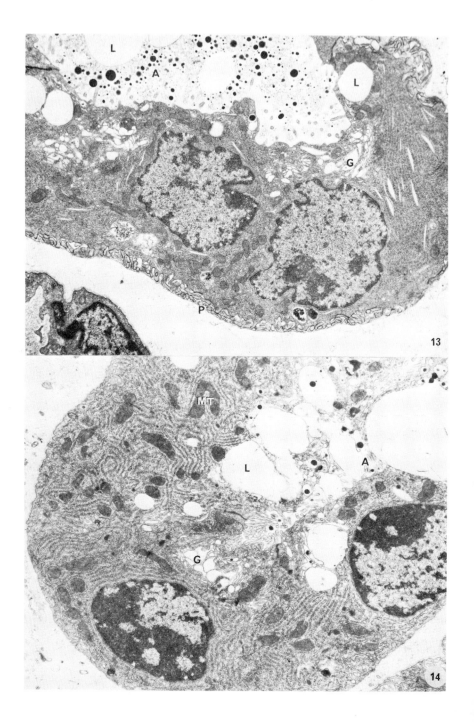

TABLE I

Percentage by volume of the organelles in the lactating mammary cell and percentage of protein and fat in the milk

Animal	Nucleus	Rough endoplasmic reticulum	Golgi membs	Golgi vesicles	Mitochondria	Lipid droplets	Cytoplasm	Plasma-lemma	Fat	Protein (in milk)*
Cow (Jersey)	12	25	5	10	13	10	24	1	4	3
Goat	13	24	8	4	12	4	27	3	5	3
Guinea-pig	17	25	6	7	11	7	24	3	4	7
Horse	15	23	8	8	9	8	28	1	2	2
Man†	22	20	10	6	9	4	28	1	4	1
Pig										
30-day pregnant	39	12	6	1	9	7	25	1		
111-day pregnant	29	9	6	4	8	21	21	2		
lactating	16	23	9	12	12	5	20	3	7	3
Rat	12	22	4	14	8	9	29	2	10	8
Sheep	21	25	4	6	9	7	25	3	7	5
Fur seal‡	18	29	11	6	4	4	27	1	53	9
Wallaby	16	25	10	8	10	4	26	1	5	4

* From Jenness (1974). † Data from blocks provided by Dr H. Tobon. ‡ Data from tissue provided by the British Antarctic Survey.

Protein Transport

The mammary epithelium plays a vital role in transporting immuno-globulin (IgG) molecules into the milk from either serum or the plasma cells in the gland itself. In primates, rats and guinea-pigs IgG is transmit-ted across the placenta to the foetus, but in cattle, sheep, goats, horses and pigs there is no placental transfer. The neonate gets its initial immunolog-ical protection only from the maternal immunoglobulins in the colostrum which are absorbed unaltered into the neonatal blood stream from the intestine (Butler, 1974). In the pig it has been shown that up to 90% of the immunoglobulin in colostrum is transferred by a selective process from the blood. Subsequent to colostrum production up to 90% of the much lower levels of immunoglobulin in the milk are produced locally by plasma cells in the mammary gland stroma (Bourne & Curtis, 1973; see the articles in this volume by Lascelles and Pumphrey, pp. 241 and 261).

The route of transfer across the mammary epithelium is still conjec-tural. It can be clearly shown by light microscope techniques that the immunoglogulins pass through the secretory cells rather than between them (Dixon, Weigle & Vasquez, 1961), but there is only one preliminary report at the electron microscopical level. This indicated that in lactating rabbits (the exact stage of lactation was not given) the immunoglobulin synthesized in the mammary-gland plasma cells was also present in the secretory epithelial cells in the Golgi body area, in small apical vesicles, and in large lysosomal-like basal vesicles (Kraehenbuhl, Galardy & Jamieson, 1974).

There are two sorts of small vesicles in the apex of the mammary cell—"coated vesicles" with a characteristic pattern of spikes around their cytoplasmic surface, and vesicles with a smooth contour (Fig. 18). It has been suggested by several workers that the smooth vesicle might be involved in transferring the contents of either the endoplasmic reticulum or the Golgi apparatus to the alveolus via a fusion with the plasmalemma, but there is no direct evidence for this (Saacke & Heald, 1974). There is some morphological evidence in other secretory systems which suggests that the coated vesicle may be involved in the transfer of excess plasma membrane back into the cell and also in the uptake of protein material from outside the cell (Heuser & Reese, 1973; Moxon, Wild & Slade, 1976).

Coated vesicles can frequently be seen at the basal (Fig. 19) and lateral plasmalemma as well as at the apical, so that it is tempting to generalize from the specific intestinal transport of immunoglobulin in the neonate by coated vesicles. In the intestine antibody is taken up at the apical

---

FIGS 15–17. Rabbit lactating mammary tissue prepared by the pyroantimonate methods of Hales *et al.*, 1974, for the demonstration of calcium localization. Note the precipitate indicative of calcium is restricted to the Golgi vesicles and membrane sacs (arrows) which frequently also contain casein micelles (double arrows).
Fig. 15, ×17 000. Fig. 16, ×31 000. Fig. 17, ×45 000.

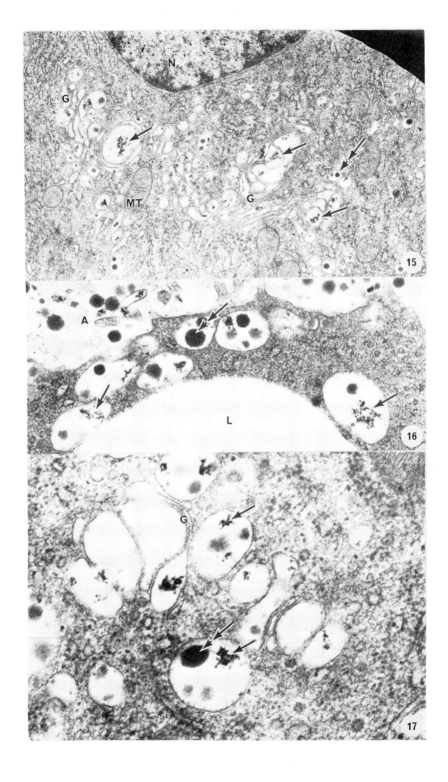

plasmalemma of the proximal cells of the small intestine and transported in coated vesicles to the laternal plasmalemma from where it can diffuse to the blood and lymph vessels (Rodewald, 1973).

A reversal of this process would provide a mechanism for the specific immunoglobulin transport carried out by the mammary gland. There is a recent morphological report that coated vesicles are involved in secretion at the apical surface of the mammary gland (Franke, Luder, Kartenbeck, Zerban & Keenan, 1976) and since some can also be found in the Golgi area it seems that the coated vesicles may be in the correct location to play an important role in the transport of immunoglobulin through the cell.

Lipid Synthesis

Lipid synthesis occurs throughout pregnancy but in the cow the proportions of fatty acids change late in pregnancy from the ratios typical of body fat to ratios unique to the milk fat (Mellenberger, Bauman & Nelson, 1973). The only electron microscopical autoradiographic study of the process is that of Stein & Stein (1967) in lactating mice. By injecting labelled palmitate or oleate intravenously they found that 90% of the label was esterified into triglyceride within one minute of injection. At this time the label was found over the rough endoplasmic reticulum and over lipid droplets of all sizes in all parts of the cell. At ten and 60 minutes there was little label left over the rough endoplasmic reticulum, but intracellular and luminal droplets were labelled.

Biochemical studies of isolated cell organelles demonstrated that enzymes for the esterification of fatty acids with glycerol were localized in the rough endoplasmic reticulum (Bauman & Davis, 1974).

Stein & Stein (1967) suggested that the start of the formation of the lipid droplet took place within the rough endoplasmic reticulum cisternae, showing pictures of cisternae with crescent-shaped dilatations of the same lack of contrast as lipid droplets. However all lipid droplets in mammary secretory cells are bounded only by a single dense line, never by a unit membrane (Figs 22–25). A droplet which originated within the endoplasmic reticulum cisternae would be surrounded, at least initially, by a unit membrane. Saacke & Heald (1974) found modified areas of rough endoplasmic reticulum which they described as smooth endoplasmic reticulum (i.e. lacking ribosomes) and which they suggested represented the site of lipid synthesis or accumulation to form lipid droplets. However if the tissue is fixed by perfusion with fixative through the blood vessels rather than by cutting into pieces in fixative (immersion fixation)

FIGS 18 and 19. Details of the apical (Fig. 18) and basal (Fig. 19) regions of the secretory epithelial cell to demonstrate "coated" inpocketings of the plasmalemma (arrows) and coated vesicles within the cytoplasm (double arrows). Note also the tight junctions (asterisks) sealing the lateral intercellular spaces.
Fig. 18, Cow ×3400. Fig. 19, Wallaby ×10 000.

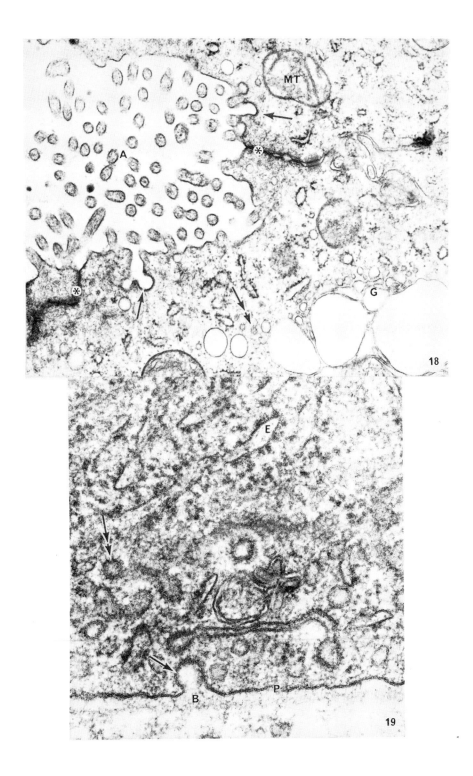

neither the dilatations nor the smooth regions of rough endoplasmic reticulum can be found (compare Fig. 4 with Figs 20 and 21).

In well perfused tissue the smallest lipid droplets do not display any consistent relationship with the rough endoplasmic reticulum (Figs 22–25). They usually have some small unit membrane-bounded vesicles close to their periphery (as found by Kurosumi, Kobayashi & Baba, 1968), and coated vesicles also if the lipid is close to the basal plasmalemma (Wooding, 1971a). The significance of this association is unknown. It is possible that such vesicles might come from the plasmalemma or rough endoplasmic reticulum and act as nucleating agents for the lipid droplets but the evidence is meagre.

Claims of continuity of lipid droplet with endoplasmic reticulum can equally well be interpreted as an unfavourable plane of section. The larger lipid droplets have no specific relationship with rough endoplasmic reticulum cisternae, but the autoradiographic evidence shows that they are actively adding triglyceride to their bulk (Stein & Stein, 1967). There is no indication of any preferred size of lipid droplet for secretion, since small and large droplets are secreted side by side and the range of size of droplets is the same inside the cell as in the alveolus (Wooding, 1971a).

Lipid Secretion

The method of lipid secretion is also uncertain. The product is generally agreed to be a unit membrane-bounded lipid droplet (milk fat globule) (inset on Fig. 34); Patton & Fowkes (1967) suggested that there are short-range molecular forces which would cause an attraction between the lipids of the droplet and of the plasmalemma, so that the plasmalemma would be adsorbed to the fat droplet surface thus drawing the droplet out of the cytoplasm and into the alveolus. The short-range molecular forces would only be effective at 1–2 nm separation, but subsequent work has shown that the plasmalemma never gets closer than 10–15 nm to the periphery of the lipid droplet (Hollmann, 1974; Saacke & Heald, 1974; Wooding, 1971a) (Figs 28 and 34). If this value is accurate, and electron microscopy of thin sections is subject to many errors at this level of resolution, then the suggestion of Patton & Fowkes in its original form would not seem tenable. However, the proponents of the 10–15 nm gap have to explain its uniformity and widespread occurrence. Since the gap can widen to contain cytoplasmic organelles (Wooding, Peaker & Linzell, 1970) (Fig. 37), and display periodic structure (after

---

FIGS 20 and 21. Rat lactating secretory epithelial cells. Tissue fixed by immersion and dissection in glutaraldehyde fixative. Note the characteristic "clefting" (arrows) in Fig. 20 which is a more frequent artifact than the short profiles of smooth endoplasmic reticulum (arrows in Fig. 21) continuous with the usual rough endoplasmic reticulum in Fig. 21.

Compare with Fig. 4 which is from tissue fixed by perfusion and which shows neither clefting nor "smooth" endoplasmic reticulum.

Fig. 20, ×29 000. Fig. 21, ×49 000.

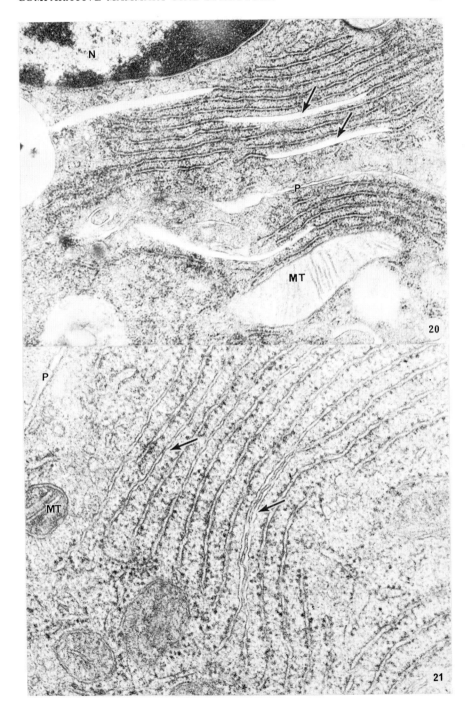

some alteration of membrane continuity of the secreted lipid droplet—
Wooding, 1975—Figs 34 and 35) and is unaffected by lipid solvents
(Wooding & Kemp, 1975) it seems most likely that a protein or glycoprotein molecule(s) is involved. Such a molecule, if attached to one surface,
either the inside of the plasmalemma or the periphery of the lipid droplet,
and interacting specifically with the other surface, could be the basis for
the association. Lipid secretion would take place because of the strength
of the association produced between the two surfaces by the intermediary
molecule.

A second suggestion for the mechanism of lipid secretion is based on
the observation that the apical lipid droplet frequently has Golgi and
other vesicles closely associated with its boundary (Figs 26–30); such an
association is maintained even if the contents of a cell are artifactually
released into the alveolus (Figs 31 and 32). The lipid droplet boundaries
are rarely accurately preserved (e.g. L1 on Fig. 31) owing to the comprehensive extraction of triglyceride, which is unavoidable in processing
tissue prior to cutting thin sections for electron microscopy. Where the
boundary is well preserved the Golgi and other vesicle membranes are
separated from the boundary line of the lipid by the same 10–15 nm gap
(Figs 28–30) that characterizes the newly-secreted milk fat globule membrane (Wooding, 1971b) (inset Fig. 34). In the goat mammary gland milk
fat globules can be produced in intracellular vacuoles by coincident fusion
of Golgi vesicles which were initially associated with an intracellular
droplet (Figs 33 and 40). In this case the plasmalemma is not involved in
the production of the milk fat droplet (Wooding, 1973).

In the normal mammary cell, lipid-associated Golgi vesicles will fuse
with the plasmalemma, as they are designed to do to release casein and
other contents. The lipid-associated Golgi membrane area now becomes
milk fat globule membrane which will then extend around the globule as
other lipid-associated Golgi vesicles fuse in turn. The speed of globule
release would depend on the availability of Golgi vesicles in the correct
orientation because there is no evidence that the release is a rapid or
continuous process. Association of Golgi vesicles with the apical lipid
droplets has been noted in a variety of mammals, but if the process of
release is not continuous there is no reason to expect many Golgi vesicles
to be associated with a lipid droplet at any one time.

FIGS 22–25. Small lipid droplets in the basal cytoplasm of secretory epithelial cells. Note the
lack of a unit membrane around the periphery and the lack of any consistent association with
any other organelle.
Fig. 22, Rat ×61 000. Fig. 23, Rat ×79 000. Fig. 24, Wallaby ×47 000. Fig. 25, Sheep ×47 000.
FIGS 26–30. Lactating secretory epithelial cells showing the association ("rosetting") between
Golgi vesicles and apical lipid droplets. Compare the membrane structure at the region of
close association (between the arrows, Fig. 28; arrow Figs 29 and 30) with that surrounding
the newly secreted milk fat globule (inset on Fig. 34).
Fig. 26, Cow ×15 000. Fig. 27, Sheep ×16 500. Fig. 28, Sheep ×91 000. Fig. 29, Horse
×45 000. Fig. 30, Rat ×32 000.

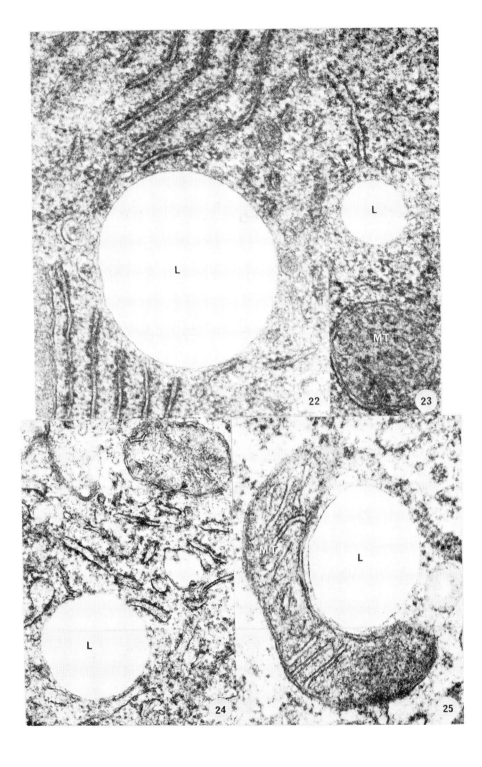

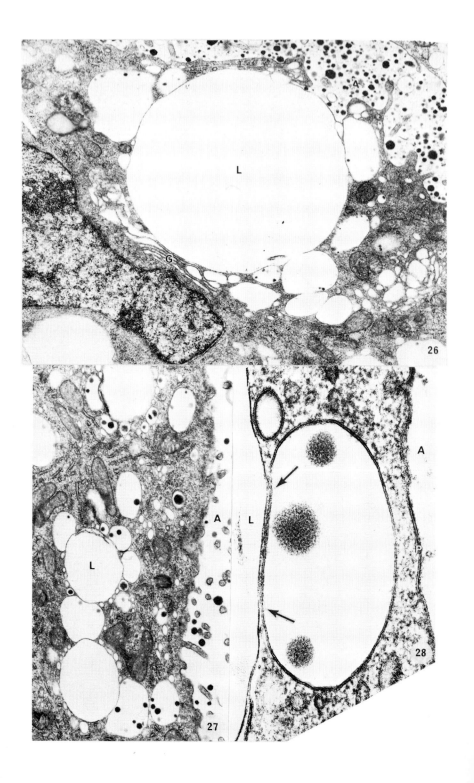

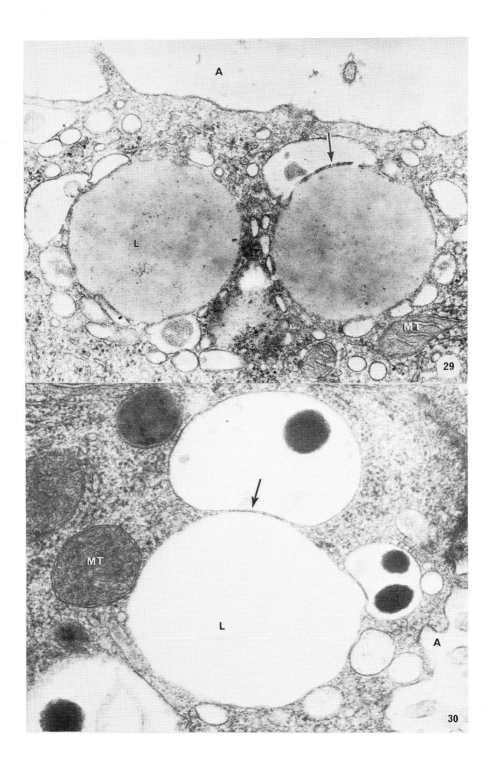

A

L

MT

29

MT

L

A

30

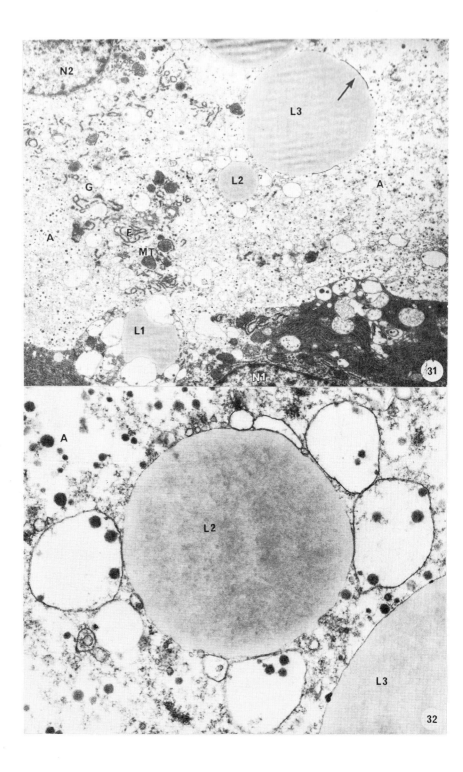

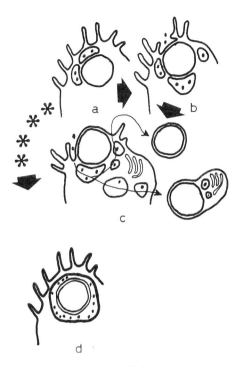

FIG. 33. Suggested method of formation of milk fat globule with unit membrane around only (see inset Fig. 34), or milk fat globule with cytoplasm and organelles under the membrane (see Fig. 37) by sequential fusion of lipid droplet associated Golgi vesicles with the plasma membrane (a → b → c). Fusion of lipid associated Golgi vesicles with each other (a → d) produces "presecretory vacuoles" (see Fig. 40) which are not necessarily continuous with the apical plasmalemma.

The relative roles of plasmalemma and Golgi vesicle membrane in the process of fat extrusion remain to be elucidated. There is some evidence that lipid secretion may be synchronized in the cells of any one alveolus, and this also seems to be true of the incorporation of labelled amino acids into protein (casein) (Saacke & Heald, 1974). It is not clear whether this synchronization reflects anything more than similar local availability of precursors and similar rates of secretion in each cell; it is certainly not universally found.

FIGS 31 and 32. Rat lactating secretory epithelium and alveolus fixed by immersion. Immersion fixation frequently produces discharge of well-fixed organelles (E, G, MT, N2) into the alveolus by rupture of the apical plasmalemma (e.g. the cell containing nucleus N1 and lipid drop L1). Despite this extrusion the association between the Golgi vesicles and the lipid droplet (L2) has persisted (Fig. 32). Lipid drop L3 has been secreted in the normal way so is a true milk fat globule as can be verified by the presence of patches of residual milk fat globule membrane on its surface (arrow).
Fig. 31, ×51 000. Fig. 32, ×34 000.

CYTOPLASMIC INCLUSIONS IN MILK

Cytoplasmic Fragments

Uninfected milk normally contains white blood cells, and various other inclusions originating from the secretory cells. The mammary secretory cell extrudes large numbers of casein micelles and lipid droplets while conserving its cellular architecture. The cell does lose a very small but definite amount of cytoplasm and plasmalemma when each lipid droplet is secreted from the cell (inset Fig. 34). In a small proportion (from 1 to 10%) of the total of the milk fat globules of all species, an appreciable amount of cytoplasm is included between the plasmalemma and fat globule boundary (Fig. 37). The amount of cytoplasm included can range from a tiny blob containing a few ribosomes up to a large fragment of cytoplasm with rough endoplasmic reticulum, mitochondria and fragments of the Golgi apparatus. The number of such cytoplasmic crescents (or "signets", Wooding et al., 1970) which are always surrounded by a membrane but do not contain a nucleus, varies with the species, being much more frequent in goats, man and rats than in cows, guinea-pigs or sheep. They are more frequent if the animals are milked at short intervals or after oxytocin administration. Their origin can be accounted for by the rapid fusion of Golgi vesicles with the plasmalemma thereby isolating an area of cytoplasm along with the lipid droplet (Fig. 33). It is difficult to see how they could be produced on the Patton & Fowkes (1967) adsorption theory of lipid droplet secretion. This rapid Golgi vesicle fusion could also isolate areas of cytoplasm without an included lipid droplet (Fig. 33); pieces of cytoplasm without a lipid droplet are found in goats but have not been reported from other animals (Wooding et al., 1970).

The mechanism of release of the lipid (with or without an area of cytoplasm) is certainly very different from the release of the casein micelles which is generally accepted as occurring by exopinocytosis. There has been considerable discussion regarding nomenclature for lipid release. Bargmann and his group (Bargmann & Welsch, 1969) point out that "apocrine secretion" is a term which was once used to describe the entire process of absorption of precursors, synthesis of product and extrusion of that product; for this reason they think it better not to use the term at all. However Kurosumi (1961) has updated its use and produced a simple but informative classification for the various forms of secretion. Secretion is defined narrowly as the way in which the material elaborated for a specific purpose leaves the cell. In holocrine secretion the whole cell becomes the secretion, in apocrine secretion part of the cell is lost with the

---

FIGS 34 and 35. Transverse (Fig. 34) and surface (Fig. 35) views of the hexagonal pattern found in the initial milk fat globule membrane. This has only been observed after breakdown of the continuous unit membrane (inset Fig. 34) which surrounds the globule immediately after it has been secreted.
Fig. 34, Cow ×111 000; inset in centre horse ×140 000. Fig. 35, Cow ×111 000.

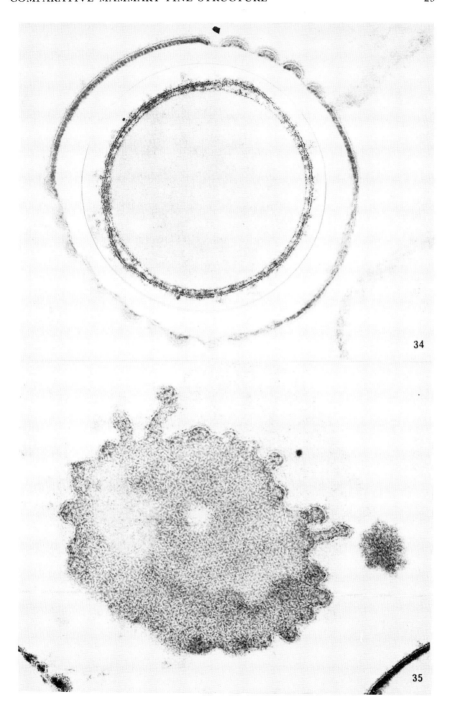

34

35

secretion and in merocrine secretion only the secretory product is released from the cell.

The importance of the secretion of cytoplasmic pieces into the milk is that they may still be biochemically active. Christie & Wooding (1975) and Christie, Vernon & Wooding (1976) have shown that, in goat milk, the lipid droplets and their associated cytoplasm can easily be isolated by virtue of their anomalous flotation properties (brought about by the included lipid). Such particles are the site of the triglyceride-synthesizing ability possessed by fresh goat milk (Figs 36, 38 and 39). The particles have most of the biosynthetic capability of intact isolated mammary cells or tissue slices, producing for example, triglycerides of the same composition as those normally present in the milk; moreover, they incorporate $[^{14}C]$-glucose into lactose and $[^3H]$-leucine into protein. This triglyceride-synthesizing ability has also been reported in fresh pig (Keenan & Colenbrander, 1969) and cow milk (McCarthy & Patton, 1964), and was considered to be due to the presence of free microsomes released into the milk from the mammary gland by rupture of the apical plasmalemma. However, there is no morphological evidence for such a release and it would seem more likely that particles similar to those found in the goat were responsible for the enzymic activity.

Membrane Material

In addition to the discrete pieces of cytoplasm, milk also contains "membrane material" or "fluff". There seem to be three possible sources for this. Firstly, by loss of membrane from the milk fat globule after its secretion into the alveolus (Wooding, 1971b, 1974); secondly, by the sloughing of microvilli and/or the apical plasmalemma from the secretory cell during normal secretion (Stewart, Puppione & Patton, 1972), and thirdly, membrane from degenerating white cells which migrate into the milk from the blood through the mammary epithelium (Anderson, Brooker, Andrews & Alichanidis, 1974). The last source is usually minor unless the udder is mastitic; when the white cell count is enormously increased its membrane contribution predominates (Anderson, Brooker, Andrews & Alichanidis, 1975).

The breakdown of the milk fat globule membrane has been well documented (Wooding, 1971b, 1972, 1974, 1975) (Fig. 34) but there is no direct morphological evidence from perfusion-fixed tissue for any sloughing or loss of the apical plasmalemma. Patton has suggested (Patton & Jensen, 1975) that a much greater area of membrane is added to the

---

FIGS 36 and 37. Fragments of goat secretory epithelial cell cytoplasm, each usually with included lipid droplet, found in freshly secreted goat milk. Fig. 37 is a sample from a goat cream layer; the swollen endoplasmic reticulum (E) is usually found; the particles in Fig. 36 were isolated by calcium precipitation (see Christie & Wooding, 1975).
Fig. 36, ×2600. Fig. 37, ×4300.

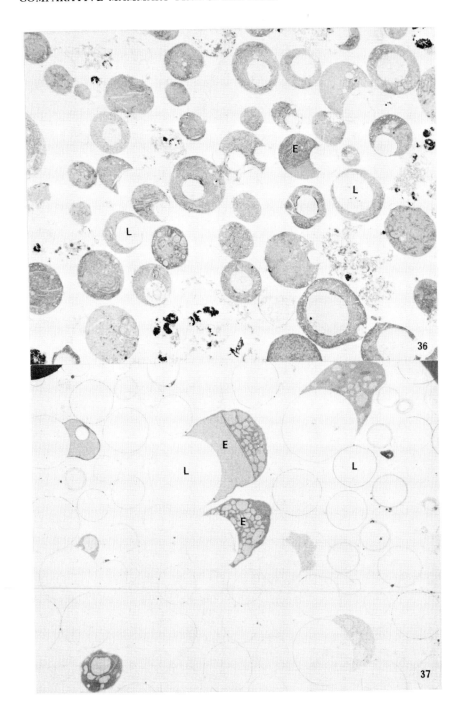

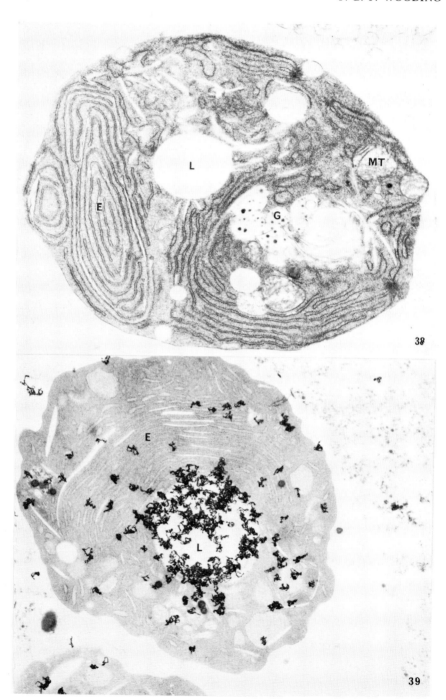

plasmalemma by Golgi vesicle fusion than is lost by lipid droplet extrusion, and that it would therefore be necessary for the excess membrane to
be sloughed off in order to maintain the balance. Patton calculated the
total Golgi vesicle membrane area from the average Golgi vesicle diame-
ter on electron micrographs and the total volume of the milk (minus
lipid). Since there is no evidence that all the aqueous phase of the milk
arises from the Golgi vesicles the calculation is of doubtful validity. In
addition, it has been shown that in other secretory systems where there is
considerable exocytosis, such as the parotid gland or pancreas, cellular
mechanisms exist for recycling the "excess" plasma membrane back into
the cell by pinocytosis.

AQUEOUS PHASE OF MILK

Although milk is isosmotic with plasma it is considerably different in
composition. Linzell & Peaker (1971) suggested that with lactose as the
major osmotic component and with the measured electrical potential and
concentration differences between secretory cell cytoplasm, blood and
milk, the cationic concentrations in milk, mammary tissue and blood
could be accounted for if there was a sodium–potassium activated ATPase
located on the basolateral surfaces of the mammary cell. This would
utilize metabolic energy to pump sodium into the intercellular space and
potassium into the cell. It is also necessary to postulate an impermeable
seal between the cells and possibly a chloride pump on the apical membrane. Kinura (1969) has demonstrated an ATPase localization by electron microscopic histochemistry equivalent to that postulated by Linzell &
Peaker; this localization has been confirmed by the use of specific sodium–
potassium ATPase inhibitors in mammary glands of both the sheep and
goat (Johnson & Wooding, unpublished observations) (Fig. 41).

In some species, like the goat, there is no evidence in the lactating
gland that any fluid passes from blood to milk via the intercellular spaces
of the alveoli or ducts (see the chapter in this volume by Peaker). The cells
are sealed around their apices by the presence of a tight junction (Figs 42
and 43). The permeability of the junction depends upon the number of
lines of fusions between the two cell plasma membranes (Claude &
Goodenough, 1973). In the lactating cow it can be seen in frozen-etched
tissue that the apical junction is of the "tight" variety (Fig. 43). It has been
shown (Pitelka, Hamamoto, Duafala & Nemanic, 1973) that in the mouse
there is a gradual increase of this junctional complexity during alveolar

FIGS 38 and 39. Detail of cytoplasmic particles from goat milk. In Fig. 38 the particle can be
seen to contain rough endoplasmic reticulum, Golgi vesicles and lipid drops, in proportions
characteristic of secretory cell cytoplasm. The particle in Fig. 39 was incubated with ^{3}H-oleic
acid and processed for electron microscope autoradiography. The majority of the label is
over the lipid droplet and it has been shown biochemically that the oleic acid is incorporated
into triglyceride by the particles (Christie & Wooding, 1975).
Fig. 38, ×15 000. Fig. 39, ×10 500.

development in pregnancy. In early pregnant and non-lactating glands there is no "blood–milk barrier", so the development of tight junction structure would correlate with the decrease in permeability that occurs at about the time of parturition (Linzell & Peaker, 1974).

INVOLUTION

On weaning or the cessation of removal of milk, the mammary gland becomes swollen with milk. Lipid droplets and casein-containing vesicles accumulate in the cytoplasm but, at least initially, during the first day there is no effect on the rate of synthesis of protein in tissue isolated from the gland (Saacke & Heald, 1974). Subsequently, there is a great increase in the enzymes of intracellular breakdown (Fig. 6k, l) correlated with an increase in lysosomal bodies in the cytoplasm (see Hollmann, 1974). The endoplasmic reticulum breaks down into vesicles and the Golgi apparatus reduces to a few flattened membranes. The cytoplasmic organelles, together with the unsecreted lipid droplets and casein micelles, are segregated and broken down in intracellular vacuoles or autophagosomes. Alternatively, the whole secretory cell may become densely staining, with a shrunken nucleus, and then eventually be lost into the alveolus. The reduction of cytoplasmic content by autodigestion in autophagic vacuoles and the death of whole cells result in a great decrease in the volume of the mammary secretory parenchyma. This process is aided by contraction of the myoepithelial cells, which have not been reported to decrease or show signs of degeneration.

Removal of the residual lipid droplets, dead cells and casein from the alveoli is accomplished by cells of the monocyte series, including macrophages, which are present at all stages of lactation within the epithelium of the mammary gland (Figs 44 and 45) in cows, goats, pigs and sheep (Wooding, unpublished results) and in rats (Helminen & Ericsson, 1968). These probably function during lactation by removing any cell debris or cells which may die and need to be eliminated. During the later phases of involution there is probably also an influx of macrophages from the blood. When they become engorged with lipid they are probably identical with the "foam cells" or corpuscles of Donné which are found at several

---

FIG. 40. Goat secretory epithelial cells with grossly swollen endoplasmic reticulum cisternae (E). There are two intracellular vacuoles (arrows) formed by fusion of Golgi vesicles. They contain casein and lipid droplets with an external membrane identical to the milk fat globule membrane. It has been demonstrated by serial sectioning that similar vacuoles are not connected to the alveolus (Wooding, 1973). The conditions necessary for such vacuole formation and endoplasmic reticulum distension have not yet been defined.
Fig. 40, Goat ×7000.

FIG. 41. Cow secretory epithelium after incubation by the Borgers, Schaper & Schaper (1971) method to demonstrate the sodium–potassium ATPase enzyme. Reaction product outlines the lateral and basal plasmalemmae. Section not stained on the grid.
Fig. 41, Cow ×4900.

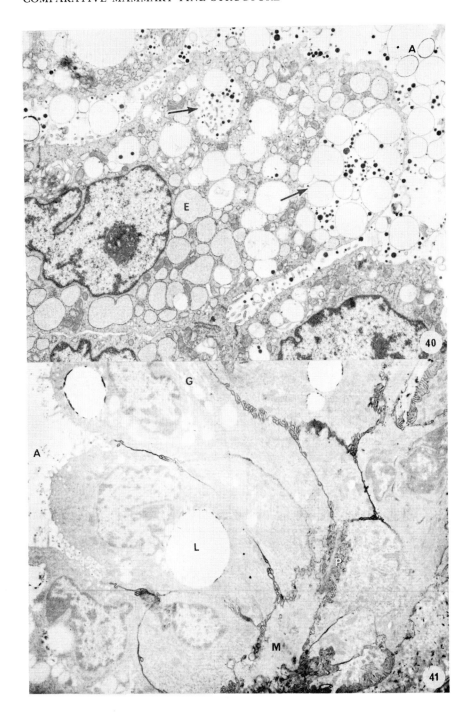

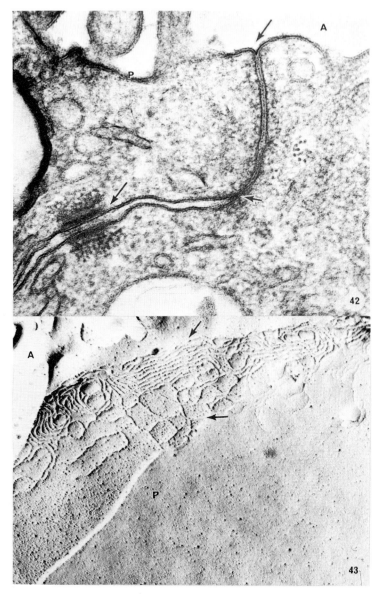

FIGS 42 and 43. Structure of the tight junction between cow secretory epithelial cells seen in section (Fig. 42) and *en face* after freeze fracturing (Fig. 43). Equivalent regions are shown between the arrow heads. A desmosome (arrow) can be seen on Fig. 42, there is no similar structure visible on Fig. 43. In the freeze-fractured tight junction, the number of horizontal lines and their closeness of packing indicates that the junction is physiologically non-leaky (see Claude & Goodenough, 1973).
Fig. 42, Cow ×58 000. Fig. 43, Cow ×41 000.

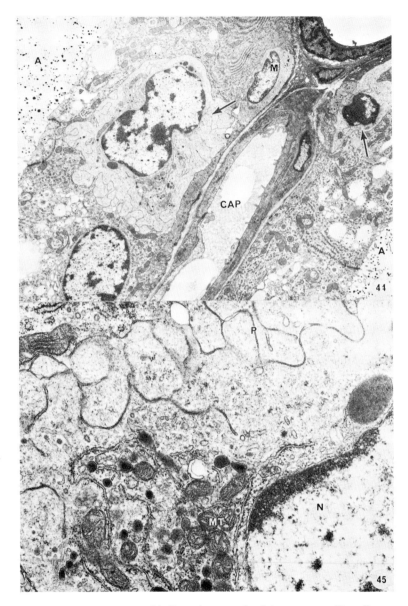

FIGS 44 and 45. Cow secretory epithelium. An example of the monocyte-like cells common
to all mammary epithelia. Large and small examples (arrows). The large cell shows typically
tortuous channels of the plasmalemma (P Fig. 45), lightly staining cytoplasm and clusters of
organelles near the nucleus. Note the capillary (CAP) between the alveoli.
Fig. 44, Cow ×4000. Fig. 45, (detail of Fig. 44) ×20 600.

stages in the mammary gland's activity. There are also suggestions that such corpuscles originate from the degeneration of secretory epithelial cells (Hollmann, 1974).

The process of involution results in very small alveoli containing a few cells; these cells have few organelles plus the occasional residual autophagosome. This is the "resting stage" and the gland shows a large increase in volume of the stroma and/or fat cells. This resting stage is found in the "dry period", which in cows is essential for the milk yield to be maintained in the next lactation. If the next pregnancy occurs after less than a six-week dry period up to 50% less milk may be produced. The reasons for this are unknown, and there is little change in the ultrastructure of the cell from the end of involution to the start of pregnancy.

ACKNOWLEDGEMENT

I would like to register my debt to Dr J. L. Linzell for his interest, stimulation and encouragement, and also for his generous friendship. Many of the micrographs in this article were produced from material obtained with his help, and any deficiencies in them are always mine and not his.

REFERENCES

Anderson, M., Brooker, B. E., Andrews, A. T. & Alichanidis, E. (1974). Membrane material in milk of mastitic and normal cows. *J. Dairy Sci.* **57**: 1448–1458.

Anderson, M., Brooker, B. E., Andrews, A. T. & Alichanidis, E. (1975). Membrane material from bovine skim milk from udder quarters infused with endotoxin and pathogenic organisms. *J. Dairy Sci.* **42**: 401–417.

Anderson, W. A., Trantalis, J. & Kang, Y. (1975). Ultrastructural localisation of endogenous mammary gland peroxidase during lactogenesis in the rat. *J. Histochem. Cytochem.* **23**: 295–302.

Baldwin, R. L. & Yang, Y. T. (1974). Enzymatic and metabolic changes in the development of lactation. In *Lactation* **1**: 399–414. Larson, B. L. & Smith, V. R. (eds). New York and London: Academic Press.

Bargmann, W. & Welsch, U. (1969). On the ultrastructure of the mammary gland. In *Lactogenesis: the initiation of milk secretion at parturition*: 43–52. Reynolds, M. & Folley, S. J. (eds). Philadelphia: University of Pennsylvania Press.

Bauman, D. E. & Davis, C. L. (1974). Biosynthesis of milk fat. In *Lactation* **2**: 31–76. Larson, B. L. & Smith, V. R. (eds). New York and London: Academic Press.

Baumrucker, C. R. & Keenan, T. W. (1975). Membranes of the mammary gland. X. ATP dependent calcium accumulation by Golgi apparatus fractions from bovine mammary gland. *Expl Cell Res.* **90**: 253–260.

Birkinshaw, M. & Falconer, I. R. (1972). The localisation of prolactin labelled with radioiodide in rabbit mammary tissue. *J. Endocr.* **55**: 323–324.

Borgers, M., Schaper, J. & Schaper, W. (1971). Localisation of specific phosphatase activities in canine coronary blood vessels and heart muscle. *J. Histochem. Cytochem.* **19**: 526–539.

Bourne, F. J. & Curtis, F. J. (1973). The transfer of immunoglobulins IgG, IgA and IgM from serum to colostrum and milk in the sow. *Immunology* **24**: 157–162.

Butler, J. E. (1974). Immunoglobulins of mammary secretions. In *Lactation* **3**: 217–257. Larson, B. L. & Smith, V. R. (eds). New York and London: Academic Press.

Ceriani, R. L. (1970). Fetal mammary gland differentiation *in vitro* in response to hormones. I: Morphological findings. *Devl Biol.* **21**: 506–529.

Chatterton, R. T., Harris, J. A. & Wynn, R. M. (1975). Lactogenesis in the rat. An ultrastructural study of the initiation of the secretory processes. *J. Reprod. Fertil.* **43**: 479–486.

Christie, W. W. & Wooding, F. B. P. (1975). The site of triglyceride biosynthesis in milk. *Experientia* **31**: 1445–1447.

Christie, W. W., Vernon, R. G. & Wooding, F. B. P. (1976). Lipid metabolism in cytoplasmic droplets from freshly secreted milk. *Biochem. Soc. Trans.* **4**: 242–243.

Claude, P. & Goodenough, D. A. (1973). Fracture faces of zonula occludens from "tight" and "leaky" epithelia. *J. Cell Biol.* **58**: 390–400.

Croom, W. J., Collier, R. J., Bauman, D. E. & Hays, R. L. (1976). Cellular studies of mammary tissue from cows hormonally induced into lactation. Histology and ultrastructure. *J. Dairy Sci.* **59**: 1232–1246.

Dixon, F. J., Weigle, W. O. & Vasquez, J. J. (1961). Metabolism and mammary secretion of serum proteins in the cow. *Lab. Invest.* **10**: 216–237.

Franke, W. W., Luder, M., Kartenbeck, J., Zerban, H. & Keenan, T. W. (1976). Involvement of vesicle coat material in casein secretion and surface regeneration. *J. Cell Biol.* **69**: 173–195.

Gardner, D. G. & Wittliff, J. L. (1973). Characterisation of a distinct glucocorticoid binding protein in the lactating mammary gland of the rat. *Biochim. biophys. Acta* **320**: 617–627.

Hales, C. N., Luzio, J. P., Chandler, A. & Herman, L. (1974). Localisation of calcium in the smooth endoplasmic reticulum of rat isolated fat cells. *J. Cell Sci.* **15**: 1–15.

Helminen, H. J. & Ericsson, J. L. E. (1968). Studies on mammary gland involution. *J. Ultrastruct. Res.* **25**: 193–252.

Heuser, J. E. & Reese, T. D. (1973). Evidence for recycling of synaptic vesicle membrane during transmitter release at the frog neuromuscular junction. *J. Cell Biol.* **57**: 315–344.

Hollmann, K. H. (1974). Cytology and fine structure of the mammary gland. In *Lactation* **1**: 3–96. Larson, B. L. & Smith, V. R. (eds). New York and London: Academic Press.

Jenness, R. (1974). The composition of milk. In *Lactation* **3**: 3–108. Larson, B. L. & Smith, V. R. (eds). New York and London: Academic Press.

Keenan, T. W. & Colenbrander, V. F. (1969). Biosynthesis of glycerides in freshly secreted sows milk. *Lipids* **4**: 168–170.

Kihm, V. (1973). Immunohistochemische Darstellung der Casein und β lactoglobulin synthese in der Rinder Milchdruse. *Histochemie* **35**: 273–281.

Kinura, T. (1969). A study of the mechanism of milk secretion. *J. Jap. Obstet. Gynaec. Soc.* **21**: 301–308.

Kraehenbuhl, J. P., Galardy, R. E. & Jamieson, J. D. (1974). Preparation and characterisation of an immunoenzyme tracer consisting of a haemeoctapeptide coupled to Fab. *J. exp. Med.* **139**: 208–223.

Kurosumi, K. (1961). An electron microscopic analysis of the secretory mechanism. *Int. Rev. Cytol.* **11**: 1–61.

Kurosumi, K., Kobayashi, Y. & Baba, N. (1968). The fine structure of mammary glands of lactating rats with special reference to the apocrine secretion. *Expl Cell Res.* **50**: 171–192.

Linzell, J. L. & Peaker, M. (1971). Mechanism of milk secretion. *Physiol. Rev.* **51**: 564–597.

Linzell, J. L. & Peaker, M. (1974). Changes in colostrum composition and in the permeability of the mammary epithelium at about the time of parturition in the goat. *J. Physiol., Lond.* **243**: 129–151.

McCarthy, R. D. & Patton, S. (1964). Biosynthesis of glycerides in freshly secreted milk. *Nature, Lond.* **202**: 347–348.

Mellenberger, R. W., Bauman, D. E. & Nelson, D. R. (1973). Metabolic adaptation during lactogenesis. Fatty acid and lactose synthesis in cow mammary tissue. *Biochem. J.* **136**: 741–748.

Mills, E. S. & Topper, Y. J. (1970). Some ultrastructural effects of insulin, hydrocortisone and prolactin on mammary gland explants. *J. Cell Biol.* **44**: 310–328.

Moxon, L. A., Wild, A. E. & Slade, B. S. (1976). Localisation of proteins in coated micropinocytotic vesicles during transport across rabbit yolk sac endoderm. *Cell Tissue Res.* **171**: 175–193.

Oka, T. & Topper, Y. J. (1972). Dynamics of insulin action on mammary epithelium. *Nature, Lond.* **239**: 216–217.

Patton, S. & Fowkes, F. M. (1967). The role of the plasmalemma in the secretion of milk fat. *J. theor. Biol.* **15**: 274–281.

Patton, S. & Jensen, R. G. (1975). Lipid metabolism and membrane functions of the mammary gland. In *Progress in the chemistry of fats and other lipids* **14** (No. 4). Holman, R. T. (ed.). Oxford: Pergamon Press.

Pitelka, D. R., Hamamoto, S. T., Duafala, J. G. & Nemanic, M. K. (1973). Cell contacts in the mouse mammary gland I. Normal gland in postnatal development and the secretory cycle. *J. Cell Biol.* **56**: 797–818.

Raynaud, A. (1971). Foetal development of the mammary gland and hormonal effects on its morphogenesis. In *Lactation*: 3–29. Falconer, I. R. (ed.). London: Butterworths.

Reid, I. M. & Chandler, R. L. (1973). Ultrastructural studies on the bovine mammary gland with particular reference to glycogen distribution. *Res. vet. Sci.* **14**: 334–340.

Rodewald, R. (1973). Intestinal transport of antibodies in the new born rat. *J. Cell Biol.* **58**: 189–211.

Saacke, R. G. & Heald, C. W. (1974). Cytological aspects of milk formation and secretion. In *Lactation* **2**: 147–190. Larson, B. L. & Smith, V. R. (eds). New York and London: Academic Press.

Stein, O. & Stein, Y. (1967). Lipid synthesis, intracellular storage and secretion. II. Electronmicroscope autoradiographic study of the mouse lactating mammary gland. *J. Cell Biol.* **34**: 251–264.

Stewart, P. S., Puppione, D. L. & Patton, S. (1972). The presence of microvilli and other membrane fragments in the nonfat phase of bovine milk. *Z. Zellforsch. mikrosk. Anat.* **123**: 161–167.

Tan, W. C., Young, S., Goldsmith, I. J. & Livingston, D. C. (1973). The biochemical and ultrastructural localisation of phosphoprotein containing particles in the lactating mammary gland of rats. *Acta Endocr.* **72**: 25–32.

Tobon, H. & Salazar, H. (1974). Ultrastructure of the human mammary gland 1. Development of the fetal gland through gestation. *J. Clin. Endocr. Metab.* **39**: 443–456.

Topper, Y. J. & Oka, T. (1974). Some aspects of mammary development in the mature mouse. In *Lactation* **1**: 327–348. Larson, B. L. & Smith, V. R. (eds). New York and London: Academic Press.

Wooding, F. B. P. (1971a). The mechanism of secretion of the milk fat globule. *J. Cell Sci.* **9**: 805–821.

Wooding, F. B. P. (1971b). The structure of the milk fat globule membrane. *J. Ultrastruct. Res.* **37**: 388–400.

Wooding, F. B P. (1972). Milk microsomes, viruses and the milk fat globule membrane. *Experientia* **28**: 1077–1079.

Wooding, F. B. P. (1973). Formation of the milk fat globule membrane without participation of the plasmalemma. *J. Cell Sci.* **13**: 221–235.

Wooding, F. B. P. (1974). Milk fat globule membrane material in skim milk. *J. Dairy Res.* **41**: 331–337.

Wooding, F. B. P. (1975). Ultrastructural aspects of the formation and fate of the milk fat globule membrane. In *Annual Bulletin, International Dairy Federation Document No. 86; proceedings of the lipolysis symposium, Cork, 5–7 March,* 1975: 7–11. Brussels.

Wooding, F. B. P. & Kemp, P. (1975). High melting point triglycerides and the milk fat globule membrane. *J. Dairy Res.* **42**: 419–426.

Wooding, F. B. P., Peaker, M. & Linzell, J. L. (1970). Theories of milk secretion: evidence from the electron microscopic examination of milk. *Nature, Lond.* **226**: 762–764.

Symp. zool. Soc. Lond. (1977) No. 41, 43–55

Comparative Aspects of Milk Fat Synthesis

R. DILS, SUSAN CLARK and J. KNUDSEN

Department of Physiology and Biochemistry, University of Reading, Whiteknights, Reading, England, Department of Biochemistry, University of Leeds, Leeds, England and Institute of Biochemistry, University of Odense, Denmark

SYNOPSIS

The fatty acid composition of milk triacylglycerols (triglycerides) varies considerably between species. Part of this variation arises from the differing pattern of fatty acids synthesized within the lactating mammary glands of these species. For example, this tissue from guinea-pig only synthesizes long-chain ($C_{14:0}$–$C_{18:1}$) fatty acids whereas rabbit and rat mammary gland synthesize predominantly medium-chain ($C_{8:0}$–$C_{12:0}$) fatty acids.

This variation can be explained, in part, by the occurrence in rabbit and rat mammary gland of a novel enzyme which terminates fatty acid elongation at medium-chain acids. An unidentified factor which is heat stable is also involved in this chain termination process. The pattern of milk fatty acids synthesized by a number of species strongly suggests that these chain termination factors may have widespread occurrence.

An attempt is made to summarize the factors which have been postulated to control the fatty acid composition of milk fat. Speculations on the correlation between these factors and the physicochemical properties of milk triacylglycerols are discussed.

INTRODUCTION

The fatty acid composition of milk triacylglycerols (triglycerides) varies considerably between species (see Morrison, 1970, for review). The fatty acid composition for a particular species is the overall result of the types of fatty acids taken up by the gland and of the types of fatty acids synthesized *de novo* within the gland. In both cases, the fatty acids may be further transformed within the gland by desaturation, elongation etc.

One of the most interesting aspects of this variation is the occurrence in many species of high proportions of medium-chain fatty acids in milk fat (Fig. 1). An extreme example is elephant milk which contains 13% of octanoic ($C_{8:0}$), 67% of deconoic ($C_{10:0}$) and 15% of dodecanoic ($C_{12:0}$) fatty acids (McCullagh, Lincoln & Southgate, 1969). Amongst laboratory animals, this species specificity is exemplified by rat milk triacylglycerols which contain up to 40% of $C_{8:0}$–$C_{12:0}$ acids and rabbit milk triacylglycerols which contain up to 70% of $C_{8:0}$ plus $C_{10:0}$ acids. By contrast, milk from guinea-pigs maintained on the same diet as rabbits contains negligible proportions of these acids (Smith, Watts & Dils, 1968; Hall,

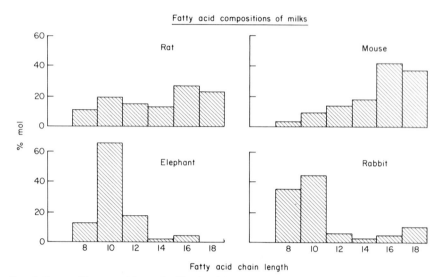

FIG. 1. Fatty acid compositions of milk triacylglycerols. The data is taken from Smith, Watts & Dils (1968) and from McCullagh *et al.* (1969).

1971). Human milk also contains about 14% of $C_{10:0}$ plus $C_{12:0}$ acids which appear between days 3 and 9 post-partum (Watts & Dils, 1968).

All the available evidence suggests that these medium-chain fatty acids are synthesized *de novo* within the lobulo-alveolar tissue of the mammary gland. It is unlikely that any significant proportion of these milk fatty acids are derived from blood lipids; medium-chain acids of dietary origin would normally be oxidized in the liver, and all other tissues of the body which are likely to contribute towards circulating lipids synthesize predominantly long-chain (i.e. $C_{16:0}$ and longer) fatty acids.

This article will therefore concentrate on the unique ability of the mammary gland to synthesize these unusual fatty acids. Though many species synthesize these milk fatty acids during lactation, there are severe problems with many species in obtaining regular supplies of mammary biopsy samples for studies *in vitro*. Amongst laboratory animals, the lactating rabbit and rat have been the most widely studied species. For convenience, the results obtained from experiments with the former will be discussed in detail and they will then be compared with those obtained using the lactating rat.

THE SYNTHESIS OF OCTANOIC $(C_{8:0})$ AND DECANOIC $(C_{10:0})$ FATTY ACIDS BY RABBIT MAMMARY GLAND

When lactating rabbits are injected with ^{14}C-acetate, the mammary gland synthesizes predominantly $C_{8:0}$ and $C_{10:0}$ fatty acids *in vivo* (Carey & Dils,

1972). Similarly, mammary slices from lactating rabbits synthesize these acids almost exclusively from ^{14}C-acetate plus glucose (Strong & Dils, 1972a). The extraordinary simplicity of this pattern of products (which incidentally shows that the gland neither elongates nor desaturates these products to any significant extent) makes this an ideal tissue for study.

An important question is: when does the mammary gland first acquire the ability to synthesize these acids? Since they are characteristic products of the differentiated mammary epithelial cells, the answer should give some insight into the time-scale of differentiation in the gland. With this in mind, the pattern of fatty acids synthesized by rabbit mammary tissue (Fig. 2) has been measured during pregnancy and early lactation (Strong

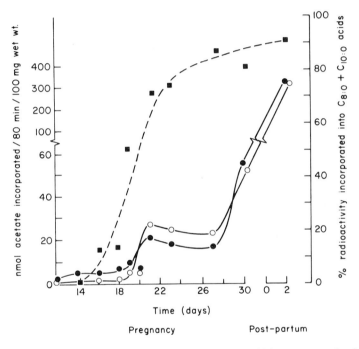

FIG. 2. The rate and pattern of fatty acids synthesized by rabbit mammary gland during pregnancy and early lactation. Mammary tissue from rabbits at different stages of pregnancy and at early lactation. (●–●), incubated with $0 \cdot 1$ mmol/l sodium $[1-{}^{14}C]$-acetate plus 1 mmol/l glucose; (○–○), incubated with 1 mmol/l sodium $[1-{}^{14}C]$-acetate plus 10 mmol/l glucose (these latter results have been divided by 10 to fit on scale); (■–■), % radioactivity incorporated into $C_{8:0}$ plus $C_{10:0}$ fatty acids. For further details, see Strong & Dils (1972b).

& Dils, 1972b). Up to day 18 of pregnancy, long-chain ($C_{14:0}$–$C_{18:1}$) fatty acids are the major products as they are in other mammalian tissues. From day 18 to day 22 of pregnancy, there is an increased rate of fatty acid synthesis due almost exclusively to the synthesis of $C_{8:0}$ and $C_{10:0}$ acids. At or around parturition, there is a second increase in lipogenesis but the pattern of fatty acids synthesized does not change. This initiation of milk fat synthesis between days 18 and 22 coincides with a decreased plasma concentration of progesterone and an increased concentration of plasma glucocorticoids (Denamur, 1971). During this period the gland undergoes intense morphological differentiation (Bousquet, Fléchon & Denamur, 1969).

HORMONAL CONTROL OF MILK FAT SYNTHESIS USING ORGAN CULTURE TECHNIQUES

Previous work has shown the usefulness of mammary-gland explants maintained in organ culture for studying the hormonal regulation of the synthesis and accumulation of milk proteins and lactose which occurs during pregnancy (see Forsyth, 1971). This technique is therefore ideal to study the induction of milk fat synthesis during this critical and co-ordinated period of differentiation in rabbit mammary gland.

Insulin, a glucocorticoid and prolactin are the minimum hormonal requirements for the stimulation of milk synthesis in mammary explants from mid-pregnant mice maintained in organ culture (see Topper, 1970). Mammary explants from 16-day pregnant rabbits (i.e. before the critical period of differentiation) have therefore been cultured with insulin (5 μg/ml), corticosterone or cortisol (1 μg/ml) and prolactin (1 μg/ml). Within 20 hours in culture, the rate of fatty acid synthesis increases 30 to 40-fold to values observed in early lactation (Fig. 3a). This is accompanied by an increased proportion of medium-chain fatty acids synthesized, though this takes longer (48 h) to reach a maximum of about 50% of the

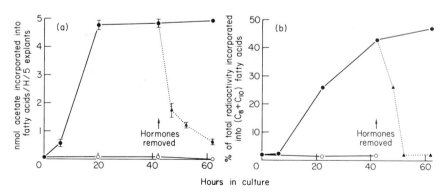

FIG. 3. The rate and pattern of fatty acid synthesis by mammary explants from 16-day pregnant rabbits cultured with insulin, cortisol and prolactin. (a) The rate of fatty acid synthesis; (b) the proportion of $C_{8:0}$ and $C_{10:0}$ fatty acid synthesized. (●–●), cultured with insulin, cortisol and prolactin; (○–○), cultured in the absence of hormones. For details, see Speake, Dils & Mayer (1975).

total fatty acids formed (Fig. 3b). These changes are not observed if prolactin is omitted from the hormone combination. Removal of the hormones from the medium after 20, 50 and 76 h in culture causes a precipitous reversal of these changes; within 5–10 h there is a decrease to very low values in the rate of fatty acid synthesis and in the proportion of medium-chain fatty acids formed (Forsyth, Strong & Dils, 1972; Speake, Dils & Mayer, 1975). These results show that the critical period of differentiation leading to the synthesis of milk fatty acids can be mimicked *in vitro* by culturing in the continuous presence of hormone combinations which contain a glucocorticoid and prolactin.

When these experiments are done with explants from rabbits beyond day 22 of pregnancy, they continue to synthesize predominantly medium-chain acids even in the absence of hormones (Speake, Dils & Mayer, 1976; Lynch & Dils, 1976). Since the explants do not revert to synthesizing long-chain fatty acids, differentiation appears to be complete by day 22 of pregnancy and to be "irreversible".

FACTORS GOVERNING THE SYNTHESIS OF MEDIUM-CHAIN FATTY ACIDS

Medium-Chain Acylthioester Hydrolase

Fatty acids are synthesized *de novo* in tissues by fatty acid synthetase. This is a multicatalytic enzyme complex which catalyses the synthesis of saturated even-numbered fatty acids from acetyl-CoA, malonyl-CoA and NADPH. The complex has been isolated from many mammalian tissues, and in all cases the products it synthesizes are short- ($C_{4:0}$) and long-chain ($C_{14:0}$ and $C_{16:0}$) fatty acids. The proportions of these products depend upon the relative concentrations of acetyl-CoA and malonyl-CoA. The rate of formation of malonyl-CoA by acetyl-CoA carboxylase (EC 6.4.1.2) may also govern the overall rate of fatty acid synthesis *de novo* in tissues.

The fatty acid synthetase complex isolated from lactating rabbit mammary gland is similar to other mammalian synthetases in that it synthesizes predominantly short- and long-chain fatty acids (Carey & Dils, 1970). This pattern of products cannot therefore explain the synthesis of medium-chain acids by the gland *in vivo* and *in vitro*. A factor (or factors) must be present in the gland which interacts with the synthetase and terminates chain lengthening of the growing fatty acid chain at $C_{8:0}$ and $C_{10:0}$ acids.

A soluble and high molecular weight factor has been found in the gland which terminates fatty acid synthesis at these chain lengths (Strong, Carey and Dils, 1973; Carey & Dils, 1973). This factor has recently been purified from the cytoplasm of lactating rabbit mammary gland by Knudsen, Clark & Dils (1976) and shown to be a novel enzyme (medium-chain acylthioester hydrolase of mol. wt. 29 000). When the purified enzyme is added to a system which synthesizes short- and long-chain fatty acids from rate-limiting amounts of malonyl-CoA, the pattern of fatty acids synthesized changes to predominantly $C_{8:0}$ and $C_{10:0}$ acids. A high proportion of these acids is synthesized when the molar ratio of medium-chain hydrolase to synthetase is $4:1$ (Table I). The medium-chain hydrolase must therefore be able to interact directly with

TABLE I

Effect of purified medium-chain hydrolase on the proportions of fatty acids synthesized by fatty acid synthetase

Additions	Mole % distribution of radioactivity in fatty acids						
	$4:0$	$6:0$	$8:0$	$10:0$	$12:0$	$14:0$	$16:0$
None	38	12	5	3	6	30	6
Albumin	18	6	0	2	6	55	13
Hydrolase + albumin	29	9	26	28	6	2	0

The molar ratio of hydrolase to fatty acid synthetase was $4:1$. For details, see Knudsen *et al.* (1976).

the synthetase and so release medium-chain acids from the acyl carrier protein of the synthetase or from other leaving sites on the synthetase. This will effectively terminate chain elongation of fatty acids on the synthetase and so prevent the synthesis of long-chain fatty acids. Recent work has shown that the medium-chain hydrolase and the fatty acid synthetase present in the cytosol from lactating rabbit gland are immunologically distinct enzymes (Chivers, Knudsen & Dils, 1977).

Immunological techniques have been used to detect the stage of mammary development at which medium-chain hydrolase appears in the gland. The enzyme cannot be detected in the cytosol of mammary tissue from 13- to 17-day pregnant rabbits, but the enzyme is clearly present in the cytosol from 22-day pregnant and from 15-day lactating rabbits (Chivers *et al.*, 1977). This strongly supports the role of this enzyme in controlling the onset of milk-fat synthesis during differentiation. In addition, the enzyme cannot be detected immunologically in the cytosol from the liver of lactating rabbits. Neither can it be detected in the cytosol from lactating guinea-pig mammary gland. These results correlate with the fact that these two tissues do not synthesize medium-chain fatty acids *in vivo* (Carey & Dils, 1972; Strong & Dils, 1972a).

It will be interesting to investigate the accumulation of the medium-chain hydrolase during the critical period of differentiation in rabbit mammary gland *in vivo*, and whether the enzyme is "induced" when mammary explants from 16-day pregnant rabbits are cultured in the presence of prolactin. Once differentiation is complete *in vivo*, the enzyme appears to terminate fatty acid synthesis even when the differentiated tissue is cultured in the absence of prolactin.

Heat-Stable Factor(s)

Though the cytosol from lactating rabbit mammary gland contains medium-chain hydrolase, it retains its ability to terminate fatty acid synthesis at $C_{8:0}$–$C_{12:0}$ acids after prolonged centrifugation (E. M. Carey, pers. comm.), and after heating for 10 min at 70°C (B. Speake & R. Dils, unpublished work). These results indicate that there may be factor(s) in the cytosol other than medium-chain hydrolase which terminate fatty acid synthesis.

TABLE II

Effects of high-speed supernatant (HSS) on the proportions of fatty acids synthesized by microsomal plus supernatant fractions

Additions	Mole % distribution of radioactivity in fatty acids						
	4:0	6:0	8:0	10:0	12:0	14:0	16:0
None	79	9	4	5	2	1	0
Lactating rabbit mammary gland HSS, heated and acid-treated	32	3	2	55	5	1	1
Lactating rabbit mammary gland HSS, heated and freeze-dried	21	1	1	64	10	1	0
Lactating rabbit liver HSS	68	6	3	7	5	7	3
Lactating guinea-pig mammary gland HSS	64	6	3	12	8	6	1

The assay contained the microsomal plus supernatant fraction from lactating rabbit mammary gland and a system designed to synthesize predominantly short-chain ($C_{4:0}$) fatty acids. Where indicated, the HSS was heated for 10 min at 70°C and treated with 5% (w/v) perchloric acid, or heated for 10 min at 70°C and then freeze-dried (Clark *et al.*, in prep.).

Evidence for this factor(s) is shown in Table II (Clark, Knudsen & Dils, in prep.). The "high-speed" supernatant (HSS) from lactating rabbit mammary gland cytosol decreases the proportion of $C_{4:0}$ and increases the proportion of $C_{10.0}$ synthesized by dilute microsomal plus cytosol fractions from lactating rabbit mammary gland or liver. The HSS can be freeze-dried, or heated at 70°C for 10 min followed by treatment with perchloric acid to remove high molecular weight proteins, and still retain its capacity to terminate fatty acid synthesis. The HSS also withstands treatment with proteolytic enzymes. The equivalent HSS fractions prepared from lactating rabbit liver or from lactating guinea-pig mammary gland have little or no ability to terminate fatty acid synthesis in this assay.

The chemical nature of this factor(s) is unknown, as is its mechanism of involvement in chain termination. Nor is its role in controlling milk-fat synthesis during mammary differentiation understood; experiments using mammary tissue in organ culture similar to those described above for the medium-chain hydrolase should throw light on its mode of action.

COMPARATIVE ASPECTS OF CHAIN TERMINATION IN MAMMARY GLAND

Little work has been done with other species on the roles of medium-chain hydrolase and of the heat-stable factor(s) in controlling chain termination in mammary gland. Libertini, Lin & Smith (1976) and Smith, Agradi, Libertini & Dilpeen (1976) have recently confirmed the presence of an enzyme of mol. wt. 32 000 in the cytosol from lactating rat mammary gland. This enzyme seems similar or identical to the medium-chain hydrolase isolated from lactating rabbit mammary gland. In addition, Smith & Abraham (1975) have found evidence for a heat-stable factor in lactating rat mammary gland which is involved in milk fat synthesis.

The high proportions of medium-chain fatty acids found in the milk of many species may indicate the widespread occurrence of these enzymes and factors in mammary gland. It is most interesting in this context that cow mammary gland (Table III) synthesizes increasing proportions of

TABLE III

Fatty acids synthesized by cow mammary gland

Days from parturition	Mole % of fatty acids synthesized							
	C_4	C_6	C_8	C_{10}	C_{12}	C_{14}	C_{16}	C_{18}
−30	7	2	3	3	4	18	55	8
−7	8	5	6	7	7	20	43	4
+7	6	6	8	13	12	21	31	3
+40	2	4	5	13	16	26	31	3

For details, see Mellenberger *et al.* (1973).

both $C_{10:0}$ and $C_{12:0}$ fatty acids during mammary development between late pregnancy and early lactation (Mellenberger, Bauman & Nelson, 1973). By contrast, goat mammary gland (Table IV) synthesizes increasing proportions of $C_{10:0}$ during this period (Lynch & Dils, 1976). Human milk on the other hand, contains increasing proportions of $C_{10:0}$–$C_{14:0}$ acids between day 3 and day 9 post-partum (Watts & Dils, 1968). Considerable work is needed to understand the subtle chain-length specificity involved in chain termination which these examples demonstrate.

Table IV

Fatty acids synthesized by goat mammary gland

Days from parturition	Mole % of fatty acids synthesized							
	C_4	C_6	C_8	C_{10}	C_{12}	C_{14}	C_{16}	C_{18}
−80	8	8	6	5	4	18	38	13
−60	6	5	4	7	7	24	38	9
−5	5	6	7	14	5	25	35	3
+10	5	5	8	20	6	20	33	3
+40	7	5	6	19	6	14	40	3
+80	6	4	6	18	8	18	36	4

For details, see Lynch & Dils (1976).

THE OCCURRENCE OF BUTYRATE IN RUMINANT MILK TRIACYLGLYCEROLS

Ruminant milk triacylglycerols are characterized by the presence of significant proportions of esterified butyric acid ($C_{4:0}$). Part of this butyrate is synthesized within the lactating ruminant mammary tissue (see Tables III and IV). Though butyryl-CoA is synthesized by purified fatty acid synthetase from both ruminant and non-ruminant lactating mammary gland (Carey & Dils, 1970; Knudsen, 1972; Strong & Dils, 1972c), non-ruminant mammary gland does not esterify the butyryl-CoA into triacylglycerols. Since non-ruminant mammary gland has a high capacity to esterify butyrate into triacylglycerol *in vitro* (Knudsen, 1976), the absence of esterification *in vivo* may be due to the butyryl-CoA being preferred as a "primer" for fatty acid synthesis *de novo* in the gland. The mechanism of esterification of butyryl-CoA into triacylglycerols in lactating ruminant mammary gland is not understood. Recent evidence indicates that the microsomal diacylglycerol transacylases of ruminant mammary gland may play a key role in this process (Marshall & Knudsen, 1977).

THE PHYSIOLOGICAL SIGNIFICANCE OF BUTYRATE AND OF MEDIUM-CHAIN FATTY ACIDS IN MILK

The lactating mammary gland needs to synthesize triacylglycerols with a sufficiently low melting point that they can be readily secreted as liquid droplets or globules. Jenness (1974) has pointed out that the melting point of milk fat is lowered by an increase in the proportion of short- and medium-chain fatty acids, by an increased degree of unsaturation of the fatty acids, and by the asymmetric positioning of fatty acids on the glycerol backbone of triacylglycerols. This may help explain why ruminant mammary gland in particular synthesizes short-chain fatty acids, desaturates the predominantly saturated fatty acids arriving from the rumen, and positions fatty acids on the glycerol backbone (Table V) so as to make the molecule as asymmetric as possible (Jenness, 1974). These "manoeuvres" may then enable the ruminant to secrete milk triacylglycerols with a lower melting point than would be the case if it only synthesized triacylglycerols from fatty acids derived from the rumen.

TABLE V

Positional distribution of fatty acids in cow and goat milk triacylglycerols

Species	Position on glycerol backbone	4:0	6:0	8:0	10:0	12:0	14:0	16:0	18:0	18 unsat.
Cow*	1	—	—	—	—	3	11	41	15	21
	2	—	—	1	4	7	23	38	4	15
	3	54	24	5	5	—	—	5	2	8
Goat†	1	—	—	2	4	4	9	43	15	17
	2	—	—	1	6	5	21	33	6	20
	3	16	13	6	12	1	2	3	7	36

* For details, see Breckenridge & Kuksis (1969).   † For details, see Marai, Breckenridge & Kuksis (1969).

Non-ruminants appear to cope with this problem in a variety of ways. For example, lactating rabbit mammary gland has very little capacity to synthesize butyrate or to desaturate fatty acids *in vivo*. It does, of course, synthesize high proportions of medium-chain fatty acids which have relatively low melting points. By contrast, lactating guinea-pig mammary gland has a considerable capacity to desaturate fatty acids, but does not synthesize short- or medium-chain fatty acids (Strong & Dils, 1972a).

This line of reasoning could lead to satisfying correlations between the physicochemical properties of secreted milk triacylglycerols and the biosynthesis and metabolism of fatty acids in the lactating gland of different species. Nevertheless, it does not necessarily lead to an understanding of whether the diverse fatty acid compositions of milk triacylglycerols are related to the needs of the suckling young of different species.

It would be interesting, for example, to know whether there are any advantages to the suckling animal in being provided with milk which contains high proportions of short- and medium-chain fatty acids. These acids may be absorbed directly into the bloodstream, rather than forming part of chylomicron triacylglycerols. The circulating fatty acids may provide a readily utilizable form of energy for the liver etc., and could even be preferred substrates for oxidation in thermogenesis by "brown" adipose tissue.

It is evident that as much work needs to be done on the metabolism of milk fatty acids in the young animal as on the synthesis of milk fatty acids by the lactating mother.

REVIEWS

As a guide to further reading, there have recently been three excellent reviews of lipid metabolism in the mammary gland. Bauman & Davis (1974) have reviewed the literature (about 230 references) to 1971–72. The composition and biosynthesis of milk triacylglycerols has also been reviewed by Smith & Abraham (1975) who cover about 120 papers up to 1974. A somewhat broader area has been dealt with by Patton & Jensen (1976) who have reviewed nearly 400 papers to 1973, and list additional references up to 1975. The parts of these reviews dealing with milk fat synthesis have been updated and extended by Dils (1977).

ACKNOWLEDGEMENTS

We thank the Science Research Council of Great Britain for providing a studentship for S. Clark, who also received a Boehringer Mannheim Travelling Fellowship. We also thank the Wellcome Trust for providing travel grants to R. Dils.

REFERENCES

Bauman, D. E. & Davis, C. L. (1974). Biosynthesis of milk. In *Lactation* 2: 31–75. Larson, B. L. & Smith, V. R. (eds). New York and London: Academic Press.

Bousquet, M., Fléchon, J. E. & Denamur, R. (1969). Aspects ultrastructuraux de la glande mammaire de lapine pendent la lactogénèse. *Z. Zellforsch. mikrosk. Anat.* **96**: 418–436.

Breckenridge, W. C. & Kuksis, A. (1969). Structure of bovine milk fat triglycerides: II. Long chain lengths. *Lipids* **4**: 197–204.

Carey, E. M. & Dils, R. (1970). Fatty acid biosynthesis V. Purification and characterization of fatty acid synthetase from lactating rabbit mammary gland. *Biochim. biophys. Acta* **210**: 371–387.

Carey, E. M. & Dils, R. (1972). The pattern of fatty acid synthesis in lactating rabbit mammary gland studied *in vivo*. *Biochem. J.* **126**: 1005–1007.

Carey, E. M. & Dils, R. (1973). Fatty acid biosynthesis X. Specificity for chain termination of fatty acid biosynthesis in cell-free extracts of lactating rabbit mammary gland. *Biochim. biophys. Acta* **306**: 156–167.

Chivers, L., Knudsen, J. & Dils, R. (1977). Immunological properties of medium-chain acyl-thioester hydrolase and fatty acid synthetase from lactating-rabbit mammary gland. *Biochim. biophys. Acta* **487**: 361–367.

Clark, S., Knudsen, J. & Dils, R. (in prep.). *Heat-stable factor(s) in lactating-rabbit mammary gland involved in milk fat synthesis.*

Denamur, R. (1971). Hormonal control of lactogenesis. *J. Dairy Res.* **38**: 237–264.

Dils, R. (1977). Lipid metabolism in mammary gland. In *Lipid metabolism in Mammals* **2**: 131–144. Snyder, F. (ed.). New York: Plenum Press.

Forsyth, I. A. (1971). Organ culture techniques and the study of hormone effects on the mammary gland. *J. Dairy Res.* **38**: 419–444.

Forsyth, I. A., Strong, C. R. & Dils, R. (1972). Interactions of insulin, corticosterone and prolactin in promoting milk-fat synthesis by mammary explants from pregnant rabbits. *Biochem. J.* **129**: 929–935.

Hall, A. J. (1971). Fatty acid composition of rabbit (*Oryctolagus cuniculus*) milk fat throughout lactation. *Int. J. Biochem.* **2**: 414–418.

Jenness, R. (1974). Biosynthesis and composition of milk. *J. Invest. Dermatol.* **63**: 109–118.

Knudsen, J. (1972). Fatty acid synthetase from cow mammary gland tissue cells. *Biochim. biophys. Acta* **280**: 408–414.

Knudsen, J. (1976). Role of particulate esterification in chain length control of de novo synthesized fatty acids in liver and mammary gland. *Comp. Biochem. Physiol.* **53B**: 3–7.

Knudsen, J., Clark, S. & Dils, R. (1976). Purification and some properties of a medium-chain acyl-thioester hydrolase from lactating-rabbit mammary gland which terminates chain elongation in fatty acid synthesis. *Biochem. J.* **160**: 683–691.

Libertini, L., Lin, C. Y. & Smith, S. (1976). Isolation and properties of two different thioesterases from rat tissues. *Fedn Proc. Fedn Am. Socs exp. Biol.* **35**: 1671.

Lynch, E. P. J. & Dils, R. (1976). Differentiation of mammary gland during pregnancy: the response to hormones of rabbit and goat mammary gland explants in organ culture. *J. Endocr.* **68**: 32P.

Marai, L., Breckenridge, W. C. & Kuksis, A. (1969). Specific distribution of fatty acids in the milk fat triglycerides of goat and sheep. *Lipids* **4**: 562–570.

Marshall, M. O. & Knudsen, J. (1977). Biosynthesis of triacylglycerols which contain short-chain fatty acids by microsomal fractions from lactating cow mammary gland. *Biochem. Soc. Trans.* **5**: 285–287.

McCullagh, K. G., Lincoln, H. G. & Southgate, D. A. T. (1969). Fatty acid composition of milk fat of the African elephant. *Nature, Lond.* **222**: 493–494.

Mellenberger, R. W., Bauman, D. E. & Nelson, D. R. (1973). Metabolic adaptations during lactogenesis. Fatty acid and lactose synthesis in cow mammary gland. *Biochem. J.* **136**: 741–748.

Morrison, W. R. (1970). Milk lipids. In *Topics in lipid research* **1**: 51–106. Gunstone, F. D. (ed.). London: Logos Press.

Patton, S. & Jensen, R. (1976). *Biomedical aspects of lactation*. Oxford: Pergamon Press.

Smith, S. & Abraham, S. (1975). The composition and biosynthesis of milk fat. In *Advances in lipid research* **13**: 195–239. Paoletti, R. & Kritchevsky, D. (eds). New York and London: Academic Press.

Smith, S., Agradi, E., Libertini, L. & Dilpeen, K. N. (1976). Specific release of the thioesterase component of the fatty acid synthetase multienzyme complex by limited trypsinization. *Proc. natn. Acad. Sci., U.S.A.* **73**: 1184–1188.

Smith, S., Watts, R. & Dils, R. (1968). Quantitative gas-liquid chromatographic analysis of rodent milk triglycerides. *J. Lipid Res.* **9**: 52–57.

Speake, B. K., Dils, R. & Mayer, R. J. (1975). Regulation of enzyme turnover during tissue differentiation: Studies on the effects of hormones on the turnover of fatty acid synthetase in rabbit mammary gland in organ culture. *Biochem. J.* **148**: 309–320.

Speake, B. K., Dils, R. & Mayer, R. J. (1976). Effect of hormones on lipogenesis in mammary explants taken from rabbits at different stages of pregnancy and lactation. *Biochem. Soc. Trans.* **4**: 238–240.

Strong, C. R., Carey, E. M. & Dils, R. (1973). The synthesis of medium-chain fatty acids by lactating-rabbit mammary gland studied *in vitro. Biochem. J.* **132**: 121–123.

Strong, C. R. & Dils, R. (1972a). Fatty acids synthesized by mammary gland slices from lactating guinea-pig and rabbit. *Comp. Biochem. Physiol.* **40B**: 643–652.

Strong, C. R. & Dils, R. (1972b). Fatty acid synthesis in rabbit mammary gland during pregnancy and lactation. *Biochem. J.* **128**: 1303–1309.

Strong, C. R. & Dils, R. (1972c). The fatty acid synthetase complex of lactating guinea-pig mammary gland. *Int. J. Biochem.* **3**: 369–377.

Topper, Y. J. (1970). Multiple hormone interactions in the development of mammary gland *in vitro. Recent Progr. Horm. Res.* **26**: 287–308.

Watts, R. & Dils, R. (1968). Human milk: quantitative gas–liquid chromatographic analysis of triglyceride and cholesterol content during lactation. *Lipids* **3**: 471–476.

Symp. zool. Soc. Lond. (1977) No. 41, 57–75

Synthesis and Secretion of Milk Proteins

T. B. MEPHAM

Department of Physiology and Environmental Studies,
University of Nottingham, Faculty of Agricultural Science,
Sutton Bonington, Loughborough, England

SYNOPSIS

Current ideas on the mechanism of milk protein synthesis and secretion are reviewed. For the purpose of discussion the total process is considered as a sequence of discrete steps. Arteriovenous difference determinations and isotope studies indicate the extent of amino acid absorption by the mammary gland in relation to milk protein secretion and suggest the possibility that the supply of certain amino acids might determine milk composition. Studies on mammary DNA and RNA indicate that synthesis of milk-specific proteins conforms to the general scheme operative in other eukaryotic cells. Gene expression is dependent on hormonal induction, and recent data suggest possible intracellular mechanisms of hormone action. Synthesized milk proteins are vectored into the cisternal lumina of the endoplasmic reticulum. It is postulated that the proteins are transported through the lumina to the Golgi apparatus, within which specific interactions with other proteins occur, e.g. caseins aggregate to form micelles, and α-lactalbumin binds to Golgi membrane-bound galactosyltransferase to form lactose synthetase. Transport of protein to the apical membrane probably involves migration of vesicles, a process apparently dependent on the activity of microtubules. Release of protein into the alveolar lumina seems to occur predominantly by reverse pinocytosis. Serum proteins pass from the extracellular fluid to the alveolar lumina by both transcellular and paracellular routes. Most of the available data derive from investigations on very few species and it does not seem possible at present to make generalizations as to the source of the observed interspecific variations in milk protein composition.

INTRODUCTION

Proteins have been detected in the milks of all of the approximately 150 species which have so far been studied. The importance of milk proteins to the neonate derives from their nutritional and immunological properties, but the recent demonstration that one of them (α-lactalbumin) plays a key role in the regulation of processes on which the secretion of milk as a whole depends, gives added significance to the study of their synthesis and secretion. Despite their ubiquity there is considerable variation between species in both the absolute and relative concentrations of the different proteins, and also in their concentrations at different stages of lactation in an individual animal (for a review of the milk proteins of a large number of species, see Jenness, 1974).

Milk proteins are classifiable into two major groups, the caseins and non-casein (whey) proteins. Caseins are milk-specific proteins, characterized by the presence in their molecules of ester-bound phosphate, a high proline content and the manner in which they aggregate, in association with colloidal calcium phosphate, in complex "micelles". Bovine caseins are precipitated at pH 4·6 (the pI) and by the action of rennin, but neither this nor the other listed characteristics constitute an adequate definition of casein, applicable to all species. In bovine milk there are three main types, α_s, β and κ, each of which comprises a number of molecular species (see Thompson & Farrell, 1974). κ-Casein is a glycoprotein which stabilizes the micelle against precipitation by calcium ions. Rennin acts by splitting the polypeptide chain of κ-casein, thereby increasing the sensitivity of the other caseins to the action of calcium. The nutritional importance of the caseins is due to their high essential amino acid content and their capacity to bind calcium and phosphorus, the elements of which bone is largely composed. The highest and lowest recorded concentrations of casein in milk are, respectively, 19·7 g/100 ml, in the whitetail jackrabbit, and 0·4 g/100 ml in women (Jenness, 1974).

The non-casein proteins, which in bovine milk remain in solution when casein is precipitated at its pI or by rennin action, comprise both milk-specific and serum proteins. Milks of some species (for example rodents) contain very few specific proteins, while others (for example carnivores) contain many. The principal milk-specific non-casein proteins are α-lactalbumin and β-lactoglobulin. The crucial role of α-lactalbumin in the biosynthesis of lactose (discussed in Jones's paper in this volume) suggests that this protein is present in the milks of all species which secrete the disaccharide. β-Lactoglobulin, however, which is the quantitatively major whey protein of bovine milk, appears to be present only, but not invariably, in the milks of artiodactyls (see Jenness, 1974). Serum proteins are also present in milk, their concentration, particularly in the case of certain of the immunoglobulins, being especially high in colostrum of those species (which include cows, sheep and pigs) in which the structure of the placental membranes prevents their transfer from the maternal to the foetal plasma. Somewhat analogously, in marsupials the increasing requirements of the young for iron are met by an increase in the concentration of transferrin in the milk in late lactation (Tyndale-Biscoe, 1973).

In the elaboration of milk protein, precursors pass from the blood to the lumina of the mammary alveoli, in most cases being involved in a process of controlled polymerization within the mammary secretory cells. For the purposes of discussion it is convenient to consider the sequence of events as a series of discrete steps viz. (i) the passage of precursors from the blood capillary to the extracellular fluid (ECF), (ii) passage from the ECF to the interior of the secretory cell, (iii) the intracellular synthesis of the milk-specific proteins, (iv) intracellular translocation of the proteins to the luminal (apical) membrane of the cell, (v) passage of the proteins across the apical membrane into the alveolar lumina. It is also necessary to consider the possibility that proteins pass from the ECF to the lumina *between* adjacent secretory cells (the paracellular route). Furthermore a full description of the process necessitates definition of the factors controlling each step.

It is the aim of the following sections to review current knowledge of the mechanisms involved in these separate steps. Since protein synthesis occupies a central position in the overall function of the secretory cell the following discussion may serve as a reference point for some of the aspects discussed in other chapters, whilst also indicating possible causes for some of the variations in milk protein secretion discussed above.

PRECURSORS OF MILK-SPECIFIC PROTEINS

Compounds absorbed by the lactating gland from the blood which are subsequently used for milk protein synthesis, have been investigated by the use of isotopic tracers and by arteriovenous difference studies. Experiments in which ^{14}C-labelled amino acids were injected into whole animals with subsequent analysis of the labelling of milk proteins suggested that casein is synthesized *de novo* from free amino acids of the plasma (see Barry, 1961), but such experiments not only lacked precision but also failed to conclusively demonstrate the form in which the label was absorbed by the glands. The use of isotopes in experiments employing isolated perfused gland preparations (see, for example, Davis & Mepham, 1976) has, however, largely confirmed earlier conclusions.

Quantitative estimation of substrate uptake in relation to milk protein output by the gland has been investigated by the arteriovenous concentration difference method in a number of species (cows: Verbeke & Peeters, 1965; Bickerstaffe, Annison & Linzell, 1974; goats: Mepham & Linzell, 1966; sows: Linzell, Mepham, Annison & West, 1969; Spincer, Rook & Towers, 1969; guinea-pigs: Davis & Mepham, 1974; ewes: S. R. Davis, R. Bickerstaffe & S. D. Hart, unpublished data). Rigorous application of the technique demands (a) the simultaneous measurement of mammary blood flow, so that concentration differences may be translated into "uptake per unit time", and (b) the adoption of procedures to ensure that the mammary venous blood sampled is not mixed with blood of non-mammary origin (see Linzell, 1974). However, these conditions have not been met in several of the investigations quoted, and in these cases the results can only be regarded as providing an indication of the relative importance of the different substrates, rather than a quantitative measure of uptake. Despite this limitation, the results, in so far as they are comparable, show a marked similarity in the pattern of amino acid absorption in relation to milk protein secretion in four of the five species (sows being the exception).

From the quantitative studies on ruminants (see Mepham, 1971, 1976) it is possible to draw the following conclusions. (i) The absorption per unit time of both amino acid carbon and nitrogen are sufficient to account for the output per unit time of milk protein carbon and nitrogen. (ii) The mammary uptake of essential amino acids is sufficient to account for the output of their corresponding residues in the protein. However, while certain amino acids are apparently transferred almost quantitatively to milk (viz. methionine, histidine, phenylalanine and tyrosine) others (valine, isoleucine, leucine, lysine, threonine and arginine) generally show uptakes in excess of the output of their residues in protein. Recent work on the guinea-pig gland indicates that tryptophan should be added to the former group (Mepham, Peters & Davis, 1976). It is suggested that the latter group are extensively catabolized in the gland, a view which is supported by the observation that even when perfused glands are deprived of amino acids in the former group, absorption of the latter group continues and the extent of their oxidation is increased (Alexandrov, Peters & Mepham, in press). (iii) The absorption of most non-essential amino acids is very variable, both between animals and in the same animal at different times, frequently being inadequate to provide all the corresponding residues of the secreted milk protein. It is deduced that these acids are synthesized in the gland, and the overall balance between the uptake of carbon and nitrogen in amino acids and their output in protein supports this view. This stoichiometric relationship does not, of course, imply that amino acids are metabolized independently of other substrates, and, indeed, a number of carbohydrates have been shown to act as carbon precursors of mammary-synthesized amino acids (for example, glucose in the goat mammary gland, Linzell & Mepham, 1968). (iv) The observation that the mammary extraction (arteriovenous difference expressed as a percentage of the arterial concentration) of certain essential amino acids is very high (e.g. about 70% for methionine and phenylalanine) suggests the possibility that the supply of these amino acids to the gland might be limiting for protein synthesis. An alternative claim, that non-essential amino acid supply may be limiting, because of a restriction on the extent of intramammary synthesis, has also been made (Halfpenny, Rook & Smith, 1969). Experiments on goats in which the supply of amino acids to the glands was increased by intramammary arterial infusion adduced no evidence in support of the latter claim, but provided some indication that methionine might be the first-limiting amino acid (Mepham & Linzell, 1974; Linzell & Mepham, 1974). These results were, however, far from conclusive and the proposition needs further testing.

It is difficult to know to what extent these conclusions are applicable to other species. The excessive mammary absorption of certain essential amino acids, high extractions of others and inadequate uptakes of most non-essential amino acids are features apparently shown by goats, cows, ewes and guinea-pigs, but since the first three are ruminants and the latter a herbivore, the mammary metabolism of which is in certain respects similar to that of ruminants (see Mepham, Davis & Humphreys, 1976), the group is hardly representative. The two studies on sows do, in fact, show a marked difference in that the mammary absorption of methionine is excessive in relation to that of other amino acids (Davis, 1974). It seems unlikely that the capillary endothelium presents a significant barrier to the passage of these substrates into the ECF since all the molecules involved are small.

In general terms, however, it is clear that blood concentrations of amino acids are likely to be an important determinant of the quantity and nature of the milk proteins secreted. Since the blood concentrations of substrates are ultimately dependent on environmental, ecological and behavioural factors it is obvious that data from many more species are required before any general scheme relating to the importance of precursor supply can be derived.

TRANSPORT ACROSS THE BASAL MEMBRANE OF THE SECRETORY CELL

The hydrophilic nature of the precursors of milk-specific proteins implies that their penetration of the predominantly hydrophobic basal membrane must be mediated by a process other than diffusion. Investigation of transport mechanisms in a large number of tissues reveals the widespread occurrence of specific carrier systems which promote amino acid accumulation by cells (see Christensen, 1975). For neutral amino acids so-called "A" (alanine preferring), "L" (leucine preferring) and "ASC" (alanine, serine and cysteine) systems have been described, as well as others for acidic and basic amino acids. The presence of such systems has been investigated by the use of non-metabolizable amino acids, e.g. aminoisobutyric acid (AIB) for the "A" system. The occurrence of these carrier systems in mammary tissue has received very little attention, but recently Anderson & Rillema (1976) showed that insulin stimulated [14C]-AIB uptake by mid-pregnant mouse mammary explants, whereas cortisol inhibited uptake and prolactin was without effect. Partington & Mepham (unpublished data) have also demonstrated a stimulatory effect of insulin on AIB uptake by slices of lactating guinea-pig mammary gland. Despite their designations, the carriers are believed to be important in the transport of all the amino acids and they clearly warrant further investigation, particularly in view of their apparent sensitivity to hormonal influences.

The basal membrane of the secretory cell shows prominent infolding to form deep clefts. It has been suggested that the cytoplasmic processes surrounding the clefts may be involved in precursor uptake by the formation of pinocytotic vesicles which enclose ECF. To what extent this is important in amino acid transport is unknown, but the deep infolding of the membrane doubtless promotes transport by increasing the surface area.

THE SYNTHESIS OF MILK-SPECIFIC PROTEINS

There is overwhelming evidence that the milk-specific proteins are synthesized in the mammary secretory cells by the same basic processes as have been described for all other eukaryotic cells. The reaction sequences have been extensively reviewed (e.g. see Larson & Jorgensen, 1974; Denamur, 1974) and the briefest of summaries here is justifiable only insofar as it provides a basis for amplifying those aspects in which milk protein synthesis exhibits noteworthy variations on the general theme.

It is envisaged that the code which determines the sequence in which the different amino acids are linked together by peptide bonds to form proteins is embodied in the molecules of DNA, which together with associated proteins, form the chromatin present in all cell nuclei. The code consists of a sequence of nucleotides which under appropriate conditions serve as a template for the synthesis of RNA molecules (transcription). Each of the types of RNA synthesized, mRNA, tRNA and rRNA, has a specific role in directing the polymerization of amino acids, which takes place on ribosomes in the cytoplasm. There are specific tRNAs for each amino acid, the two molecules combining to form a complex. Molecules of mRNA are present as long strands, composed of nucleotides; groups of three nucleotides form a "codon", which is recognized by the appropriate nucleotide triplet ("anticodon") on a tRNA-amino acid complex. However, it is only when mRNA is associated with rRNA in the ribosome that binding of codon to anticodon occurs and the attached amino acids are aligned in sufficient proximity for peptide-bond formation to occur (translation). Each ribosome has binding sites for only two tRNA molecules at a time, so that for the code on the mRNA to be "read" the mRNA must move with respect to the ribosome. In view of its length a single mRNA molecule may be read simultaneously by more than one ribosome, the multiple structure being termed a "polysome" (see Fig. 1).

In those few species (chiefly rats, mice, guinea-pigs, rabbits, ewes and cows) in which it has been studied, conclusive evidence has been obtained indicating that milk protein synthesis conforms to this general scheme. Thus, for example (i) the RNA/DNA ratio, which indicates protein synthetic potential per cell, is much higher in lactating than in non-lactating tissue; (ii) actinomycin D, which specifically inhibits DNA-dependent RNA synthesis, also inhibits milk protein synthesis; (iii) polysome aggregation is markedly greater in lactating than in non-lactating tissue (for references see Denamur, 1974). More recently species of mRNA coding for milk proteins have been isolated and partially characterized from lactating tissue of rats (Rosen, Woo & Comstock, 1975), guinea-pigs (Craig, Brown, Harrison, McIlreavy & Campbell, 1976) and

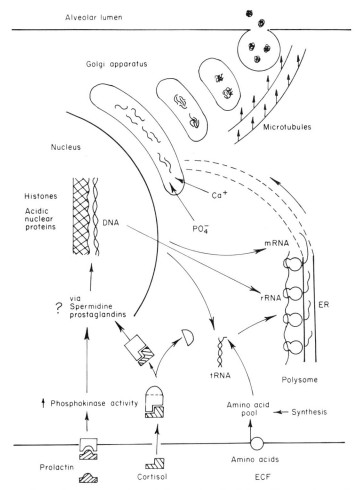

FIG. 1. A schematic representation of factors involved in milk protein synthesis and secretion. Gene derepression by the action of prolactin (membrane receptor) and cortisol (mobile receptor) is believed to be mediated by spermidine and/or prostaglandins. Proteins synthesized on the endoplasmic reticulum (ER) are conveyed to the Golgi apparatus, where they enter associations with other proteins (e.g. caseins aggregate in micelles). Golgi vesicle migration is thought to involve microtubules, while release from the cell is predominantly by reverse pinocytosis. For details see text.

ewes (Gaye & Houdebine, 1975), and shown to be capable of directing milk protein synthesis in heterologous cell-free systems.

In fact, lactating mammary tissue has proved to be very good material in which to examine protein synthesis, not only because of the high rates of synthesis exhibited, but also because the intermittent activity of the secretory cells provides an ideal subject for the study of factors controlling gene expression. Thus, the genetic material present (in nuclei and mitochondria) in all the cells of the body presumably contains genes necessary for directing the synthesis of all the proteins the body is capable of synthesizing. In lactation the secretory cells use only part of the encoded information, i.e. that related to the synthesis of milk proteins and of enzymes necessary for the functioning of the cell. Resorting to teleology, what is required at parturition is that signals generated outside the mammary gland be conveyed to the secretory cells to derepress the appropriate portions of the genome.

Interspecific variation in the composition of the complex of hormones which initiates lactogenesis is discussed in Kuhn's chapter in this volume, but whatever the precise composition of the complex in any particular species, it seems inevitable that the intracellular effects are partially, if not indeed entirely, concerned with gene expression. The studies of Topper and co-workers (see Topper & Oka, 1974) showed that explants of mid-pregnant mouse mammary tissue could be stimulated to synthesize milk proteins, characteristic of mouse milk, by *in vitro* culture in media containing insulin, cortisol and prolactin. These hormones exemplify two distinct mechanisms of hormonal modulation of gene expression (see Tepperman, 1973). Cortisol and other steroids are believed to penetrate cell membranes and then combine with a mobile cytoplasmic receptor which is transported to the nucleus where it interacts with chromosomal proteins. By contrast, polypeptide hormones such as insulin and prolactin are thought not to enter the cell, but by binding to membrane receptors to induce changes in the cytoplasm. In some cases the synthesis of cyclic nucleotides (e.g. cyclic AMP) is stimulated, which in turn activates cytoplasmic phosphokinases, but in the case of prolactin it is the synthesis of these phosphokinases themselves which is apparently stimulated rather than their activation by cyclic AMP (Majumder & Turkington, 1971). The precise nature of the interaction between the steroid mobile receptors and phosphokinases on the one hand and the chromatin on the other is unknown, but some recent results have suggested a general mechanism of action. The chromosomal proteins, which by the nature of their binding to the DNA strands are believed to repress gene expression, are of two major types, histones and acidic nuclear proteins. Recently it has been shown that in mouse mammary cultures, combinations of insulin, cortisol and prolactin markedly stimulate phosphorylation of both acidic nuclear proteins and histones (McCarty & McCarty, 1975), while insulin alone stimulates acetylation of histones (Marzluff & McCarty, 1970). Moreover, phosphorylation is closely associated with increased RNA synthesis (Turkington, 1972). McCarty & McCarty (1975) have postulated that phosphorylation of the acidic nuclear proteins may, by rendering them negatively charged, cause them to be repelled by the histones and hence lead to unfolding of specific areas of the genome. This allosteric effect would thus make available for transcription certain genes which were previously masked.

In the case of cortisol, effects may be mediated by the polyamine, spermidine, which is known to interact with chromatin in modifying DNA-dependent RNA polymerase activity (McCarty & McCarty, 1975). In mouse mammary cells cultured *in vitro*, spermidine has been shown capable of substituting for cortisol in the lactogenic complex (Oka, 1974) and the mammary concentration of spermidine increases sharply at lactogenesis in the rat (Russell & McVicker, 1972). The observation that spermidine inhibits cortisol uptake and stimulates trytophan uptake by the isolated perfused lactating guinea-pig mammary gland (Peters & Mepham, 1976) may be indicative of a role for this amino acid in derepression. Of considerable interest in this context is the recent report that prostaglandins B_2, E_2 or $F_{2\alpha}$ in combination with spermidine stimulated casein synthesis in mouse explants in a prolactin-like manner (Rillema, 1976). Furthermore, addition of an inhibitor of polyamine synthesis abolished the stimulatory effect of prolactin, but not the effect of spermidine plus a prostaglandin.

GENETIC POLYMORPHISM

The occurence in milk of genetic polymorphs of milk proteins derives from mutations, i.e. changes in the sequence of the base pairs of DNA. Genetic variants of the proteins have been studied most extensively in bovine milk. They arise as a result of (i) a simple change in one of the DNA bases, which causes the substitution of a single amino acid for another in the protein molecule, and (ii) the fracture of chromosomes during meiosis, followed by the formation of recombination products in which a segment of DNA (containing up to 39 base pairs) is deleted (Thompson & Farrell, 1974). In consequence it has been found possible to distinguish, for example, four variants of α_{s1}-casein on the basis of electrophoretic mobility. Frequencies of these variants show a breed-specificity in cows. Moreover, the observation that in a large number of breeds only certain variants of α_{s1}- and of β-casein occurred together led to the conclusion that the genes controlling the synthesis of the proteins are linked. The possible implications of such mutations for subsequent micelle formation are discussed below.

INTRACELLULAR TRANSLOCATION

Insight into the intracellular site of synthesis of milk proteins, and the route by which they traverse the cell, is provided by high-resolution autoradiographic studies. In this technique the intracellular localization of a radioactive protein precursor at different times is determined by electron microscopical examination of samples of mammary tissue taken at intervals after isotope injection into the animal. Studies such as those of Stein & Stein (1967) indicated that 20 min after the intravenous injection of [^{3}H]-leucine most of the radioactivity was detected in the region of the rough endoplasmic reticulum (RER). Lactating mammary tissue has abundant RER, 80% of the ribosomes being membrane-bound (Denamur, 1974). Moreover synthesis of milk proteins apparently occurs exclusively on bound ribosomes.

The autoradiographic studies show that about 30 min after injection of [^{3}H]-leucine the radioactivity is primarily associated with the Golgi apparatus. The question of how protein synthesized on bound ribosomes reaches the vacuoles of the Golgi apparatus is not peculiar to mammary tissue. Ribosomes are composed of two subunits, and it is known that it is the larger of these two units which is attached to the ER membrane. As in other tissues it seems clear that newly-synthesized protein passes into the cisternal lumen of the ER through a channel in the larger subunit (Redman & Sabatini, 1966). A hypothesis for the secretion of proteins across the ER membranes has recently been proposed by Blobel & Dobberstein (1975). This, so-called, "signal" hypothesis envisages that mRNAs of proteins destined for secretion contain a unique sequence of codons which results in the synthesis of a corresponding "signal" sequence of amino acids at the amino terminal end of the nascent polypeptide chain. As the signal sequence emerges from the large ribosomal subunit it is believed both to order the loose association of certain membrane proteins in such a manner that they form a "tunnel" in the membrane and also to promote the binding of the large subunit to these membrane proteins so that the emerging nascent chain is vectored through the tunnel into the cisternal lumen. It is postulated that the signal sequence in the peptide chain is removed by an endopeptidase during chain elongation. After release of the nascent chain the detachment of the ribosome from the membrane would cause disaggregation of the membrane proteins, and hence elimination of the tunnel. The probability that such a mechanism is involved in transmembrane transport of milk proteins is indicated by the recent finding of Craig et al. (1976) that mRNA from guinea-pig mammary tissue directs synthesis of a "pre-α-lactalbumin", i.e. α-lactalbumin modified at its N terminus by the addition of an extra seven to ten amino acid residues; this was demonstrated by its synthesis in a heterologous cell-free system deficient in the endopeptidase which would normally remove the signal sequence.

While it has been suggested that protein is conveyed from the ER to the Golgi apparatus by a budding-off of vesicles from the former, followed by their migration through the cytoplasm (Kurosumi, Kobayashi & Baba, 1968), it is more widely believed that protein passes directly to the Golgi through the lumen of the ER, which is thought to be continuous with it (for example, see Hollmann, 1974). According to Patton & Jensen (1976) the continuity of the ER and Golgi membranes is an illustration of the fact that the former becomes transformed into the latter, i.e. that there is a continuous process of membrane flow and modification within the cell. Electron microscopic studies reveal the presence, in several species, of Golgi vacuoles containing proteinaceous material, some of which is only loosely associated, and some, particularly in vacuoles nearer the cell apex, more tightly bound (see Hollmann, 1974). It is presumed that this aggregation process gives rise to a quaternary structure (the micelle), which is peculiarly dependent on the binding properties of the different casein molecules present. In most milks studied the arrangement of the micellar subunits is too complex to allow visualization by electron microscopy, but some of the casein micelles in milk of the monotreme, the echidna, show a prominent internal structure consisting of layers of concentric rings (Griffiths, Elliott, Leckie & Schoefl, 1973).

The association characteristics of the different casein molecules are dependent on a number of modifications which they undergo in the Golgi apparatus. Turkington & Topper (1966) indicated the presence of a pool of non-phosphorylated casein in mammary tissue, and subsequently Bingham, Farrell & Basch (1972) showed that a kinase for the phosphorylation of casein is concentrated in the Golgi. According to Mercier, Grosclaude & Ribadeau-Dumas (1971) phosphorylation of the serine and threonine residues of the polypeptide chain is dependent on the presence of specific residues in neighbouring segments of the chain. Differences in the extent of phosphorylation of different genetic variants of a protein may thus appear disproportionately large, and lead to profound differences in the quaternary structure of the micelle. Various models for the structure of the bovine casein micelle have been proposed (see Thompson & Farrell, 1974). Two factors of undoubted importance in the formation of the micelle are the glycosylation of the peptide chain in the formation of κ-casein; and the incorporation of calcium phosphate which is thought to stabilize the micellar structure. Baumrucker & Keenan (1975) have recently reported the presence in bovine mammary Golgi fractions of a magnesium-dependent ATPase (distinct from a sodium–potassium activated ATPase) which is markedly stimulated by calcium ions. These authors suggested that this enzyme may be important both in providing phosphate for casein phosphorylation and in sequestering calcium ions for subsequent micellar aggregation.

There is growing evidence that in many tissues the Golgi is the site of synthesis of the carbohydrate moieties of glycoproteins (see Whaley, Dauwalder & Kephart, 1972). It is also the site of lactose synthesis, a situation which derives from the temporary association of α-lactalbumin with a Golgi membrane-bound galactosyltransferase and the consequent change in the reactivity of the enzyme such that lactose is synthesized. The osmotic pressure exerted by lactose would seem to be a key factor controlling the secretion of the aqueous phase of milk (Linzell & Peaker, 1971a).

TRANSLOCATION FROM THE GOLGI APPARATUS TO THE ALVEOLAR LUMEN

There is considerable electron microscopical evidence that the mammary-synthesized proteins are conveyed from the Golgi to the apical membrane by a process which involves the budding of vesicles from the apical surface of this organelle (Patton & Jensen, 1976). However, because of the difficulties inherent in attempting to describe a dynamic process on the basis of "still" photographs, the precise way in which the translocation is achieved remains a matter of conjecture, as is indicated by the numerous theories which have been proposed. Thus (i) Hollmann (1974) suggested that microvesicles have a "shuttle" function, those containing protein granules being in transit to the apical membrane, while others, apparently empty, being assumed to be returning to the Golgi to be re-filled; (ii) Saacke & Heald (1974), on the basis of the apposition of contiguous vesicles in a "ball and socket" configuration, suggested that the contents of one vesicle might be injected into its neighbour following merger of the two vesicles; (iii) Patton & Jensen (1976) speculated that the vesicular membrane system might behave as a "flexible lattice", "through which milk fluid percolates". However, the apparently most widely held view is that vesicles containing protein move to the apex of the cell and then fuse with the apical plasmalemma, releasing their contents by "reverse pinocytosis". This concept is consistent with the findings of studies in which the lipid and protein composition of membranes has been studied, and lends further support to the theory that there is a continuous transformation of ER membrane into apical plasmalemma, Golgi membrane representing an intermediate state (Patton & Jensen, 1976).

The process of reverse pinocytosis would lead, in the absence of any compensatory mechanism. to a net gain of plasma membrane. This tendency is offset by the process, discussed elsewhere in this volume, by which fat has been shown to be secreted i.e. the "pinching off", from the cell apex, of fat globules surrounded by plasmalemma. It has been suggested that the two processes ("pinching off" and reverse pinocytosis) are balanced, and the dimensions of the apical membrane thereby maintained (Patton & Jensen, 1976). In fact Wooding (1973), on the basis of the prevalence in goat, cow and guinea-pig mammary tissue, of lipid droplets in the apical cytoplasm which are surrounded by Golgi vesicles, has suggested that the interaction between the Golgi vesicle membrane and the lipid droplet might be instrumental in the process of extrusion of the latter.

The presence in the alveolar lumina of fat droplets with attached crescents of cytoplasm (signets), in some cases containing protein granules (Kurosumi et al., 1968), has been adduced as evidence of an alternative mechanism of protein secretion, protein being thought to be released from the signet within the lumen. The criticism that the signets were artifacts, representing detached portions of neighbouring cells, was disproved by the observation that they occur in secreted milk (Wooding, Peaker & Linzell, 1970). However, only 1–5% of the lipid globules in goat and guinea-pig milk had attached crescents of cytoplasm and since, in sections of guinea-pig mammary tissue (Tomlinson, Davis & Mepham, unpublished data), the number does not appear greatly in excess of this, it seems unlikely that this mechanism represents a major route of protein secretion. The situation may be different in other species since Wooding (pers. comm.) has found that in wallaby milk about 8% of lipid globules had attached crescents of cytoplasm.

In other protein-secreting tissues the secretory process has been shown to be dependent on the activity of the microtubules (for example, in the β cells of the pancreas: Lacy & Malaisse, 1973; in thyroid cells: Wolff & Williams, 1973). The basic structural unit of microtubules is a protein, tubulin, which aggregates in a specific way to form helical structures. Aggregation may be specifically inhibited by the plant alkaloid colchicine, which binds to the subunits of the quaternary structure, and a similar inhibition is produced by the vinca alkaloids, vincristine and vinblastine. Administration of these antimicrotubular agents also inhibits secretion of insulin (Lacy & Malaisse, 1973), and it has been deduced that this is due to inhibition of the conformational changes in the microtubules which normally result in displacement of the secretory vesicles towards the cell apex. These authors also suggest that the supply of calcium ions may constitute the intracellular trigger which initiates contraction or changes in configuration of the microtubules. I am aware of only one study of the effect of colchicine and vincristine on milk secretion (Patton, 1976), but the results obtained strongly suggest that Golgi vesicle transport in the mammary cell conforms to the above scheme. Thus, unilateral intracisternal injection of both alkaloids into glands of goats caused a marked inhibition of milk yield from the injected gland, but had virtually no effect on the control gland. Milk obtained from both glands during the period of inhibition (2–3 days) was of normal composition in respect of lactose, fat and protein, except that there was an increase in the concentration of immunoglobulins in the milk from the injected glands. Ultrastructural studies on the mammary tissue indicated that the epithelial cells were engorged with secretory vesicles and fluid, which suggests that the inhibition was primarily acting on the secretory process rather than on synthesis. It would seem that recovery from the effects of antimicrotubular agents represents a situation not unlike that occurring at lactogenesis. In the parturient gland synthesis of milk components begins well before the onset of copious milk secretion, the latter being heralded by a rapid rise in the milk concentration of citrate (Peaker & Linzell, 1975). The recent evidence of Linzell, Mepham & Peaker (1976) that citrate is secreted from the cell by the same route as protein and lactose suggests that the factors controlling microtubular activity might, through their effects on Golgi vesicle transport, be instrumental in initiating lactation.

In concluding this section on the secretion of milk-specific proteins it seems appropriate to consider the significance of α-lactalbumin secretion. Reference has been made above both to the role of this protein in altering the substrate specificity of a Golgi-membrane bound enzyme, such that lactose is synthesized, and to the vital function of lactose in regulating water movement by osmosis. At first sight it might appear puzzling that such an important enzyme component is lost from the cell by secretion into milk, but, as Brew (1970) has pointed out, the rapid turnover of α-lactalbumin provides a means of fine control over lactation as a whole: factors which control α-lactalbumin secretion ultimately control milk secretion.

Some of the factors discussed in this and earlier sections, relating to the synthesis and secretion of milk-specific proteins, are summarized in Fig. 1.

TRANSFER OF PROTEINS FROM THE ECF TO THE ALVEOLAR LUMEN

In addition to the milk-specific proteins, serum proteins also occur in milk, their concentrations, especially in the case of immunoglobulins, being particularly high in colostrum. The question of the origin of these proteins, whether from the blood plasma or by synthesis in plasma cells situated in the mammary gland, is discussed in Lascelles' chapter in this volume, and the following brief comments relate solely to the means by which these proteins traverse the mammary secretory epithelium. There would appear to be two possible means of transport, viz. through the secretory cells (the transcellular route) and between contiguous secretory cells (the paracellular route).

There is considerable evidence that immunoglobulins (Ig) are transported through the secretory cells by a process which is capable of discriminating between molecules of very similar molecular weight. For example, IgG_1 and IgG_2 are present in bovine blood (and mammary lymph) at approximately equal concentrations, but in colostrum the concentration of IgG_1 greatly exceeds that of IgG_2 (Brandon, Watson & Lascelles, 1971). The selective mechanism is believed to depend on the presence of specific receptor sites on the basal membrane of the secretory cell (Brandon et al., 1971) and to be regulated by the blood concentrations of oestrogen and progesterone (Smith, 1971). Peaker & Linzell (1974) showed that intravenous infusion of oestrogen into goats, at a rate designed to match the endogenous production rate in late pregnancy, led to an increased Ig content of milk. The changes in the concentration of other milk components were not compatible with the proteins reaching the alveolar lumina by the paracellular route, and it was suggested that the immunoglobulins traversed the cell within pinocytotic vesicles previously formed at the basal membrane, a view which receives some support from electron microscopical studies of mammary tissue during pregnancy (e.g. in the rat, Murad, 1970).

Apart from the transcellular route there is clear evidence that proteins, along with other components of the ECF, might reach the alveolar lumina by passage between epithelial cells. Thus, conformational changes of the alveolus associated with oxytocin-induced contraction of the myoepithelial cells may cause disruption of some tight junctions between neighbouring cells, and thus allow partial equilibration of milk and ECF (see Linzell & Peaker, 1971b). However, it appears that in some species (the rabbit, for example, Peaker & Taylor, 1975) the paracellular route is prominent even at peak lactation, becoming even more pronounced in the declining phase. Whilst administration of prolactin to rabbits in late lactation significantly increased milk yield and altered its sodium and potassium concentrations, the protein content of the milk was not significantly altered (Linzell, Peaker & Taylor, 1975).

Clearly, interspecies variation in the extent to which the paracellular route is operative may be an important determinant of differences in milk protein concentration.

CONCLUSIONS

Although it seems likely that the basic mechanism of milk protein synthesis is identical in all species, there would appear to be a number of key steps, variations in the activity of which would be expressed in terms of differing milk protein compositions. Thus, genetic factors will determine (i) the precise molecular structure of the milk-specific proteins, (ii) the sensitivity of the transcriptional and translational process to hormonal modulation, and (iii), consequently, the tertiary and quaternary structure of the micellar aggregates. But, apart from such considerations, the extent of protein synthesis in the mammary gland may be critically dependent on (a) the supply of precursors to the gland, itself a consequence of dietary factors and of inter-organ competition for substrates, and (b) the partitioning of absorbed substrates between different metabolic pathways within the secretory cell. The process of secretion, as opposed to synthesis, may be controlled by factors such as microtubular activity, or by some necessary interaction with fat globule extrusion. Moreover, the total protein content of milk will clearly be highly dependent on the extent to which serum proteins enter the milk, by either the paracellular or transcellular routes.

It is hardly surprising that the wide diversity in protein content of milks of different species has stimulated attempts to formulate simple generalizations defining the source of these variations. Such attempts are unfortunately hampered by the multiplicity of variables encountered. Not only, for example, are the young of different species born at greatly differing stages of maturity, but factors such as frequency of suckling and length of lactation complicate any comparison based simply on relative concentrations of protein in milks. Blaxter (1971), comparing milks of 20 species, showed that, on average, milk protein content increased with decreasing weight of the species, but data from individual species showed large deviations from the mean.

In this chapter, I have attempted to summarize current views on factors which might control milk protein secretion. Most of the available data, however, derive from investigations of, at best, a mere handful of species. It seems clear that data from many more species are required before attempts can profitably be made to define the factors determining milk protein composition.

ACKNOWLEDGEMENTS

I am most grateful to Mrs Carol Stanton and Miss Kathryn Canovan for their help in preparing the manuscript.

REFERENCES

Alexandrov, S., Peters, A. R. & Mepham, T. B. (in press). [Effects of methionine and leucine depletion on amino acid uptake by the isolated perfused guinea-pig mammary gland.] *Anim. Sci.* (Sofia). [In Bulgarian.]

Anderson, L. D. & Rillema, J. A. (1976). Effects of hormones on protein and amino acid metabolism in mammary gland explants of mice. *Biochem. J.* **158**: 355–359.

Barry, J. M. (1961). Protein metabolism. In *Milk: The mammary gland and its secretion*: **2**: 389–419. Kon, S. K. & Cowie, A. T. (eds). London and New York: Academic Press.

Baumrucker, C. R. & Keenan, T. W. (1975). Membranes of mammary gland. Adenosine triphosphate-dependent calcium accumulation by Golgi apparatus rich fractions of bovine mammary glands. *Expl Cell Res.* **90**: 253–290.

Bickerstaffe, R., Annison, E. F. & Linzell, J. L. (1974). The metabolism of glucose, acetate, lipids and amino acids in lactating dairy cows. *J. Agric. Sci., Camb.* **82**: 71–85.

Bingham, E. W., Farrell, H. M. & Basch, I. J. (1972). Phosphorylation of casein. Role of the Golgi apparatus. *J. biol. Chem.* **247**: 8193–8194.

Blaxter, K. L. (1971). The comparative biology of lactation. In *Lactation*: 51–69. Falconer, I. R. (ed.). London: Butterworths.

Blobel, G. & Dobberstein, B. (1975). Transfer of proteins across membranes. *J. Cell Biol.* **67**: 835–862.

Brandon, M. R., Watson, D. L. & Lascelles, A. K. (1971). The mechanism of transfer of immunoglobulin into mammary secretion of cows. *Aust. J. Exp. Biol. Med. Sci.* **49**: 613–623.

Brew, K. (1970). Lactose synthetase: Evolutionary origins structure and control. In *Essays in biochemistry* **6**: 93–117. Campbell, P. N. & Dickens, F. (eds). London and New York: Academic Press.

Christensen, H. N. (1975). *Biological transport.* 2nd edn. Reading, Massachusetts: W. A. Benjamin Inc.

Craig, R. K., Brown, P. A., Harrison, O. S., McIlreavy, D. & Campbell, P. N. (1976). Guinea-pig milk protein synthesis. Isolation and characterisation of messenger ribonucleic acids from lactating mammary gland and identification of caseins and pre-α-lactalbumin as translation products in heterologous cell-free systems. *Biochem. J.* **160**: 57–74.

Davis, S. R. (1974). Amino acid metabolism in the isolated perfused guinea-pig mammary gland. Ph.D. thesis: Nottingham University.

Davis, S. R. & Mepham, T. B. (1974). Amino acid and glucose uptake by the isolated perfused guinea-pig mammary gland. *Q. Jl exp. Physiol.* **59**: 113–130.

Davis, S. R. & Mepham, T. B. (1976). Metabolism of L-[U-^{14}C] valine, L-[U-^{14}C] leucine, L-[U-^{14}C] histidine and L-[U-^{14}C] phenylalanine by the isolated perfused lactating guinea-pig mammary gland. *Biochem. J.* **156**: 553–560.

Denamur, R. (1974). Ribonucleic acids and ribonucleoprotein particles of the mammary gland. In *Lactation* **1**: 413–465. Larson, B. L. & Smith, V. R. (eds) New York and London: Academic Press.

Gaye, P. & Houdebine, L. M. (1975). Isolation and characterisation of casein and RNAs from lactating ewe mammary glands. *Nucleic Acids Res.* **2**: 707–722.

Griffiths, M., Elliott, M. A., Leckie, R. M. C. & Schoefl, G. I. (1973). Observations of the comparative anatomy and ultrastructure of mammary glands and on the fatty acids of the triglycerides in platypus and echidna milk fats. *J. Zool., Lond.* **169**: 255–279.

Halfpenny, A. F., Rook, J. A. F. & Smith, G. H. (1969). Variations with energy nutrition in the concentrations of amino acids of the blood plasma in the dairy cow. *Br. J. Nutr.* **23**: 547–557.

Hollmann, K. H. (1974). Cytology and fine structure of the mammary gland. In *Lactation* **1**: 3–95. Larson, B. L. & Smith, V. R. (eds). New York and London: Academic Press.

Jenness, R. (1974). The composition of milk. In *Lactation* **3**: 3–107. Larson, B. L. & Smith, V. R. (eds). New York and London: Academic Press.

Kurosumi, K., Kobayashi, Y. & Baba, N. (1968). The fine structure of the mammary glands of lactating rats, with special reference to the apocrine secretion. *Expl Cell Res.* **50**: 177–192.

Lacy, P. E. & Malaisse, W. J. (1973). Microtubules and beta-cell secretion. *Recent Prog. Horm. Res.* **29**: 199–221.

Larson, B. L. & Jorgensen, G. N. (1974). Biosynthesis of the milk proteins. In *Lactation* **2**: 115–146. Larson, B. L. & Smith, V. R. (eds). New York and London: Academic Press.

Linzell, J. L. (1974). Mammary blood flow and methods of identifying and measuring precursors of milk. In *Lactation* **1**: 143–225. Larson, B. L. & Smith, V. R. (eds). New York and London: Academic Press.

Linzell, J. L. & Mepham, T. B. (1968). Mammary synthesis of amino acids in the lactating goat. *Biochem. J.* **107**: 18–19 P.

Linzell, J. L. & Mepham, T. B. (1974). Effects of intramammary arterial infusion of essential amino acids in the lactating goat. *J. Dairy Res.* **41**: 101–109.

Linzell, J. L., Mepham, T. B., Annison, E. F. & West, C. E. (1969). Mammary metabolism in lactating sows: arteriovenous differences of milk precursors and the mammary metabolism of [^{14}C] glucose and [^{14}C] acetate. *Br. J. Nutr.* **23**: 319–332.

Linzell, J. L., Mepham, T. B. & Peaker, M. (1976). The secretion of citrate into milk. *J. Physiol., Lond.* **260**: 739–750.

Linzell, J. L. & Peaker, M. (1971a). Mechanism of milk secretion. *Physiol. Rev.* **51**: 564–597.

Linzell, J. L. & Peaker, M. (1971b). The effects of oxytocin and milk removal on milk secretion in the goat. *J. Physiol., Lond.* **216**: 717–734.

Linzell, J. L., Peaker, M. & Taylor, J. C. (1975). The effects of prolactin and oxytocin on milk secretion and on the permeability of the mammary epithelium in the rabbit. *J. Physiol., Lond.* **253**: 547–563.

McCarty, K. S. & McCarty, K. S. (1975). Early mammary gland responses to hormones. *J. Dairy Sci.* **58**: 1022–1032.

Majumder, G. C. & Turkington, R. W. (1971). Adenosine 3′, 5′-monophosphate-dependent and independent protein phosphokinase isoenzymes from mammary gland. *J. biol. Chem.* **246**: 2650–2657.

Marzluff, W. F. & McCarty, K. S. (1970). Two classes of histone acetylation in developing mouse mammary gland. *J. biol. Chem.* **245**: 5635–5642.

Mepham, T. B. (1971). Amino acid utilisation by the lactating mammary gland. In *Lactation*: 297–315. Falconer, I. R. (ed.). London: Butterworths.

Mepham, T. B. (1976). Amino acid supply as a limiting factor in milk and muscle synthesis. In *Principles of cattle production*: 201–219. Swan, H. & Broster, W. H. (eds). London: Butterworths.

Mepham, T. B., Davis, S. R. & Humphreys, J. R. (1976). Acetate utilization by the isolated perfused guinea-pig mammary gland. *J. Dairy Res.* **43**: 197–203.

Mepham, T. B. & Linzell, J. L. (1966). A quantitative assessment of the contribution of individual amino acids to the synthesis of milk proteins by the goat mammary gland. *Biochem. J.* **101**: 76–83.

Mepham, T. B. & Linzell, J. L. (1974). Effects of intramammary arterial infusion of non-essential amino acids and glucose in the lactating goat. *J. Dairy Res.* **41**: 111–121.

Mepham, T. B., Peters, A. R. & Davis, S. R. (1976). Uptake and metabolism of tryptophan by the isolated perfused guinea-pig mammary gland. *Biochem. J.* **158**: 659–662.

Mercier, J. C., Grosclaude, F. & Ribadeau-Dumas, B. (1971). [The primary structure of bovine α_{s1}-casein.] *Eur. J. Biochem.* **23**: 41–51. [In French.]

Murad, T. M. (1970). Ultrastructural study of rat mammary gland during pregnancy. *Anat. Rec.* **167**: 17–36.

Oka, T. (1974). Spermidine in hormone-dependent differentiation of mammary gland *in vitro*. *Science, N.Y.* **184**: 78–80.

Patton, S. (1976). Mechanisms of secretion: Effects of colchicine and vincristine on composition and flow of milk in the goat. *J. Dairy Sci.* **59**: 1414–1419.

Patton, S. & Jensen, R. G. (1976). *Biomedical aspects of lactation*. Oxford: Pergamon Press.

Peaker, M. & Linzell, J. L. (1974). The effects of oestrus and exogenous oestrogens on milk secretion in the goat. *J. Endocr.* **61**: 231–240.

Peaker, M. & Linzell, J. L. (1975). Citrate in milk: a harbinger of lactogenesis. *Nature, Lond.* **253**: 464.

Peaker, M. & Taylor, J. C. (1975). Milk secretion in the rabbit: changes during lactation and the mechanisms of ion transport. *J. Physiol., Lond.* **253**: 527–546.

Peters, A. R. & Mepham, T. B. (1976). The effect of spermidine on the uptake of tryptophan and corticosteroids by the isolated perfused guinea-pig mammary gland. *Biochem. Soc. Trans.* **4**: 118–119.

Redman, C. M. & Sabatini, D. D. (1966). Vectorial discharge of peptides released by puromycin from attached ribosomes. *Proc. Nat. Acad. Sci. U.S.A.* **56**: 608–615.

Rillema, J. A. (1976). Activation of casein synthesis by prostaglandins plus spermidine in mammary gland explants of mice. *Biochem. Biophys. Res. Commun.* **70**: 45–49.

Rosen, M., Woo, S. L. C. & Comstock, J. P. (1975). Regulation of casein messenger RNA during the development of the rat mammary gland. *Biochem.* **14**: 2895–2903.

Russell, D. H. & McVicker, T. A. (1972). Polyamine biogenesis in the rat mammary gland during pregnancy and lactation. *Biochem. J.* **130**: 71–76.

Saacke, R. G. & Heald, C. W. (1974). Cytological aspects of milk formation and secretion. In *Lactation* **2**: 147–189. Larson, B. L. & Smith, V. R. (eds). New York and London: Academic Press.

Smith, K. L. (1971). Role of oestrogen in the selective transport of IgGl into the mammary gland. *J. Dairy Sci.* **54**: 1322–1323.

Spincer, J., Rook, J. A. F. & Towers, K. G. (1969). The uptake of plasma constituents by the mammary gland of the sow. *Biochem. J.* **11**: 727–732.

Stein, O. & Stein, Y. (1967). Lipid synthesis, intracellular transport, and secretion. *J. Cell Biol.* **34**: 251–263.

Tepperman, J. (1973). *Metabolic and endocrine physiology.* 3rd Edn. Chicago: Year Book Medical Publishers Incs.

Thompson, M. P. & Farrell, H. M. (1974). Genetic variants of the milk proteins. In *Lactation* **3**: 109–134. Larson, B. L. & Smith, V. R. (eds). New York and London: Academic Press.

Topper, Y. J. & Oka, T. (1974). Some aspects of mammary gland development in the mature mouse. In *Lactation* **1**: 327–348. Larson, B. L. & Smith, V. R. (eds). New York and London: Academic Press.

Turkington, R. W. (1972). Molecular biological aspects of prolactin. In *Lactogenic hormones*: 111–127. Wolstenholme, G. E. W. & Knight, J. (eds). Edinburgh: Churchill Livingstone.

Turkington, R. W. & Topper, Y. J. (1966). Casein biosynthesis: evidence for phosphorylation of precursor proteins. *Biochim. biophys. Acta* **127**: 366–372.

Tyndale-Biscoe, H. (1973). *Life of marsupials.* London: Edward Arnold.

Verbeke, R. & Peeters, G. (1965). Uptake of free plasma amino acids by the lactating cow's udder and amino acid composition of udder lymph. *Biochem. J.* **94**: 183–189.

Whaley, W. G., Dauwalder, M. & Kephart, J. (1972). Golgi apparatus: Influence on cell surfaces. *Science, N.Y.* **175**: 596–599.

Wolff, J. & Williams, J. A. (1973). The role of microtubules and microfilaments in thyroid secretion. *Recent Prog. Horm. Res.* **29**: 229–278.

Wooding, F. B. P. (1973). Formation of the milk fat globule membrane without participation of the plasmalemma. *J. Cell Sci.* **13**: 221–235.

Wooding, F. B. P., Peaker, M. & Linzell, J. L. (1970). Theories of milk secretion: evidence from the electron microscopic examination of milk. *Nature, Lond.* **226**: 762–764.

Symp. zool. Soc. Lond. (1977) No. 41, 77–94

Synthesis and Secretion of Milk Sugars

E. A. JONES

National Institute for Research in Dairying, Shinfield, Reading, England

SYNOPSIS

The milks of most mammals contain significant amounts of carbohydrate and in the majority of species studied the predominant carbohydrate is lactose, though oligosaccharides derived from lactose are also important in some milks. Lactose is formed in the Golgi apparatus of the mammary epithelial cells by a complex of the enzyme galactosyltransferase and the whey protein α-lactalbumin. It is the presence of α-lactalbumin which confers on the mammary gland its unique ability to produce lactose.

The concentration of carbohydrate in milk is characteristic of the particular species but how this concentration is controlled is not understood. There is an inverse relationship between the carbohydrate and fat contents of milks. The aquatic mammals have high fat and low carbohydrate while the horse is an example of the opposite condition. However, there is no simple relationship between the lactose content of milks and environmental and developmental factors.

The evolution of lactose synthesis has probably involved the development of α-lactalbumin from the widely distributed enzyme lysozyme following gene duplication. In the milk of the monotreme, the echidna, there occurs a protein which has the characteristics of both lysozyme and α-lactalbumin. The milks of monotremes are characterized by low concentrations of lactose and high concentrations of oligosaccharides derived from lactose so the emergence of lactose as the principal milk sugar must have involved, in addition to the evolution of α-lactalbumin, the suppression of those enzymes producing more complex carbohydrates.

INTRODUCTION

As the role of milk is to provide a balanced diet for the young it is not surprising that in most species it contains significant concentrations of fats, proteins and carbohydrates. With the exception of some seals all milks contain detectable amounts of carbohydrates at concentrations up to 10 g/100 ml. It is widely assumed in comparisons of the milks of different species that lactose [galactosyl-(β1,4)-glucose] is the predominant sugar though normally the assay methods used are not specific. Lactose certainly is the major sugar component in the milk of humans and the domesticated ruminants, the species which have been most intensively studied, but more complex oligosaccharides are important in some other species and perhaps a more extensive survey of the milks of wild animals would reveal that the substitution of lactose by other oligosaccharides is widespread. However, the lactose moiety forms the reducing end of the great majority of these other sugars and is presumably an intermediate in their synthesis, so it is the ability to produce lactose that makes the mammary gland unique among mammalian organs. In this review our current knowledge of lactose synthesis will be described before proceeding to consideration of other oligosaccharides and variations in milk composition. With the space available it is impossible to give all the primary references for the data in this chapter and readers are referred to the reviews mentioned in the various sections for a more comprehensive coverage.

LACTOSE SYNTHESIS

The only major natural source of lactose is the mammary gland, though traces occur in a few plants. Its synthesis is of great interest not simply from the viewpoint of lactational physiology but also to the enzymologist because of unique features in the reaction mechanism. It has been subject to intensive study in recent years and has been reviewed in detail by Ebner & Schanbacher (1974) and Brew & Hill (1975a,b).

Lactose is synthesized from blood glucose in the epithelial cells lining the mammary alveoli, the same cells which synthesize the other major milk constituents. The final reaction of the synthetic pathway (Equation 1) is typical of reactions involving the synthesis of a glycosidic bond in that the glycosyl residue is present in an activated form as a nucleotide derivative in order to shift the equilibrium of the reaction in the synthetic direction.

$$\text{UDP-galactose} + \text{glucose} \rightarrow \text{lactose} + \text{UDP} \tag{1}$$

The crucial observation concerning this reaction was made by Brodbeck & Ebner (1966) while studying the soluble lactose synthetase found in bovine milk, who showed that for lactose synthesis to occur at physiological glucose concentrations two components were required:

(a) an enzyme component, galactosyltransferase, which transfers the galactose moiety of UDP-galactose to a variety of acceptors including N-acetylglucosamine, oligosaccharides and glycoproteins with a suitable terminal sugar residue but for which glucose is a very poor substrate, having a K_m in the region of 1 mol/l;

(b) the whey protein, α-lactalbumin, which forms a bimolecular complex with galactosyltransferase which has a greatly increased affinity for glucose. It is the occurrence of α-lactalbumin which confers upon the mammary gland its unique ability to synthesize lactose and its derivatives.

Galactosyltransferase

In the mammary epithelial cells galactosyltransferase occurs firmly bound to the membranes of the Golgi apparatus (Keenan, Morré & Cheetham,

1970) and is, in fact, the best marker for Golgi membranes in cell fractionation studies. However, far more is known of the properties of the soluble form of the enzyme which is found in milk and is presumed to be derived from the particulate form. The enzyme, which has been studied mainly in bovine and human milks, is a glycoprotein which can be isolated relatively simply by affinity chromatography using Sepharose-α-lactalbumin (Andrews, 1970) or Sepharose derivatives of other ligands (Barker, Olsen, Shaper & Hill, 1972). In bovine milk it exists in a number of forms in the molecular weight range 40 000 to 60 000 (Powell & Brew, 1974a; Magee, Mawal & Ebner, 1974), all of which are enzymically active and which probably result from proteolytic degradation of the highest molecular weight form. This undegraded form is found in colostrum in which proteolysis appears to be inhibited.

The various species of galactosyltransferase have similar though not identical kinetic properties which can be studied in the absence of α-lactalbumin using N-acetylglucosamine as acceptor. The reaction mechanism has been investigated by the steady-state kinetic approach (Morrison & Ebner, 1971a; Khatra, Herries & Brew, 1974) and it has been shown that the substrates add in the obligatory order UDP-galactose; N-acetylglucosamine, followed by the release of the product, N-acetyllactosamine. It is known that the reaction requires the presence of a divalent cation and until recently it was believed that manganese was the only effective activator. This was puzzling as the manganese content of mammary tissue is low, in the order of 50–100 μmol/l (Meyer & Lemmer, 1974). However, recent studies (Powell & Brew, 1976) have shown that the role played by metal ions in the reaction is more complex than had been realized. Two sites are present, one of high affinity which is saturated by 10 μmol/l manganese and one of low affinity which can accept a number of divalent cations. *In vivo*, galactosyltransferase is probably activated by a mixture of cations though the situation requires further investigation.

Though the studies outlined above have been carried out using the soluble enzyme, from the limited data available it seems that the particulate Golgi enzyme has broadly similar properties. The enzyme is tightly bound to Golgi membranes and is thus presumably in contact with the phospholipid bilayer which is a feature of all biological membranes. It can be solubilized using the detergent Triton X100 (Fraser & Mookerjea, 1976) to give a hydrophobic high molecular weight form. It is not known how the water-soluble form found in milk is derived.

α-Lactalbumin

The α-lactalbumins are a group of soluble proteins, found in the milks of all lactose-synthesizing mammals, which appear to have the specific function of facilitating the synthesis of lactose within the mammary gland. All those studied have molecular weights close to 14 000 and those that have had their amino acid sequence determined [human, bovine, guinea-pig and kangaroo (partially)] show good structural homology with about 50% of amino acid residues identical and many of the replacements involving amino acids of similar properties (see Brew & Hill, 1975a,b). The most interesting observation concerning α-lactalbumin is that it has a significant sequence homology with both human and egg-white lysozyme. Lysozymes are enzymes found in various body fluids which are concerned with the hydrolysis of bacterial cell wall mucopolysaccharides. The homology is close enough for a three-dimensional structure of α-lactalbumin to have been proposed by analogy with the known structure of lysozyme, and various physical and chemical studies have tended to confirm this structure.

As lysozyme has an oligosaccharide binding site as part of its enzyme function and the role of α-lactalbumin in lactose synthesis is concerned with changing affinities for carbohydrates, it might be expected that α-lactalbumin itself would be able to bind sugars but extensive investigations have not produced any evidence of this.

Galactosyltransferase-α-Lactalbumin Interaction

It is well established that α-lactalbumin and soluble galactosyltransferase interact to form a bimolecular complex in the presence of UDP-galactose and manganese or a carbohydrate acceptor (Klee & Klee, 1972; Challand & Rosemayer, 1974; Powell & Brew, 1975). This complex can be stabilized by the introduction of a covalent link (Powell & Brew, 1975; Brew, Shaper, Olsen, Trayer & Hill, 1975) to give an enzyme with similar properties but greatly reduced specific activity. Some of the kinetic properties of this complex, both for soluble and particulate galactosyltransferase, are presented in Table I. The values for the various Michaelis constants (i.e. the ligand concentration giving half the maximum velocity) in this table are the apparent values obtained under normal assay conditions, in some cases deduced from data presented in the original papers, rather than absolute values calculated by computer programmes from these data. Apparent values are more useful when considering *in vivo* metabolic control than true values which, in fact, can never be realized under attainable conditions. It will be seen that the properties of the soluble and membrane-bound enzymes are quite similar, especially when one considers that the values for the membrane enzyme were obtained in crude systems and subject to large error.

The effect of α-lactalbumin on galactosyltransferase is to increase its affinity for certain galactosyl acceptors (see Ebner & Schanbacher, 1974, for more detail) but this effect is complicated by various forms of substrate inhibition, so that in any given situation the effect might be an activation or an inhibition, depending on acceptor concentration (Brew, Vanaman & Hill, 1968), temperature (Andrews, 1972) and pH (Osborne & Steiner, 1974). However in practical terms as far as lactose production in the

TABLE I

Apparent Michaelis constants for substrates and activators of galactosyltransferase

Ligand	Soluble enzyme		Membrane-bound enzyme	
	Apparent K_m*	Comment	Apparent K_m*	Comment
Glucose	2 mmol/l (human)* 5 mmol/l (bovine)†	Deduced from data in papers	2 mmol/l (mouse)‡	At saturating α-lactalbumin
UDP-Galactose	0·1 mmol/l (human)* 0·1 mmol/l (bovine)†	Deduced from data in papers	0·20 mmol/l (sheep)§ 0·06 mmol/l (rat)‖	Values approximate because of assay difficulties
α-Lactalbumin	7 μmol/l (human)* 5–8 μmol/l (bovine)†	Deduced from data in papers Bovine galactosyltransferase with various α-lactalbumins	2 μmol/l (mouse)‡	Using bovine α-lactalbumin
Manganese	1·4 mmol/l (human)* 1·2 mmol/l (bovine)†	At low affinity site	0·5 mmol (mouse)**	In absence of other cations

* These approximate values for the concentrations of the various ligands which give half maximum reaction velocities at physiological concentrations of the other reactants. Manganese is a special problem (see text) and an adequate concentration of divalent cations is assumed.
† Khatra, Herries & Brew (1974). ‡ Morrison & Ebner (1971b). § Jones (1972). ‖ Smith, Powell & Brew (1975). ** E. A. Jones, unpublished result.

mammary gland is concerned the effect under physiological conditions is one of great activation.

The mechanism of the α-lactalbumin effect is still the subject of much controversy. Chemical modification of α-lactalbumin has implicated tryptophan and histidine residues in the binding to galactosyltransferase (Barman & Bagshaw, 1972; Schindler, Sharon & Prieels, 1976) but, though these modified forms bind less well, the bimolecular complex when formed still has a high affinity for glucose. Current attempts to elucidate the mechanism of this reaction are based largely on the steady-state kinetics approach and it seems that in a reaction of this complexity involving at least five ligands this method cannot yield unambiguous results. Further progress probably depends on isolating galactosyltransferase in larger amounts and determining its sequence and structure. One intriguing observation is that α-lactalbumin can not only modify the properties of a wide range of mammalian galactosyltransferases but also induce the transferase from onions to synthesize lactose (Powell & Brew, 1974b). Whatever structural feature of galactosyltransferase confers its responsiveness to α-lactalbumin has thus been preserved over a large evolutionary span.

SYNTHESIS OF OTHER MILK SUGARS

Milk is a very rich source of oligosaccharides. Human milk contains over 50 chromatographically-identifiable oligosaccharides (Grimmonprez & Montreuil, 1975) of which the principal components are glucose, galactose, fucose, N-acetylglucosamine and sialic acids, and which may contain up to 20 monosaccharide residues. A common feature of almost all these sugars is the presence of a lactose residue at the reducing end and it can be presumed that they result from the action of glycosyltransferases associated with the Golgi apparatus on lactose. Some of these compounds are closely related to the blood group substances and give positive responses in blood group immunological tests (see Ebner & Schanbacher, 1974). Their presence in milk is indicative of the presence of the appropriate transferase which forms the glycosidic bond characteristic of that particular blood group. Though human milk is a rich source of these sugars they also occur in bovine milk, and the milks of a wide range of species contain uncharacterized oligosaccharides detectable by chromatography (Jenness, Regehr & Sloan, 1964).

In addition to the lactose derivatives, an inositol derivative, 6-β-galactinol, is found in rat milk (Naccarato, Ray & Wells, 1975) and appears to be unique to the mammary gland.

One of the most studied oligosaccharides is sialyllactose (several isomers exist) which occurs in large amounts in the milks of rat and mouse (Kuhn, 1972). On the fourth day of lactation in the rat sialyllactose accounts for 38% of the carbohydrate content of milk but falls to 2% by day 20. In contrast, guinea-pig mammary gland contains no sialyllactose. The enzyme responsible for the synthesis of sialyllactose from lactose and CMP-sialic acid, sialyltransferase, is found in a particulate fraction of rat mammary gland (Barra, Cumar & Caputto, 1969) and isolates with lactose synthetase in density gradient centrifugation. As mammary gland has the ability to form sialic acid from glucose (Carubelli, Taha, Trucco & Caputto, 1964) the whole enzyme complement necessary for the synthesis of this sugar is present in the gland. Rat mammary tissue also contains an active neuraminidase (Taha & Carubelli, 1967) capable of producing lactose from sialyllactose but it is not known how the balance between synthesis and hydrolysis is maintained.

It seems that the quantity of oligosaccharides appearing in a particular milk depends on the balance between lactose escaping further modification and lactose which acts as a substrate for glycosyltransferase reactions. It is possible that some modification of lactose occurs in milk as several glycosyltransferases occur there in soluble forms (see Ebner & Schanbacher, 1974) and the milk of some species contains appreciable quantities of sugar nucleotides which could act as donors. Bovine milk is a particularly rich source containing $0 \cdot 2$–$0 \cdot 6$ mmol/l UDP-hexoses (Johke, 1963) in early lactation, and goat milk is also a good source. However, the relative importance of post-secretion oligosaccharide synthesis in milk has not, to my knowledge, been determined.

Whether these milk oligosaccharides are merely a biochemical accident or have some nutritional significance is not known. According to Kobata, Tsuda & Ginsburg (1969) human milk contains 1–5 mmol/l lactose equivalents of oligosaccharides compared to about 200 mmol/l lactose so their contribution to the calorific value of the milk is trivial.

SECRETION OF MILK SUGARS

Milk sugars are formed within the vesicles of the Golgi apparatus and secreted into the alveolar lumen without coming into contact with the cytosol (Fig. 1). Brew (1969) first suggested this form of spatial organization and postulated that the flux of α-lactalbumin through the Golgi apparatus was the factor determining lactose output. Though there is much circumstantial evidence for this scheme direct evidence is difficult to acquire, as in cell fractionation studies some damage inevitably occurs to cell organelles, and vesicles can be disrupted and then reform (Ehrenreich, Bergeron, Siekevitz & Palade, 1973). However some of the Golgi contents are retained, as well-washed particulate fractions from lactating mammary glands are able to synthesize lactose in the absence of added α-lactalbumin, an ability which is lost when the structure is destroyed by detergent treatment (Jones, 1972). Kuhn & White (1975a, 1976) have studied the properties of a particulate fraction from rat mammary gland in detail and established several properties which are consistent with the

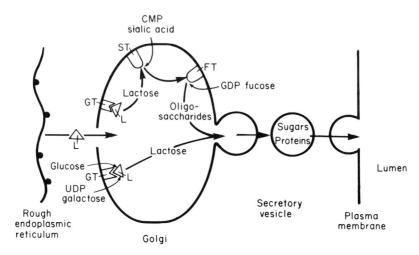

FIG. 1. Diagram of milk sugar synthesis and secretion in the mammary epithelial cell. Abbreviations: L, α-lactalbumin; GT, galactosyltransferase; ST, sialyltransferase; FT, fucosyltransferase.

hypothesis that lactose is synthesized and retained within the Golgi apparatus. They showed that, under appropriate conditions, lactose synthesized *in vitro* by these particles was retained and that galactosyl-transferase was situated largely on the inner walls of the vesicles. They also demonstrated the existence of a glucose transport system similar to that found in the plasma membrane, which could be inhibited by phlorid-zin, but which could not accumulate glucose against a concentration gradient. In addition the particles are freely permeable to UDP-galactose but less so to the competitive inhibitors UDP-glucose and UDP-glucuronate. Thus all the necessary metabolites for lactose production can enter the Golgi apparatus and, if their concentrations are adequate, lactose will be produced.

The subsequent transport of milk sugars and proteins across the cell and their discharge into the alveolar lumen are not fully understood. Presumably the secretory vacuoles are directed towards the correct cell surface by the microtubule system and it has been shown that colchicine, which disrupts microtubule structure, inhibits lactose secretion by guinea-pig mammary gland slices (Loizzi, De Pont & Bonting, 1975). Once discharged into the lumen, milk sugars are kept there by the impermea-bility of the plasma membrane of the epithelial cells though in some species leakage back into the blood stream occurs by an intercellular pathway (Linzell & Peaker, 1974; Peaker & Taylor, 1975). Kuhn & Linzell (1970) calculated that in the goat 2–3% of total lactose output passes into the plasma and if the intracellular junctions are tight this implies a limited transport through the plasma membrane or a misdirection of some secretory vacuoles towards the basal face of the cell. In the rat during lactation the urinary excretion of sialyllactose equals the rate of secretion in milk (Maury, 1972) which represents a major loss of carbohydrate, but the route taken by this sugar is not known. Until more has been learnt about the driving force behind the movement of secretory vacuoles across the cell it is impossible to speculate on the importance of the transport and secretion phase in determining the composition of milk.

Another obscure point is the origin of the soluble galactosyltransferase in milk. If one assumes that it is derived from Golgi galactosyltransferase, and there does not seem to be any other possible source, it must be solubilized during the secretory process. The origin of the milk fat globule membrane is still controversial but it may contain some Golgi components. However in the cow it has no galactosyltransferase activity (Keenan, 1974) though it has a limited amount in man (Martel, Dubois & Got, 1973). Thus at some point during the membrane transformation the transferase becomes detached and loses the hydrophobic component which attaches it to the membrane. Freshly-synthesized milk proteins take about 40 minutes to reach the alveolar lumen (Heald & Saacke, 1972) and if the Golgi membranes traverse the cell at the same rate it would be expected that the galactosyltransferase content of the mammary gland would turn over in this time. However, it can be calculated from the transferase activity of bovine mammary tissue and milk (Trayer & Hill, 1971; Klee & Klee, 1972; Mellenberger, Bauman & Nelson, 1973) that the enzyme content of the gland is transferred to the milk about once every two days. Thus there is either a big loss of activity during the solubilization process or some mechanism exists for the intracellular re-cycling of the enzyme.

CONTROL OF MILK SUGAR SYNTHESIS

This section is concerned with factors determining the rate of lactose and oligosaccharide production in a normal lactating gland and not with the more dramatic changes which occur at the onset of lactation. Comparison of the actual lactose output of the mammary gland with its measured galactosyltransferase activity reveals that synthetic capacity is stretched to the limit. In the cow it can be calculated from the data of Mellenberger *et al.* (1973) that 1 g of udder tissue has the capacity to produce 150 μmol lactose per day while the actual output is in the order of 200 μmol per day. Comparable figures for the rabbit (Jones & Cowie, 1972) are 1200 for synthetic capacity and 1500 for output. Of course enzyme activities measured *in vitro* may be underestimated, especially when one is dealing with a membrane-bound enzyme, but even so it is evident that there is little spare capacity. The problem becomes even greater if the metabolite concentrations in the Golgi vesicles are sub-optimal.

The factor most likely to be limiting lactose production is the glucose concentration. As shown in Table I, in the presence of excess α-lactalbumin the K_m of the transferase for glucose is 2mmol/l, i.e. a concentration of 10 mmol/l would give an 80% maximum reaction rate. Though the measurement of intracellular glucose concentrations is difficult because of the presence of glucose in extracellular fluids, esti-mates suggest that it is well below the plasma glucose concentration. Kuhn & White (1975b) argued that intracellular glucose probably equilibrates across the apical surface of the mammary epithelial cells with the glucose in milk and that the glucose concentration of 0·29 mmol/l in rat milk is a reasonable estimate of the intracellular glucose concentration. This is obviously far too low for efficient lactose synthesis so we are faced with a dilemma which can only be resolved by more data. In practice the extracellular glucose concentration does not seem to influence lactose production as Rao, Hegarty & Larson (1975) showed that isolated bovine mammary cells produce lactose at the same rate over a range of external glucose concentrations of 3–20 mmol/l.

The intracellular concentration of UDP-galactose in the rat mam-mary gland is 60–80 μmol/l (Murphy, Ariyanayagam & Kuhn, 1973) which is in the same order as the K_m values for UDP-galactose of the particulate enzyme (Table I). Thus the concentration of UDP-galactose may also limit lactose production but more needs to be known about the properties of the particulate enzyme before firm conclusions can be drawn. The synthesis of milk oligosaccharides depends on the availability of other sugar nucleotides such as CMP-sialic acid and GDP-fucose but, as far as I am aware, nothing is known of their intracellular distribution or transport.

α-Lactalbumin is present in milk in the concentration range 70–480 μmol/l in the species studied by Ley & Jenness (1970), which is in large excess of the requirement for lactose synthesis. Concentration may take place during transport and secretion but it seems probable that α-lactalbumin concentration is not limiting in the mature mammary gland. Jones (1972) estimated that the effective α-lactalbumin concentra-tion in particles isolated from mouse mammary gland was 3–7 μmol/l but considerable losses probably occur during preparation. As mentioned on p. 79, it appeared at one time that the availability of manganese would severely limit lactose production but now that the metal ion requirements of the synthetase are known to be less stringent this problem is not so great although which ions are actually effective *in vivo* remains to be dis-covered.

From the foregoing it will be seen that our knowledge of the control of lactose synthesis is extremely scanty. The difficulties of investigating metabolic control in a system in a small subcellular conpartment and involving a membrane-bound enzyme are formidable and some resource-ful research will be needed to overcome them.

SPECIES DIFFERENCES

The milks of different species have widely varying compositions though there is marked consistency within each of the mammalian orders. The most extensive and reliable compilation of milk composition data was made by Jenness & Sloan (1970) and most of the data in this section is from that source. For most wild species values are based on a few samples opportunely obtained or from animals in zoos. As Jenness & Sloan point out, milk obtained from animals in captivity can differ from that of the same species in the wild and when only a small number of samples have been analysed the results may not be typical. In addition values expressed as lactose are usually total carbohydrate or total reducing sugar and may consist largely of other sugars.

TABLE II

Average carbohydrate and fat content and calorific value of the milks of some mammalian orders and families

Order or family	Carbohydrate (g/100 ml)	Fat (g/100 ml)	Calorific value kcal/ 100 ml	Carbohydrate contribution (%)
Pinnipedia (seals)	0·8	43·9	465	1
Cetacea (whales)	0·9	29·6	328	1
Lagomorpha (rabbits)	1·4	17·4	266	2
Muridae (mice)	1·8	14·7	206	4
Marsupialia (kangaroos etc.)	3·9	4·8	94	16
Felidae (cats)	4·1	10·5	169	9
Proboscidea (elephants)	4·2	10·5	141	12
Bovidae (cows etc.)	4·7	5·2	99	19
Camelidae (camels)	5·4	4·1	87	24
Equidae (horses)	6·8	1·7	54	50
Primates	6·9	4·1	79	34

Calculated from the data compiled by Jenness & Sloan (1970) using the calorific factors of Perrin (1958).

Table II presents data on some mammalian orders and families ranked in order of carbohydrate content with the contribution of the carbohydrate to the total calorific value of the milk also listed. The most striking fact to emerge is that the aquatic mammals, whales and seals, have milks with little carbohydrate but a high fat content. As pointed out by Palmiter (1969) there is good inverse correlation between lactose and fat contents for different species. As fat is secreted in the non-aqueous phase of milk there is an obvious advantage for water conservation in replacing lactose by fat and this may be one factor favouring the high fat content in the milk of sea-living mammals. However a group such as the camels which normally live under arid conditions, has an average carbohydrate/fat ratio very similar to that of the other artiodactyls with no evidence of particular specialization.

Another point to emerge from Table II is that, in most milks, carbohydrates provide only a small proportion of the calorific value, horses and primates being exceptions. In many species milk carbohydrate is probably less important as an energy source than as a raw material for the synthesis of essential tissue components which would otherwise have to be produced from protein by gluconeogenesis. As the carbohydrate content of milk rises its calorific value falls so there is no iso-calorific replacement of fat by carbohydrate.

As detailed on pp. 85–86, our knowledge of the control of milk carbohydrate synthesis is meagre, so it is not surprising that the factors determining the composition of the milk of a particular species are not understood. The constancy of milk composition within a particular order suggests that genetic control predominates. For example, the Dall sheep (*Ovis dalli*) is an arctic species found in Alaska but its milk lactose content of 4·8% (Cook, Pearson, Simmons & Baker, 1970) is virtually identical with the 4·6% characteristic of sheep in less extreme environments. It might be expected that the galactosyltransferase content of the epithelial cells would be important but this enzyme is widely distributed and has functions other than lactose production. Though the Californian sea-lion (*Zalophus californianus*) produces no lactose the galactosyltransferase content of its mammary glands is similar to that of the rat (Johnson, Christiansen & Kretchmer, 1972). The reason the sea-lion produces no lactose is that it has no α-lactalbumin. Ley & Jenness (1970) showed that there was a good correlation between the α-lactalbumin and lactose contents of milks of different species, a fact which suggests that the rate of α-lactalbumin synthesis may be the main factor determining the lactose content of milk. However there are objections to this hypothesis. As mentioned on p. 86, there is no firm evidence that lactose production is limited by the α-lactalbumin concentration in the Golgi vesicles. Also, as shown by Quarforth & Jenness (1975) and Khatra *et al.* (1974), there is no evidence that α-lactalbumin has evolved so as to combine with maximum efficiency with galactosyltransferase from the same species. For example, bovine α-lactalbumin is a better activator of pig galactosyltransferase than pig α-lactalbumin itself. If α-lactalbumin were limiting lactose production it might be expected that a mutation reducing the affinity with galactosyltransferase would be highly disadvantageous. However, if α-lactalbumin is normally present in excess, mutations would be more tolerable which would explain the rapid rate of evolution it has undergone (Brew, Steinman & Hill, 1973).

EVOLUTIONARY ASPECTS

The most reasonable assumption is that the mechanism for the synthesis of milk sugars evolved from the synthesis of the oligosaccharide side-chains of glycoprotein. Almost all proteins secreted by the cell contain a carbohydrate component which is added during transit through the Golgi apparatus by the action of various glycosyltransferases. A monosaccharide such as glucose would have serious disadvantages as a major milk component. Cell membranes are permeable to glucose and thus the maintenance of a concentration of 150 mmol/i glucose (the equivalent of 5% lactose) would require energy expenditure to support a hundredfold concentration difference across the apical surface. If the lactose content of the carbohydrate-rich milks were present as a monosaccharide they would be hypertonic with respect to intracellular fluid and thus place an even greater energy drain on the cells.

It is difficult to conceive how the lactose synthetase system with its dependence on a specialized α-lactalbumin evolved. As galactosyltransferase by itself can employ N-acetylglucosamine as acceptor it would seem that N-acetyllactosamine was a more probable candidate for a milk disaccharide. However, there are probably advantages in the ability to switch lactose synthesis on and off, which the involvement of α-lactalbumin confers, and N-acetyllactosamine is a relatively poor substrate for hydrolysis by β-galactosidase and hence might be a poor food for the young.

As α-lactalbumin has a structural affinity with lysozyme from various sources and as both molecules co-exist in the same animal it presumably originated by gene duplication (Brew, Steinman et al., 1973). From the evolutionary point of view the milks of the monotremes, the echidnas and platypus, are of great interest but it is difficult to obtain them in reasonable quantities. The milks of both contain mixtures of oligosaccharides; echidna milk sugar consists of 50% sialyl-lactose, 26% fucosyl-lactose, 13% difucosyl-lactose and 8% lactose (Messer, 1974); platypus milk sugars contain only 1% lactose and the major carbohydrate is difucosyl-lactose (Messer & Kerry, 1973). These results are based on a limited number of samples and it is possible that the composition changes with stage of lactation. Hopper & McKenzie (1974) investigated the protein composition of echidna and platypus milks and obtained highly interesting results though because of the limited material available they require amplification. The milks of both species have the ability to synthesize lactose from added UDP-galactose and glucose and thus presumably contain a galactosyltransferase. The echidna galactosyltransferase activity is stimulated by bovine α-lactalbumin but the milk itself does not contain a typical α-lactalbumin. Instead it contains a protein with the iso-electric and enzymatic characteristics of lysozyme but the ability to stimulate lactose synthesis by bovine galactosyltransferase. If further research confirms that this protein doubles as lysozyme and α-lactalbumin, its amino acid sequence and conformation are of extreme interest but at the moment some caution is necessary as it can be calculated from the data of Hopper & McKenzie that its apparent K_m in the lactose synthetase reaction is at least 70 μmol/l compared with values in the 5–10 μmol/l range for normal α-lactalbumins. The possibility of the presence of an active impurity must be excluded. If the milks of the monotremes are genuinely primitive they suggest that the emergence of lactose as the principal milk sugar involved not only the evolution of α-lactalbumin but also some spatial or biochemical re-organization to prevent the lactose becoming an acceptor for further sugar residues. Assuming the oligosaccharides have no nutritional significance their synthesis represents a waste of high-energy intermediates.

Our knowledge of the composition of marsupial milk is still incomplete. Jenness *et al.* (1964) found that the milk of the red kangaroo contains a complex mixture of sugars among which lactose was only a minor component, and monosaccharides with the chromatographic properties of glucose and galactose predominated. The milks of other marsupials studied also contained only small amounts of lactose. If these were typical samples they suggest that in some way marsupials are able to cope with the problems associated with high monosaccharide concentrations in milk. A puzzling observation (Stephens, Irvine, Mutton, Gupta & Harley, 1974) is that the young of kangaroos are galactose intolerant, lacking the enzymes necessary for the metabolism of this compound, and develop cataracts if fed on milk containing lactose. However as the milk of the grey kangaroo contains a characteristic and functional α-lactalbumin (Brew, Steinman *et al.*, 1973) it must be presumed that some lactose is synthesized in the mammary gland and appears in the milk. It is evident that there are contradictions here, the resolution of which is important for the understanding of the evolution of milk sugar synthesis. The other fascinating aspect of lactation in the kangaroo which requires further investigation is its ability to suckle simultaneously young of different ages with milks of different composition. This presents a unique opportunity for studying the control of milk composition independent of endocrine control.

CONCLUSION AND FUTURE PROSPECTS

The discovery of the involvement of α-lactalbumin in lactose synthesis introduced a period of intense research on the properties of milk galactosyltransferase which still continues but which has already provided much valuable information. However, knowledge of the synthesis of lactose and other milk sugars *in vivo* and its control remains scanty and we are unable to say why a particular species produces lactose at its characteristic rate. What is needed initially is a more detailed study of the membrane-bound glycosyltransferases to determine in which ways, if any, they differ from the soluble forms. Then it will be necessary to tackle the more difficult task of determining metabolite concentrations within the Golgi apparatus *in vivo* in order to understand how synthesis is controlled. Evidently more work on the milks of monotremes and marsupials might help to reveal how lactose synthesis has evolved and more careful analysis of the milks of more exotic species might reveal that lactose derivatives rather than lactose itself are more important than has been realized. The knowledge of milk sugar synthesis which has been gained so far, though very valuable, has largely served to reveal the large tracts of ignorance which still remain.

REFERENCES

Andrews, P. (1970). Purification of lactose synthetase A protein from human milk and demonstration of its interaction with α-lactalbumin. *FEBS Lett.* **9**: 297–300.

Andrews, P. (1972). The effect of temperature on a reaction catalysed by lactose synthetase A protein. *FEBS Lett.* **26**: 333–335.

Barker, R., Olsen, K., Shaper, J. H. & Hill, R. L. (1972). Agarose derivatives of uridine diphosphate galactose and N-acetylglucosamine for the purification of galactosyltransferase. *J. biol. Chem.* **247**: 7135–7147.

Barman, T. E. & Bagshaw, W. (1972). The modification of tryptophan residues of bovine α-lactalbumin with 2-hydroxy-5-nitrobenzyl bromide and with dimethyl-(2-hydroxy-5-nitrobenzyl)-sulphonium bromide. II. Effect on the specifier protein activity. *Biochim. biophys. Acta* **278**: 491–450.

Barra, H. S., Cumar, F. A. & Caputto, R. (1969). The synthesis of neuramin lactose by preparation of the mammary gland and its relation to the synthesis of lactose. *J. biol. Chem.* **244**: 6233–6240.

Brew, K. (1969). Secretion of α-lactalbumin into milk and its relevance to the organization and control of lactose synthetase. *Nature, Lond.* **222**: 671–672.

Brew, K. & Hill, R. L. (1975a). Lactose biosynthesis. *Rev. Physiol. Biochem. Pharmacol.* **72**: 105–158.

Brew, K. & Hill, R. L. (1975b). Lactose synthetase. *Adv. Enzymol.* **43**: 411–490.

Brew, K., Shaper, J. H., Olsen, K. W., Trayer, I. P. & Hill, R. L. (1975). Cross-linking of the components of lactose synthetase with dimethyl-pimelimidate. *J. biol. Chem.* **250**: 1434–1444.

Brew, K., Steinman, H. M. & Hill, R. L. (1973). A partial amino acid sequence of α-lactalbumin-I of the grey kangaroo. *J. biol. Chem.* **248**: 4739–4742.

Brew, K., Vanaman, T. C. & Hill. R. L. (1968). The role of α-lactalbumin and the A protein in lactose synthetase: a unique mechanism for the control of a biological reaction. *Proc. natn. Acad. Sci. U.S.A.* **59**: 491–497.

Brodbeck, U. & Ebner, K. E. (1966). Resolution of a soluble lactose synthetase into two protein components and solubilization of microsomal lactose synthetase. *J. biol. Chem.* **241**: 762–764.

Carubelli, R., Taha, B., Trucco, R. E. & Caputto, R. (1964). Incorporation of [^{14}C]glucose into lactose and neuraminlactose by rat mammary glands. *Biochim. biophys. Acta* **83**: 224–230.

Challand, G. S. & Rosemeyer, M. A. (1974). The correlation between the apparent molecular weight and the enzyme activity of lactose synthetase. *FEBS Lett.* **47**: 94–97.

Cook, H. W., Pearson, A. M., Simmons, N. M. & Baker, B. E. (1970). Dall sheep milk. I. Effects of stage of lactation on the composition of the milk. *Can. J. Zool.* **48**: 629–633.

Ebner, K. E. & Schanbacher, F. L. (1974). Biochemistry of lactose and related carbohydrates. In *Lactation* 2: 77–113. Larson, B. L. & Smith, V. R. (eds). New York and London: Academic Press.

Ehrenreich, J. H., Bergeron, J. J. M., Siekevitz, P. & Palade, G. E. (1973). Golgi fractions prepared from rat liver homogentates. I. Isolation procedure and morphological characterization. *J. Cell Biol.* **59**: 45–72.

Fraser, I. H. & Mookerjea, S. (1976). Studies on the purification and properties of uridine diphosphate galactose-glycoprotein galactosyltransferase from rat liver and serum. *Biochem. J.* **156**: 347–355.

Grimmonprez, L. & Montreuil, J. (1975). Isolation and study of the physico-chemical properties of human milk oligosaccharides. *Biochimie* **57**: 695–701.

Heald, C. W. & Saacke, R. G. (1972). Cytological comparison of milk protein synthesis of rat mammary tissue *in vivo* and *in vitro*. *J. Dairy Sci.* **55**: 621–628.

Hopper, K. E. & McKenzie, H. A. (1974). Comparative studies of α-lactalbumin and lysozyme: echidna lysozyme. *Molec. Cell. Biochem.* **3**: 93–108.

Jenness, R., Regehr, E. A. & Sloan, R. E. (1964). Comparative biochemical studies of milk. II. Dialyzable carbohydrates. *Comp. Biochem. Physiol.* **13**: 339–352.

Jenness, R. & Sloan, R. E. (1970). The composition of milks of various species: a review. *Dairy Sci. Abstr.* **32**: 599–612.

Johke, T. (1963). Acid-soluble nucleotides of colostrum, milk and mammary gland. *J. Biochem. (Tokyo)* **54**: 388–397.

Johnson, J. D., Christiansen, R. O. & Kretchmer, N. (1972). Lactose synthetase in the mammary gland of the California sea-lion. *Biochem. Biophys. Res. Commun.* **47**: 393–397.

Jones, E. A. (1972). Studies on the particulate lactose synthetase of mouse mammary gland. *Biochem. J.* **126**: 67–78.

Jones, E. A. & Cowie, A. T. (1972). The effect of hypophysectomy and subsequent replacement therapy with sheep prolactin or bovine growth hormone on the lactose synthetase activity of rabbit mammary gland. *Biochem. J.* **130**: 997–1002.

Keenan, T. W. (1974). Membranes of mammary gland. IX. Concentration of glycosphingolipid galactosyl and sialyltransferase in Golgi apparatus from bovine mammary gland. *J. Dairy Sci.* **57**: 187–192.

Keenan, T. W., Morré, D. J. & Cheetham, R. D. (1970). Lactose synthesis by a Golgi apparatus fraction from rat mammary gland. *Nature, Lond.* **228**: 1105.

Khatra, B. S., Herries, D. G. & Brew, K. (1974). Some kinetic properties of human milk galactosyltransferase. *Eur. J. Biochem.* **44**: 537–560.

Klee, W. A. & Klee, C. B. (1972). The interaction of α-lactalbumin and the A protein of lactose synthetase. *J. biol. Chem.* **247**: 2336–2344.

Kobata, A., Tsuda, M. & Ginsburg, V. (1969). Oligosaccharides of human milk. I. Isolation and characterization. *Arch. Biochem. Biophys.* **130**: 509–513.

Kuhn, N. J. (1972). The lactose and neuraminlactose content of rat milk and mammary tissue. *Biochem. J.* **130**: 177–180.

Kuhn, N. J. & Linzell, J. L. (1970). Measurement of the quantity of lactose passing into mammary venous plasma and lymph in goats and in a cow. *J. Dairy Res.* **37**: 203–208.

Kuhn, N. J. & White, A. (1975a). The topography of lactose synthesis. *Biochem. J.* **148**: 77–84.

Kuhn, N. J. & White, A. (1975b). Milk glucose as an index of the intracellular glucose concentration of rat mammary gland. *Biochem. J.* **152**: 153–155.

Kuhn, N. J. & White, A. (1976). Evidence for specific transport of uridine diphosphate galactose across the Golgi membrane of rat mammary gland. *Biochem. J.* **154**: 243–244.

Ley, J. M. & Jenness, R. (1970). Lactose synthetase activity of α-lactalbumin from several species. *Arch. Biochem. Biophys.* **138**: 464–469.

Linzell, J. L. & Peaker, M. (1974). Changes in colostrum composition and in the permeability of mammary epithelium at about the time of parturition in the goat. *J. Physiol., Lond.* **243**: 129–151.

Loizzi, R. F., De Pont, J. J. H. H. M. & Bonting, S. L. (1975). Inhibition by cyclic adenosine monophosphate of lactose production in lactating guinea pig mammary gland slices. *Biochim. biophys. Acta* **392**: 20–25.

Magee, S. C., Mawal, R. & Ebner, K. E. (1974). Multiple forms of galactosyltransferase from bovine milk. *Biochemistry* **13**: 99–102.

Martel, M. B., Dubois, P. & Got, R. (1973). Membranes des globules lipidiques du lait humain. *Biochim. biophys. Acta* **311**: 565–575.

Maury, P. (1972). Increased excretion of N-acetylneuramin (2–3) lactose in the urine of pregnant and lactating rats. *J. biol. Chem.* **247**: 3153–3158.

Mellenberger, R. W., Bauman, D. E. & Nelson, D. R. (1973). Metabolic adaptions during lactogenesis: fatty acid and lactose synthesis in cow mammary tissue. *Biochem. J.* **136**: 741–748.

Messer, M. (1974). Identification of N-acetyl-4-O-acetylneuraminyl lactose in Echidna milk. *Biochem. J.* **139**: 415–420.

Messer, M. & Kerry, K. R. (1973). Milk carbohydrates of the Echidna and the Platypus. *Science, N.Y.* **180**: 203–205.

Meyer, H. & Lemmer, U. (1974). Nutrient contents of by-products (forestomach, udder, lung) of slaughter cattle. *Kleintier Praxis* **19**: 44-45.

Morrison, J. F. & Ebner, K. E. (1971a). Studies on galactosyltransferase. Kinetic investigations with N-acetylglucosamine as the galactosyl group acceptor. *J. biol. Chem.* **246**: 3977–3984.

Morrison, J. F. & Ebner, K. E. (1971b). Studies on galactosyltransferase. Kinetic effects of α-lactalbumin with N-acetylglucosamine and glucose as galactosyl group acceptors. *J. biol. Chem.* **246**: 3992–3998.

Murphy, G., Ariyanayagam, A. D. & Kuhn, N. J. (1973). Progesterone and the metabolic control of the lactose biosynthetic pathway during lactogenesis in the rat. *Biochem. J.* **136**: 1105–1116.

Naccarato, W. F., Ray, R. E. & Wells, W. W. (1975). Characterization and tissue distribution of 6-O-β-D-galactopyranosyl myo-inositol in the rat. *J. biol. Chem.* **250**: 1872–1876.

Osborne, J. C. & Steiner, R. F. (1974). Interaction of the components of the lactose synthetase system. *Arch. Biochem. Biophys.* **165**: 615–627.

Palmiter, R. D. (1969). What regulates lactose content in milk? *Nature, Lond.* **221**: 912–914.

Peaker, M. & Taylor, J. C. (1975). Milk secretion in the rabbit: changes during lactation and the mechanism of ion transport. *J. Physiol., Lond.* **253**: 527–547.

Perrin, D. (1958). Factors for calorific values of milk components. *J. Dairy Res.* **25**: 215–220.

Powell, J. T. & Brew, K. (1974a). The preparation and characterization of two forms of bovine galactosyltransferase. *Eur. J. Biochem.* **48**: 217–228.

Powell, J. T. & Brew, K. (1974b). Glycosyltransferase in the Golgi membranes of onion stem. *Biochem. J.* **142**: 203–209.

Powell, J. T. & Brew, K. (1975). Interaction of α-lactalbumin and galactosyltransferase during lactose synthesis. *J. biol. Chem.* **250**: 6337–6344.

Powell, J. T. & Brew, K. (1976). Metal ion activation of galactosyltransferase. *J. biol. Chem.* **251**: 3645–3652.

Quarforth, G. J. & Jenness, R. (1975). Isolation, composition and functional properties of α-lactalbumin from several species. *Biochem. biophys. Acta* **379**: 476–487.

Rao, D. R., Hegarty, H. M. & Larson, B. L. (1975). Effect of cell density on lactose synthesis in bovine mammary cell cultures. *J. Dairy Sci.* **58**: 159–164.

Schindler, M., Sharon, N. & Prieels, J.-P. (1976). Reversible inactivation of lactose synthetase by the modification of histidine 32 in human α-lactalbumin. *Biochem. biophys. Res. Commun.* **69**: 167–173.

Smith, C. A., Powell, J. T. & Brew, K. (1975). Puromycin does not inactivate the galactosyltransferase of Golgi membrane. *Biochim. biophys. Res. Commun.* **62**: 621–626.

Stephens, T., Irvine, S., Mutton, P., Gupta, J. D. & Harley, J. D. (1974). The case of the cataractous kangaroo. *Med. J. Aust.* **2**: 910–911.

Taha, B. H. & Carubelli, R. (1967). Mammalian neuraminidase: intracellular distribution and changes of enzyme activity during lactation. *Arch. Biochem. Biophys.* **119**: 55–61.

Trayer, I. P. & Hill, R. L. (1971). The purification and properties of the A protein of lactose synthetase. *J. biol. Chem.* **246**: 6666–6675.

Symp. zool. Soc. Lond. (1977) No. 41, 95–111

Mammary Energy Metabolism

G. H. SMITH and D. J. TAYLOR

Department of Animal Physiology and Nutrition,
University of Leeds, Leeds, England

SYNOPSIS

In addition to a supply of substrates for the synthesis of milk constituents, the mammary gland needs a supply of materials for oxidation in order to provide the energy, in the form of adenosine triphosphate (ATP) and reducing equivalents, required by biosynthetic pathways. In ruminants there is a large demand upon the supply of glucose for lactose synthesis, and the availability of alternative substrates in such animals raises the question of the extent to which these may replace the oxidation of glucose for the provision of both energy and of reducing equivalents.

Evidence which has been produced for the existence of pathways which may spare glucose in the ruminant mammary gland as compared with the non-ruminant is reviewed. It is difficult, however, to assess the extent to which these pathways operate, and various experimental approaches are described.

Measurements of the relative oxidation of different substrates, including glucose and acetate, have been made by comparing their contribution to milk citrate. The results provide what is felt to be reliable evidence of the participation of glucose in the tricarboxylic acid cycle, but indicate that the contribution of different substrates varies according to the nutritional state of the animal.

The conclusion drawn is that there is no marked block upon the oxidation of glucose in the tricarboxylic acid cycle such as the low activity of ATP-citrate lyase imposes upon the entry of glucose into fatty acid synthesis, and the concept that the mammary gland of the ruminant always operates so as to spare glucose for lactose synthesis is probably an oversimplification. The existence of pathways which may spare glucose and which appear to be unique to the ruminant mammary gland suggests, however, that the gland has this capability, and the extent of oxidation of glucose and other substrates needs to be investigated in a variety of nutritional and physiological states.

INTRODUCTION: SATELLITE PATHWAYS

The biochemical pathways, aptly described by Gumaa, Greenbaum & McLean (1973) as the "satellite systems", whose activity provides the major biosynthetic pathways with their requirements for energy in the form of adenosine triphosphate (ATP) and reducing equivalents for fatty acid synthesis, are discussed in this paper.

The relationship of the satellite systems to the major pathways is illustrated in Fig. 1. The pathways for lactose, fat and protein synthesis require a supply of ATP, the major source of which is obtained through the oxidation of materials in the tricarboxylic acid cycle and related pathways. The synthesis of fatty acids requires a supply of reducing

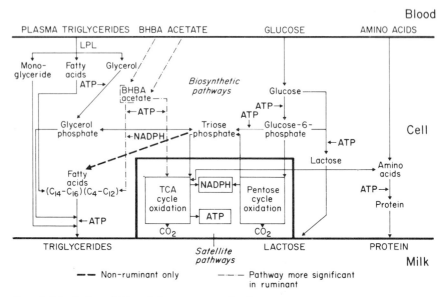

FIG. 1. Metabolic activities of the mammary gland.

equivalents which is also met by oxidation of substrates, but in this case the hydrogen atoms removed from the substrate by oxidizing enzymes are used for reduction rather than for oxidation in the mitochondria to generate ATP. The pathways generating reducing equivalents and ATP are thus related since they represent alternative fates of oxidizable substrates, and should therefore be considered together when discussing the energy metabolism of the mammary gland. Materials for oxidation represent a demand made by the gland in addition to that for the carbon precursors of the milk constituents, and the internal economy of the gland may be expected to be adjusted according to the type and amount of substrates available.

In this connection ruminants are of particular interest. The great advantage of the ability to utilize β-linked glucose polymers (the celluloses) in the diet, gained through the activities of the micro-organisms in the rumen, has entailed the sacrifice of the ready availablity of glucose from the small intestine, since the starches as well as the celluloses are fermented. The hepatic portal blood of ruminants is supplied with acetic, propionic and butyric acids from the rumen. Of these three, only propionic acid is glucogenic, being converted to glucose mainly in the liver; the others appear in the systemic blood as acetate and 3-hydroxybutyrate. Therefore, the ruminant mammary gland has alternative substrates, which, as well as being available for oxidation, may be used for fatty acid synthesis; conversely the supply of glucose may be limited. Nevertheless, the mammary gland does use considerable quantities of glucose. Indeed

Annison & Linzell (1964) have shown in lactating goats that uptake by the mammary glands accounts for 60–85% of the entire glucose production. Therefore, the glucose metabolism of the lactating animal is dominated by the demands of the udder. Much of the glucose removed by the udder (60–70%) is used for lactose production, for which glucose is effectively the sole substrate. One might suppose therefore that the ruminant mammary gland has developed so as to conserve glucose for this essential function, utilizing alternative substrates for other purposes where possible. One such restriction on the metabolism of glucose is well recognized: glucose carbon does not contribute to fatty acid synthesis, as it does in the non-ruminant, and is replaced for that purpose by acetate and 3-hydroxybutyrate. This hypothesis may serve as a standpoint from which to compare the operation of satellite pathways in ruminants and non-ruminants, and raises the question as to whether there are other such well marked differences in glucose metabolism.

SATELLITE PATHWAYS IN RUMINANTS AND NON-RUMINANTS

Fatty acid synthesis, whether from glucose or other substrates, occurs in the cytosol of cells and has a specific requirement for reducing equivalents in the form of reduced nicotinamide adenine dinucleotide phosphate (NADPH). Reducing equivalents from the oxidation of substrates in the tricarboxylic acid cycle, however, are produced in the mitochondria (whose membranes are impermeable to the nictotinamide co-enzymes) largely in the form of reduced nicotinamide adenine dinucleotide (NADH). Attention has therefore been drawn to the pentose phosphate pathway for the oxidation of glucose, which occurs in the cytosol and specifically produces NADPH as an important source of these reducing equivalents. There is much evidence which confirms the importance of the pathway in both ruminant and non-ruminant species. The activity of the enzymes concerned (particularly glucose-6-phosphate dehydrogenase and phosphogluconate dehydrogenase) increases in mammary tissues with the onset of lactation (see, for example, McLean, 1958; Gumaa *et al.*, 1973, and the comprehensive review by Bauman & Davis, 1974); there is also more direct evidence from isotope work using tritium-labelled compounds and measurements of carbon fluxes (Abraham & Chaikoff, 1959; Abraham, Katz, Bartley & Chaikoff, 1963; Lowenstein, 1961). The pentose pathway therefore represents a further demand upon the glucose supply to the mammary gland cell; this point will be considered further below.

The pentose cycle is not, however, the only pathway for the regeneration of NADPH (Fig. 2). In those tissues which synthesize fatty acids from glucose, the path of carbon involves the conversion of glucose to pyruvate in the cytosol of the cell, and the passage of pyruvate into the mitochondria where it is converted to acetyl coenzyme A (CoA). Since the

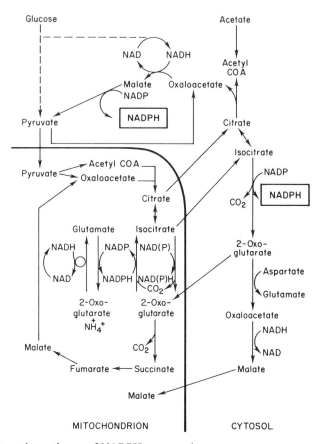

FIG. 2. Alternative pathways of NADPH regeneration.

mitochondrial membrane is impermeable to acetyl-CoA, the efflux of
carbon units for fatty acid synthesis in the cytosol occurs as citrate, which is
split in the cytosol into oxaloacetate and acetyl-CoA. The activity of the
enzyme catalysing this last reaction (ATP-citrate lyase) is very low in
ruminant tissue, and this accounts for the failure of glucose carbon to
contribute to fatty acid synthesis in these animals (Hardwick, 1966). The
conversion of glucose into pyruvate is accompanied by the production of
NADH in the cytosol. This may be used for the production of NADPH
through the coupled operation of NAD-malate dehydrogenase and
NADP-malate dehydrogenase, which at the same time disposes of the
oxaloacetate produced by citrate cleavage (Wise & Ball, 1964). The
oxaloacetate is re-formed in the mitochondria by pyruvate carboxylase, an
enzyme which in some species also occurs in the cytosol (Gul & Dils, 1969).
Evidence for the operation of the pyruvate–malate cycle comes from the

great increase in NADP-malate dehydrogenase in rat mammary tissue at the onset of lactation, and also from experiments with tritium-labelled lactate which showed that NADH produced in the cytosol could contribute to fatty acid synthesis (Lowenstein, 1961). The complete malic cycle is unlikely to operate to a significant extent in ruminant tissues owing to the low activity of ATP-citrate lyase which is paralleled by the low activity of NADP-malate dehydrogenase (Bauman, Brown & Davis, 1970).

A third source of NADPH in the cytosol involves the enzyme isocitrate dehydrogenase, and Bauman *et al.* (1970) have suggested that this is an important alternative pathway for NADPH regeneration, particularly for ruminants. Their suggestion is based upon the observation that ruminant tissue *in vitro* can synthesize fatty acids from acetate even in the absence of glucose (although it should be noted that glucose stimulated this synthesis thus indicating the importance of the pentose cycle in normal circumstances), and upon the high activity of the enzyme in ruminant mammary tissue compared with non-ruminant tissue. This scheme was later extended by Gumaa *et al.* (1973), who observed a similar large activity for malate dehydrogenase and glutamate dehydrogenase, to include the reduction of the 2-oxoglutarate, produced inside the mitochondria, back to isocitrate. In the former scheme only one mole of NADPH could be produced per mole of acetate oxidized, while the latter scheme, which allows the production of two, accords better with the observation that there was a one-to-one relationship between the amount of acetate oxidized and that incorporated into lipid in slices of mammary tissue incubated with acetate in the absence of glucose (two moles of NADPH are required to incorporate each two-carbon unit as fatty acids are synthesized) (Bauman *et al.*, 1970; Gumaa *et al.*, 1973). A further advantage of the latter scheme is the greater degree of flexibility it allows the metabolic system. In the scheme of Bauman *et al.* (1970), only one of the four sites where reducing equivalents are produced in the tricarboxylic acid cycle is coupled to NADPH production. The concomitant formation of ATP from the other sites might limit the availability of NADPH to the extent to which ATP is required, unless there is to be wastage of energy through uncoupling. In theory, all of the NADH formed in the mitochondria might be linked with NADPH production by a further extension of the scheme of Gumaa *et al.* (1973). For example, the oxidation of one mole of acetate may yield either two moles of NADPH with a concomitant yield of four moles of ATP (two moles of ATP are consumed in order to produce acetyl-CoA) or three moles of NADPH with one mole of ATP, if 2-oxoglutarate dehydrogenase is also involved.

There appear to be differences between species among non-ruminants, as well as the differences when compared with ruminants already mentioned, in the relative extent of the operation of the three pathways. For example, rat and mouse mammary glands have a high NADP-malate dehydrogenase activity, but isocitrate dehydrogenase is relatively low, whereas in guinea-pigs, pigs and rabbits the activity of isocitrate dehydrogenase is high, and (in pigs and rabbits) the activity of malate dehydrogenase is low (see Bauman & Davis, 1974).

The schemes involving isocitrate dehydrogenase allow the provision of reducing equivalents from any material which is finally oxidized in the tricarboxylic acid cycle. There is, however, no evidence for extensive oxidation of materials other than acetate or glucose in normal circumstances. Although 3-hydroxybutyrate would appear to be an alternative fuel for ruminants, little appears to be oxidized. For example, only 1·4% of the carbon dioxide produced by a perfused goat udder arose from this source (Linzell, Annison, Fazakerley & Leng, 1967). The behaviour of 3-hydroxybutyrate in this respect is a little puzzling since 3-hydroxybutyrate is used as a carbon precursor for fatty acid biosynthesis in ruminant mammary gland *in vivo* (Smith, McCarthy & Rook, 1974), where it contributes predominantly as a four-carbon unit. It has been suggested by Bauman *et al.* (1970) that the location of 3-hydroxybutyrate dehydrogenase (the first enzyme in the pathway for cleavage to two-carbon units) in the mitochondria will result in the acetyl-CoA units from 3-hydroxybutyrate being confined, like those of glucose, inside the mitochondria. This would not explain the low contribution of 3-hydroxybutyrate to mammary carbon dioxide production. However, a substantial contribution of 3-hydroxybutyrate as two-carbon units to fatty acid synthesis has been observed (Smith *et al.*, 1974; Linzell, Annison *et al.*, 1967). Since it is possible that the mitochondrial membrane is permeable to 3-hydroxybutyrate and acetoacetate (Devlin & Bedell, 1959) it may be that the location of enzymes later in the pathway (acetoacetate-succinyl-CoA transferase or acetoacetyl-CoA thiolase) whose products, as esters of CoA, will be confined to the part of the cell where they are formed in the absence of a translocation mechanism, determines the site of metabolism of 3-hydroxybutyrate. There is at present little information about this. A major contribution to oxidation from amino acids appears to be ruled out, since the amino acid uptake by the gland appears in general to balance the output in milk proteins (Linzell & Mepham, 1968). If anything, there is evidence for a slight deficiency of amino acids, which would require a net efflux of carbon from the tricarboxylic acid cycle. The existence of the scheme proposed by Gumaa *et al.* (1973) which involves glutamate dehydrogenase, would increase the exchange of carbon atoms between 2-oxoglutarate and glutamate. This possibility makes it difficult to assess the mammary oxidation of amino acids from the results of tracer studies which investigated the origin of carbon dioxide.

Fatty acids, from plasma triglycerides or as plasma non-esterified fatty acids, are a further potential source of oxidizable materials, but make only a very small contribution ($< 1\%$) to the carbon dioxide produced by the ruminant mammary gland, except in starvation when their contribution increases to 17% (see Linzell, 1974). The capacity to oxidize fatty acids seems, however, to have an active role in lactation, since the enzyme concerned in transporting the CoA esters into the mitochondria for oxidation (carnitine palmitoyl transferase) is present in mammary tissue of the sheep (Snoswell & Linzell, 1975) and in preliminary studies we have found that the mammary glands of several species contain this enzyme, and that it increases in activity in the rat from pregnancy to lactation.

RELATIVE CONTRIBUTIONS OF THE SATELLITE SYSTEMS

The conclusion from the foregoing summary is that there is a considerable amount of evidence, much of it based on enzyme profile studies, for a variety of possibilities in the operation of the satellite systems. There is much less direct evidence for the relative participation of different substrates in the pathways. There are several possible approaches to this problem which have been made.

One such approach is the measurement of the carbon dioxide output from radioactive substrates in the mammary gland as a measure of their contribution to oxidative metabolism, and much information on this has been collected by Annison, Linzell and their colleagues (see for example Annison & Linzell, 1964; Annison, Linzell & West, 1968; Bickerstaffe, Annison & Linzell, 1974); some of these data are summarized in Table I. There is little information about the non-ruminant. Considerable quantities of both glucose and acetate are oxidized, and in the goat these substrates at least account for the larger part of the total carbon dioxide output. The deficit may represent the oxidation of unidentified sub-

TABLE I

Oxidation of substrates by the lactating mammary gland

Animal	Substrate	Substrate oxidized (%)	Carbon dioxide from substrate (%)
Goat (fed)*	Glucose	$25\cdot4\pm3\cdot3$	$38\cdot8\pm4\cdot2$
	Acetate	$47\cdot0\pm8\cdot7$	$26\cdot0\pm2\cdot0$
	Fatty acid	0	0
Goat (fasted)*	Glucose	$7\cdot5\pm2\cdot5$	$9\cdot0\pm1\cdot0$
	Acetate	$62\cdot5\pm17\cdot5$	$11\cdot0\pm2\cdot0$
	Fatty acid	$13\cdot0\pm6\cdot0$	$17\cdot5$
Cow (fed)†	Glucose	$11\cdot0\pm2\cdot3$	$25\cdot0\pm4\cdot8$
	Acetate	$29\cdot0\pm9\cdot7$	$30\cdot0\pm6\cdot0$
Sow (fed)‡	Glucose	$34\cdot0$	$54\cdot0$
	Acetate		$2\cdot0$

* Data from Linzell (1974).
† Data from Bickerstaffe *et al.* (1974).
‡ Data from Linzell, Mepham, Annison & West (1969).

strates, but at least some of it might arise through dilution of tricarboxylic acid cycle intermediates by exchange reactions, for example between 2-oxoglutarate and glutamate, rather than by net oxidation of other materials. In the case of acetate or other precursors of acetyl-CoA, their contribution to carbon dioxide may be a reliable indication of their oxidation in the tricarboxylic acid cycle, since two carbon atoms are lost as carbon dioxide for every two which enter as acetyl-CoA (although one may participate in glutamate synthesis). The interpretation of carbon dioxide production from glucose is less straightforward owing to the operation of the pentose cycle as an alternative source of carbon dioxide. Several methods have been devised for assessing the relative contribution of the pentose cycle based upon the use of specifically-labelled glucose, and the methods have been extensively reviewed (Katz & Wood, 1960; Landau & Katz, 1965). What these methods determine is the proportion of triose phosphates which arise from glucose through the operation of the pentose cycle as compared with the operation of the Embden-Meyerhof-Needham pathway (Fig. 3), and do not directly measure the proportion of carbon dioxide produced by the pathway, since the subsequent fate of the triose phosphate is not determined. The method illustrated in Fig. 3, which depends upon the randomization of carbon atom from position two of glucose into positions one and three, is that which appears less affected by other pathways, and was used by Wood, Peeters, Verbeke, Lauryssens & Jacobson (1965) to conclude that in the isolated perfused cow udder 62% of the glucose was used for lactose synthesis, 30% in the pentose pathway and 8% in the Embden-Meyerhof-Needham pathway. This would indicate that 15% of the glucose had been metabolized to carbon dioxide by the pentose pathway. In the non-ruminant, the proportions of glucose oxidized in the Embden-Meyerhof-Needham pathway and in the pentose pathway appear to be approximately equal (McClean, 1964; Abraham & Chaikoff, 1959). The greater participation of the Embden-Meyerhof-Needham pathway in non-ruminants probably reflects the flux of carbon from glucose for fatty acid synthesis.

In the pentose cycle carbon dioxide appears only from carbon atoms in positions one, two and three and this has led to the use of the amount of carbon dioxide from the carbon atom in position six of glucose as a measure of its oxidation in the tricarboxylic acid cycle. However, in the ruminant mammary gland, but not in the non-ruminant, there is a considerable activity of fructose-diphosphatase (Baird, 1969). This enzyme is absent from adipose tissue, in which the methods of investigating the contribution of the pentose pathways were developed. The presence of the enzyme allows the possibility of recombination of the triose phosphates from the pentose cycle to produce glucose-6-phosphate again, so that in the extreme case the whole of the glucose molecule can be oxidized to carbon dioxide by the pentose cycle and measurements of the relative participation of the Embden-Meyerhof-Needham pathway

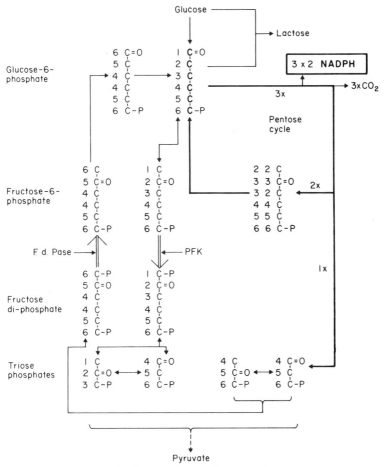

FIG. 3. The pentose cycle and related pathways. For simplicity, some possible randomizations of carbon atoms are not shown.

become rather meaningless. The activity of the enzyme is sufficient to account for extensive re-cycling. In the experiments of Wood *et al.* (1965), label from carbon in position six of glucose was transferred to position one of the galactose moiety of lactose (which is derived from glucose-6-phosphate), providing direct evidence for re-cycling of triose phosphates. The label from the six-position, in entering the one-position, is in competition with unlabelled carbon atoms appearing as fresh glucose-6-phosphate enters the system for lactose synthesis, and with carbon atoms returned from the pentose cycle itself. The appearance of label in the one position (which was one-fifth as active as the six-position in the experiments of Wood *et al.*, 1965) suggests that the re-cycling of triose phosphates may be extensive, but we have so far been unable to devise a practical method of determining the extent.

The re-cycling of glucose in the manner outlined in the ruminant mammary gland might represent a further means of economy in glucose utilization. In the pentose cycle itself only one-half of the glucose is used for NADPH regeneration and the remainder appears as a triose phosphate. In the ruminant, this cannot be used for fatty acid synthesis, and although it might contribute to the glycerol of triglycerides there would be an excess produced as compared with the amount of fatty acid synthesis made possible by the accompanying generation of NADPH. Another means of disposal is oxidation in the tricarboxylic acid cycle, which might be a wasteful use of glucose carbon. Lactate production is a further possibility but there is in fact a net uptake of lactate by the gland (see Linzell, 1974). The possibility of re-cycling glucose thus confers greater flexibility on the ruminant mammary gland.

Several investigations into the relative participation of glucose in the alternative oxidative pathways have assumed that the appearance of carbon dioxide from the six-position of glucose represents oxidation in the tricarboxylic acid cycle. The possibility of extensive re-cycling clearly questions the validity of these conclusions. Even if the true contribution of the pentose cycle can be assessed, the remaining carbon dioxide from the glucose does not represent net oxidation, since glucose may enter the tricarboxylic acid cycle as oxaloacetate or as acetyl-CoA. Only the latter route represents net oxidation, but carbon dioxide will become labelled by either route if labelled glucose is used. Indeed in an earlier review (Smith, 1971) it was suggested that there was little reliable evidence for the oxidation of glucose in the tricarboxylic acid cycle.

A further approach to the problem of assessing the satellite pathways depends upon the assumption that the flux through both the satellite and the main metabolic pathways will be so balanced as to ensure that each meets the demands of the other, without creating surpluses or deficits of metabolites which cannot be accounted for in changes in the composition of blood passing through the gland. The operation of satellite pathways may thus be predicted from the calculated requirements of ATP and NADPH to synthesize the constituents of milk of normal composition. The most comprehensive example of such an approach is that of Baldwin & Smith (1971), who have collated much of the available data together with the calculated energy, reducing equivalent and carbon requirements. However, as these authors point out, such an approach can only be tentative, particularly where there are alternative pathways producing the same products, and where greater flexibility is introduced by the separation of pathways, as the recycled pentose cycle and the pathway of Gumaa et al. (1973) separate the production of NADPH from the production of ATP. Nevertheless, the approach at least gives an indication of the order of activity of satellite pathways which may be expected to accompany a given rate of synthesis of milk constituents. A simple version of

Table II

Calculated requirements for the synthesis of 100 *g milk solids in the goat* (38 *g lactose*, 29 *g protein*, 33 *g fat*)

| | Oxidation to provide ATP (2·2 mol) and NADPH (0·66 mol) | | | |
| | Glucose | | Acetate | |
Pathway	wt. (g)	%	wt. (g)	%
Pentose cycle*	20	34	—	—
Re-cycled pentose cycle	10	21	12	38
Scheme of Bauman *et al.* (1970)	0	0	40†	67
Scheme of Gumaa *et al.* (1973)	0	0	26‡	56

* Including oxidation of triose phosphate by glycolytic pathway and tricarboxylic acid cycle.
† To produce NADPH—also produces 4·6 mol ATP.
‡ 20 g for NADPH, 6 g for ATP.

such an approach is shown in Table II. The ATP requirement shown must be regarded as a minimum since there is insufficient information available about the requirements for the maintenance of the metabolic machinery, or for more direct requirements for milk secretion such as the ion pumps described by Linzell & Peaker (1971).

In Table II, the amount of substrate oxidized to provide the ATP and NADPH requirement by the alternative systems has been assessed. If glucose is oxidized in the pentose cycle without re-cycling then sufficient energy is available from the oxidation of the triose phosphates produced. If the pentose cycle is re-cycled and glucose is conserved, then more acetate (or other substrate) must be oxidized to meet the ATP requirement. The consequences of assuming that all the reducing equivalents are supplied by oxidation in the tricarboxylic acid cycle in the absence of glucose are also shown. The relative oxidations of glucose and acetate predicted by this calculation show encouraging similarities with actual observations (Table I)—that is they are at least of the right order—but the information is not sufficiently precise to allow an accurate estimation of the relative contributions of the different possible strategies. Comparison of the figures obtained with the available data shown in Table I seems to suggest that no pathway operates exclusively.

Another approach is through the examination of enzyme profiles. This method is often only able to give information on the existence of possible pathways rather than on their quantitative importance, since the maximum activities of many enzymes are in excess of the flux through the pathways. This is because the actual activity *in vivo* is reduced by low concentrations of substrates and the presence of significant concentrations of products, as well as by other regulating mechanisms, allosteric effectors for example. There are, however, certain enzymes whose maximum activities do seem to be similar to the maximum fluxes (measured by more direct means) through the pathways concerned. If an enzyme shows this characteristic in several tissues, it is reasonable to assume that its activity may also be similar to the flux in tissues where the flux cannot be directly measured. Unfortunately, by definition, such enzymes have low activities (especially if the flux is low) and somewhat complex kinetics; their activities are therefore not often measured because of the technical difficulties involved. One such enzyme is 2-oxoglutarate dehydrogenase which in a variety of tissues reflects tricarboxylic acid cycle activity (Read, Crabtree & Smith, 1977). We have examined the activity of this enzyme in mammary tissue, and in Table III the activity can be seen to be similar to the tricarboxylic acid cycle flux predicted from the oxygen uptake. Unfortunately, there is no information which allows the selection of an enzyme reflecting the activity of the glycolytic pathway in mammary tissue, but included in Table III is the activity of pyruvate dehydrogenase, which in adipose tissue seems to be similar to the maximum flux of pyruvate to acetyl-CoA (Wieland, Siess, Löffler, Potzelt, Portenhauser, Hartmann & Schirmann, 1973). In the rat the activity is greater than that of 2-oxoglutarate dehydrogenase, possibly representing the extra flux of carbon through citrate for fatty acid synthesis. The activity of fructose-diphosphatase, compared with the calculated entry of glucose into pathways other than lactose synthesis, suggests that this enzyme would allow extensive re-cycling in the pentose cycle. The fluxes in Table III also show a correlation between the activity of the tricarboxylic acid cycle and the rate of lactose formation of the same order as that predicted by the calculations shown in Table II. This suggests that the estimates of ATP requirements based upon the requirements for biosynthesis are not widely in error. Again, however, they are not sufficiently precise to allow the calculation of the relative contributions of the different pathways.

We have made an attempt to obtain a more direct measure of the relative contribution of different substrates to the tricarboxylic acid cycle by measuring their contribution to the small amount of citrate which appears in milk. This method was first employed by Hardwick (1966), and we also have assumed that the citrate in milk represents that in the mitochondria of the secretory cells. To avoid some of the difficulties of interpretation caused by interchange of carbon atoms in the tricarboxylic acid cycle referred to earlier, it is necessary to determine the part of the citrate molecule to which different substrates contribute, and this must be done by splitting the citrate with the same stereochemical specificity as that shown by the synthetase. We have used the citratase of *Aerobacter aerogenes* for this purpose. The yield and specific activity of citrate is low, and we have collaborated with Dr M. Peaker who has made available goats with autotransplanted mammary glands, allowing close-arterial infusion of labelled substrates. The results of some experiments are summarized in Table IV. Label from both acetate and glucose entered both the acetyl and oxaloacetyl portions of the citrate. The label in the oxaloacetyl portion from acetate could only have come from the operation of the

TABLE III

Metabolic rate of mammary tissue ($\mu mol/min/g$ tissue at $37°C$)

	Tricarboxylic acid cycle (from oxygen uptake)	2-Oxoglutarate dehydrogenase (measured)	Glycolysis* + pentose cycle	Pyruvate dehydrogenase (measured)	Fructose‖ diphosphatase (measured)	Acetate oxidation	Lactose production (as glucose)
Goat	0·4†	0·24	0·15–0·68§	0·26	—	0·4–0·6 §	0·07–0·22§
Cow	0·1–0·5‡	0·34	0·1–0·2‡	0·18	1·2	0·08–0·44‡	0·14–0·32‡
Rat	—	0·22	—	1·1	0·15	—	—

* Difference between glucose uptake and lactose production: as triose phosphate.   † Data from Linzell (1960).   ‡ Data from Bickerstaffe *et al.* (1974).   § Data from Annison & Linzell (1964).   ‖ Data from Baird (1969).

TABLE IV

Contribution of substrates to milk citrate

		Acetyl portion	Oxaloacetyl portion
Goat (fed)	Glucose	37	37
	Acetate	40	11
Goat (fasted)	Glucose	14	14
	Acetate	20	7

The contribution is calculated from the ratio (expressed as a percentage) of the area under the specific activity/time curve of milk citrate to the plasma glucose or acetate following infusion of either $[U-^{14}C]$-glucose or $[1-^{14}C]$-acetate. The specific activity of the oxaloacetyl portion following acetate infusion is corrected to allow direct comparison with that following glucose infusion.

tricarboxylic acid cycle, which suggests, although not conclusively, that the assumption that milk citrate arises from the mitochondria of cells is justified. Further work must be done before these results are interpreted too intensively, but the observation that the oxaloacetyl portion of the citrate did not reach the specific activity (on a per mole basis) of the acetyl portion from acetate suggests a dilution of carbon atoms through exchange, or as a result of influx to balance losses from the cycle. The specific activity of the oxaloacetyl portion from glucose was not so diluted, indicating that at least some of this entry into the cycle is from glucose— possibly to make up for losses from amino acid synthesis in addition to the loss of citrate in milk. The labelling of the acetyl portion from both substrates indicates that both contribute to oxidation in the tricarboxylic acid cycle in the animals we examined, and contrast with earlier observations in other animals in which the entry of glucose appeared limited (Chesworth & Smith, 1971). In the fed animal, the oxidation of glucose and acetate together appear to account for about 80% of the total oxidation in the cycle. In the fasted animal, entry of both substrates was greatly reduced, but that from glucose was more markedly affected. This is compatible with the observations of reduced oxidation obtained by other methods (Table I). Relative entry to the cycle appears to change with different circumstances. These experiments on their own only allow assessment of the comparative entry of different substrates into the tricarboxylic acid cycle. In order to determine the actual flux they must be put together with measurements of the total flux through the cycle, as assessed by measurements of oxygen uptake, carbon dioxide production (from acetate) or enzyme measurement. It is hoped that extension of these experiments may allow the assessment of tricarboxylic acid cycle oxidation of different substrates in different circumstances, and therefore allow the assessment of the operation of the pentose cycle to be calculated by difference.

CONCLUSIONS

In conclusion, one may ask how far the hypothesis that the ruminant mammary gland operates to conserve glucose is supported by the experiments and information which have been reviewed. Certainly it appears that the gland has this capability; the "re-cycling" of the pentose cycle, and the possibility of replacing the pentose cycle with acetate oxidation seems to be established. The experiments with milk citrate, however, suggest that there is no invariable block upon glucose oxidation as there is upon the entry of glucose into fatty acids, and the picture which emerges is of a flexible system adaptable to changes in the supply of substrates. For example, if glucose oxidation is avoided or confined to a re-cycled pentose phosphate pathway then the demands upon acetate are increased (Table II), and this may become the limiting substrate. In the experiments with fasted goats (Table IV) acetate, as well as glucose, would have been in short supply. In turn, this flexibility suggests the difficulty of obtaining some part of the picture from a certain type of experiment and other parts from different experiments with other animals. For example, in the cow, approximately equal amounts of acetate and glucose were oxidized in the experiments of Bickerstaffe *et al.* (1974) (Table I). The supposition that equal amounts of these materials were oxidized in the tricarboxylic acid cycle would mean that the pentose cycle was not operating in these animals.

Full information about the operation of satellite systems can only be obtained by simultaneous measurements on the same animals, and the degree of flexibility which exists assessed by repeating these measurements in different physiological circumstances.

ACKNOWLEDGEMENTS

The authors acknowledge with thanks the assistance of Dr B. Crabtree and G. Read in the preparation of this paper.

REFERENCES

Abraham, S. & Chaikoff, I. L. (1959). Glycolytic pathways and lipogenesis in mammary glands of lactating and non-lactating normal rats. *J. biol. Chem.* **234**: 2246–2253.

Abraham, S., Katz, J., Bartley, J. & Chaikoff, I. L. (1963). The origin of hydrogen in fatty acids formed by lactating rat mammary gland. *Biochim. biophys. Acta* **70**: 690–693.

Annison, E. F. & Linzell, J. L. (1964). The oxidation and utilization of glucose and acetate by the mammary gland of the goat in relation to their overall metabolism and to milk formation. *J. Physiol., Lond.* **175**: 372–385.

Annison, E. F., Linzell, J. L. & West, C. E. (1968). Mammary and whole animal metabolism of glucose and fatty acids in fasting and lactating goats. *J. Physiol., Lond.* **197**: 445–459.

Baird, G. D. (1969). Fructose-1, 6-diphosphatase and phosphopyruvate carboxy-kinase in bovine lactating mammary gland. *Biochim. biophys. Acta* **177**: 343–345.

Baldwin, R. L. & Smith, N. E. (1971). Intermediary aspects and tissue interactions of ruminant fat metabolism. *J. Dairy Sci.* **54**: 583–595.

Bauman, D. E., Brown, R. E. & Davis, C. L. (1970). Pathways of fatty acids synthesis and reducing equivalent generation in mammary gland of rat, sow and cow. *Archs Biochem. Biophys.* **140**: 237–244.

Bauman, D. E. & Davis, C. L. (1974). Biosynthesis of milk fat. In *Lactation* **2**: 31–69. Larson, B. L. & Smith, V. R. (eds). New York and London: Academic Press.

Bickerstaffe, R., Annison, E. F. & Linzell, J. L. (1974). The metabolism of glucose, acetate, lipids and amino acids in lactating dairy cows. *J. agric. Sci., Camb.* **82**: 71–85.

Chesworth, J. M. & Smith, G. H. (1971). Factors limiting the utilization of glucose for milk fat synthesis in the ruminant. *Proc. Nutr. Soc.* **30**: 47–48A.

Devlin, T. M. & Bedell, B. H. (1959). Stimulation by acetoacetate of DPNH oxidation by liver mitochondria. *Biochim. biophys. Acta* **36**: 564–566.

Gul, B. & Dils, R. (1969). Pyruvate carboxylase in lactating rat and rabbit mammary gland. *Biochem. J.* **111**: 263–271.

Gumaa, K. A., Greenbaum, A. L. & McLean, P. (1973). Adaptive changes in satellite systems related to lipogenesis in rat and sheep mammary gland and in adipose tissue. *Eur. J. Biochem.* **34**: 188–198.

Hardwick, D. C. (1966). The fate of acetyl groups derived from glucose in the isolated perfused goat udder. *Biochem. J.* **99**: 228–231.

Katz, J. & Wood, H. G. (1960). The use of glucose $-C^{14}$ for the evaluation of pathways of glucose metabolism. *J. biol. Chem.* **235**: 2165–2177.

Landau, B. R. & Katz, J. (1965). Pathways of glucose metabolism. In *Handbook of physiology*, Sect. 5 Adipose tissue: 253–271. Renold, A. E. & Cahill, G. F. (eds). Washington: American Physiological Society.

Linzell, J. L. (1960). Mammary gland blood flow and oxygen, glucose and volatile fatty acid uptake in the conscious goat. *J. Physiol., Lond.* **153**: 492–509.

Linzell, J. L. (1974). Mammary blood flow and methods of identifying and measuring precursors of milk. In *Lactation* **1**: 143–225. Larson, B. L. & Smith, V. R. (eds). New York and London: Academic Press.

Linzell, J. L., Annison, E. F., Fazakerley, S. & Leng, R. A. (1967). The incorpora-tion of acetate, stearate and $D(-)\beta$-hydroxybutyrate into milk fat by the isolated perfused mammary gland of the goat. *Biochem. J.* **104**: 34–42.

Linzell, J. L. & Mepham, T. B. (1968). Mammary synthesis of amino acids in the lactating goat. *Biochem. J.* **107**: 18–19P.

Linzell, J. L., Mepham, T. B., Annison, E. F. & West, C. E. (1969). Mammary metabolism in lactating sows: arteriovenous differences of milk precursors and mammary metabolism of [^{14}C] glucose and [^{14}C] acetate. *Br. J. Nutr.* **23**: 319–332.

Linzell, J. L. & Peaker, M. (1971). Mechanism of milk secretion. *Physiol. Rev.* **51**: 564–597.

Lowenstein, J. M. (1961). The pathway of hydrogen in biosynthesis: 1. Experi-ments with glucose-1-H^3 and lactate-2-H^3. *J. biol. Chem.* **236**: 1213–1216.

McLean, P. (1958). Carbohydrate metabolism of mammary tissue: 1. Pathways of glucose catabolism in the mammary gland. *Biochim. biophys. Acta* **301**: 303–315.

McLean, P. (1964). Interrelationship of carbohydrate and fat metabolism in the involuting mammary gland. *Biochem. J.* **90**: 271–278.

Read, G., Crabtree, B. & Smith, G. H. (1977). The activities of 2-oxoglutarate-dehydrogenase and pyruvate dehydrogenase in hearts and mammary glands of ruminants and non-ruminants. *Biochem. J.* **164**: 349–355.

Smith, G. H. (1971). Glucose metabolism in the ruminant. *Proc. Nutr. Soc.* **30**: 265–272.

Smith, G. H., McCarthy, J. & Rook, J. A. F. (1974). Synthesis of milk fat from B hydroxybutyrate and acetate in lactating goats. *J. Dairy Res.* **41**: 175–191.

Snoswell, A. M. & Linzell, J. L. (1975). Carnitine secretion into milk of ruminants. *J. Dairy Res.* **42**: 371–380.

Wieland, O. H., Siess, E. A., Loffler, G., Potzelt, C., Portenhauser. R., Hartmann, U. & Schirmann, A. (1973). Regulation of the mammalian pyruvate dehydrogenase complex by covalent modification. *Symp. Soc. exp. Biol.* **27**: 371–400.

Wise, E. M. & Ball, E. G. (1964). Malic enzyme and lipogenesis. *Proc. natn. Acad. Sci. U.S.A.* **52**: 1255–1263.

Wood, H. G., Peeters, G. J., Verbeke, R., Lauryssens, H. & Jacobson, B. (1965). Estimation of the pentose cycle in the perfused cow's udder. *Biochem. J.* **96**: 607–615.

Symp. zool. Soc. Lond. (1977) No. 41, 113–134

The Aqueous Phase of Milk: Ion and Water Transport

M. PEAKER

ARC Institute of Animal Physiology,
Babraham, Cambridge, England

SYNOPSIS

The ionic composition of the aqueous phase of milk and variations in different species are described; the difficulties involved in such studies are stressed. Relationships are evident between the concentrations of lactose and the major ions, potassium, sodium and chloride. Recent evidence on the mechanisms of ion transport across the secretory cell in relation to water and lactose secretion is considered and a basic scheme for the secretory mechanism described. Quantitative differences in the operation of this mechanism can probably account for the ionic composition of milk in many species. A simple modification to the basic scheme involving movement of substances from milk to blood between cells—the paracellular pathway—can account for milk composition in some other species. Unresolved questions on milk composition and secretory mechanism in some groups of animals are also pointed out.

INTRODUCTION

In addition to substances like fat, protein and carbohydrates which are of obvious nutritional importance, milk contains salts, and in this contribution the secretion of ions and water, which together with lactose form the bulk of the aqueous phase of milk, will be considered. The major ions in milk are potassium, sodium and chloride although other substances, for example citrate, phosphate and calcium, contribute to the total ionic concentration (Fig. 1).

The basis for our current understanding of the transport of the major ions present in milk across the mammary epithelium has been derived mainly from physiological experiments involving permeability studies and the determination of concentration and electrical gradients between extracellular fluid, the inside of the secretory cell and milk (see Linzell & Peaker, 1971a; Peaker, in press a). In order to attempt to explain the relations between the concentrations of lactose and ions in milk, a hypothesis has recently been outlined to suggest a mechanism by which lactose secretion, water movement into milk and ion transport are linked (Peaker, in press a,b). In this article, the salient features of the proposed pathways and mechanisms of ion transport in the mammary gland will be described in relation to milk composition in various species

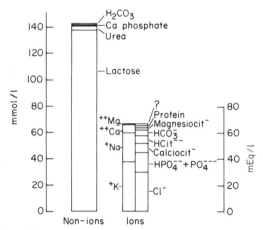

FIG. 1. Composition of the aqueous phase of cows' milk (pH 6·8). Calculated from the data of Davies & White (1960).

in order to consider the extent to which basic mechanisms are common and the strategies employed by different animals to achieve differences in the composition of the aqueous phase of milk. While it will be seen that simple modifications to basic mechanisms can explain milk composition in some species, it will also be evident that there are groups of animals in which the mechanism of secretion of water and ions remains unknown and largely unexplored.

SOME GENERAL CONSIDERATIONS

Osmotic Concentration of Milk

One constant feature of the milk of all species so far examined is that the osmolality is virtually the same as blood plasma. This means that we are concerned with the manner in which the mammary epithelium establishes concentration gradients for ions and small molecules, but not an osmotic gradient, between extracellular fluid and milk.

Milk Storage

The mammary gland is unlike the majority of exocrine glands in that the rate of secretion is slow (1–2 ml/g mammary tissue/day) and the secretion is stored in the lumen of the gland until it is removed by suckling or, in dairy animals, at milking. Therefore the gland must not only establish concentration gradients and produce a fluid markedly different from that of plasma but it must also maintain that difference during the storage of milk. In the case of milk stored in the alveoli this is achieved either by impermeability of the epithelium to secreted products, lactose for example, and some plasma constituents or, in the case of sodium and potassium for example, by maintaining a dynamic equilibrium between gain and loss across the epithelium (see Linzell & Peaker, 1971a). By contrast, Linzell & Peaker (1971c) found that the epithelium lining the ducts (Fig. 2) is

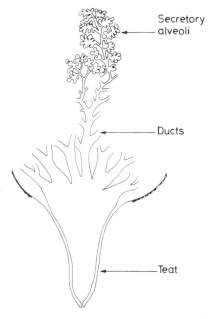

FIG. 2. Diagram showing the arrangement of the ducts and alveoli in the goat mammary gland. Modified from Linzell & Peaker (1971c).

impermeable to ions as well as to lactose, but not to water. Therefore energy need not be expended in maintaining the ionic composition of milk stored in the duct system. The ultrastructure of the duct epithelium is similar in all the species that have been examined and it seems highly probable that the duct system acts as an impermeable storage region.

Difficulties in Determining Milk Composition

Although there are many reports of milk composition in various species in the literature, many are unsuitable for deriving reliable information on the concentrations of components of the aqueous phase because the apparently simple process of milk sampling is fraught with difficulties. In addition, numerous investigators have only analysed their sample by the standard techniques of dairy chemistry, which are virtually meaningless in a physiological context.

Some of the difficulties involved can be illustrated by considering two examples of typical situations in which milk samples are taken. In the first a female in the wild is restrained and a sample taken from the teat; oxytocin may be injected to aid milking. In the second, an animal in captivity is separated from its young to allow milk to accumulate, and oxytocin is injected; this procedure may be repeated at relatively frequent intervals. The problems such procedures raise will be evident from the following account of factors which can influence the composition of the aqueous phase.

Stage of lactation

There are major changes in milk composition at the onset of lactation and during involution of the mammary gland, and it is important to ensure that an animal is in full lactation before a sample is claimed to be milk. In fact there is the possibility that some samples have given an erroneous impression of milk composition because they may have been taken from "dry" animals which had previously lactated; particularly in large animals the fluid in the glands may retain a milky appearance for some time.

In some species, the rabbit being an excellent example (see below), the composition of the aqueous phase varies markedly during lactation, and it is therefore impossible to quote the typical composition of milk in such animals without reference to the date of parturition.

Differences between glands

In an animal suckling fewer young than the number of teats, some glands may involute and therefore contain a fluid quite unlike that from the lactating glands.

Milk accumulation

When milk is allowed to accumulate for periods longer than the normal interval between successive bouts of suckling or milking, milk composition may change. In the guinea-pig separation from the young for more than four hours may lead to a change (unpublished observations, see also Mepham & Beck, 1973).

Oxytocin

In many studies the dose of oxytocin employed has been enormous in relation to the normal amounts released by the posterior pituitary in response to suckling. Such doses may alter the ionic composition of milk removed immediately afterwards (Linzell, Peaker & Taylor, 1975). However, doses within the physiological range have been found to be satisfactory in the short term; nevertheless if such treatment is repeated during the day the composition of the milk may alter (Linzell, 1967; Linzell & Peaker, 1971c; Linzell, Peaker et al., 1975).

Apart from the immediate effects of large doses of oxytocin, changes may also be evident in the long term. For example, Linzell, Peaker et al.

(1975) found that following the administration of 1 iu oxytocin intravenously in the rabbit, the immediate change was a rise in the concentrations
of sodium and chloride in milk. By contrast, 24 hours later milk sodium
and chloride were lower, and potassium higher, than in untreated animals. In other words long-term effects were the reverse of the short-term
changes.

Disease

Infection of the mammary glands and some systemic conditions can
markedly affect the composition of the aqueous phase of milk (see Linzell
& Peaker, 1975a).

Changes after death

In some studies animals have been shot and milk samples taken post-
mortem. It is important that the samples should be taken as quickly as
possible after death because active transport mechanisms, which maintain
the ionic composition of milk in the alveoli, cease to operate. Although the
ducts probably remain impermeable in such circumstances milk stored
there may be affected by the diffusion of substances between them and the
alveoli.

It is clear that a milk sampling regime must take these factors into
account for reliable information to be obtained on the composition of the
aqueous phase. For non-domesticated species there are likely to be few
animals in which information can be obtained under natural conditions,
the exception being ungulates and some other large animals where under
field conditions it is possible by the use of tagged individuals and close
observation to determine the stage of lactation and to obtain milk samples
relatively easily following restraint. For other animals, the establishment
of a breeding colony in captivity seems the most satisfactory solution.

INTERSPECIFIC VARIATION IN IONIC COMPOSITION OF MILK

From published and unpublished data of milk composition in various
species, and taking into account the difficulties in interpretation outlined
above, I consider that reliable figures exist on the sodium, potassium,
chloride and lactose concentrations during established lactation in the 11
species shown in Table I. It is evident that the species can be divided into
two groups: (i) those with a sodium: potassium ratio of approximately 1:3,
and (ii) those in which this ratio is considerably less than 1:3. The species
falling into the latter category will be considered on pp. 122–128.

In the nine species with a sodium: potassium ratio in the order of 1:3,
the milk lactose concentration varies from 130 to 204 mmol/l fat-free
milk, sodium from 5·5 to 16, potassium from 14 to 45 and chloride from
10·3 to 43. When the concentration of lactose is plotted against those for
sodium, potassium and chloride, significant correlations are apparent

TABLE I

Milk lactose, sodium, potassium and chloride concentrations in some eutherian mammals

| | Lactose | mmol/l fat-free milk | | | K/Na | Source |
		Na	K	Cl		
(i) Species with milk sodium: potassium ratio of about 1:3 (2·5–3·3)						
Man (*Homo sapiens*)	204	6·5	18·1	12·1	2·8	Macy, Kelly & Sloan (1953) and unpublished*
White rhinoceros (*Diceros simus*)	200	5·5	14·0	15·5	2·6	unpublished*
Horse (*Equus caballus*)	182	6·5	21·7	10·3	3·3	unpublished
Cow (*Bos taurus*) (Jersey breed)	140	15·0	43·0	24·0	2·9	unpublished†
Sheep (*Ovis aries*) (Friesland breed)	140	13·0	38·0	26·0	2·9	unpublished
Barbary sheep (*Ammotragus lervia*)	140	13·5	45·0	20·0	3·3	unpublished*
Goat (*Capra hircus*) (Saanen breed)	139	16·0	45·0	43·0	2·8	Linzell & Peaker (1974)
Guinea-pig (*Cavia porcellus*)	138	10·5	32·0	27·0	3·0	Peaker, Jones, Goode & Linzell (1975)
Red deer (*Cervus elaphus*)	130	13·5	34·0	20·0	2·5	Arman, Kay, Goodall & Sharman (1974)
(ii) Species with milk sodium: potassium ratio of less than 1:2·5						
Rabbit (*Oryctolagus cuniculus*) (Dutch breed)	110	56·0	90·0	50·0	1·6	Peaker & Taylor (1975)
Plains viscacha (*Lagostomus maximus*)	143	18·0	38·0	40·0	2·1	M. Peaker, B. J. Weir & J. A. Goode, unpublished

* Samples taken by Mr D. M. Jones, The Zoological Society of London, Whipsnade Park.
† Established to be free from sub-clinical mastitis.

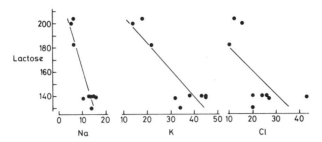

FIG. 3. Relations between the concentrations of lactose, sodium, potassium and chloride in the milks of the species listed in Table I, part (i). All concentrations are expressed as mmol/l fat-free milk. Details of the regressions: lactose = 231–6·61 Na ($r = -0·907$, $P < 0·001$); lactose = 227–2·19 K($r = -0·800$, $P < 0·01$); lactose = 200–1·97 Cl⁻($r = -0·65$, n.s.).

(Fig. 3). Both sodium and potassium are significantly and inversely correlated with lactose, whereas the negative correlation between lactose and chloride is not statistically significant. In other words, the higher the lactose concentration, the lower the concentration of ions, with the sodium:potassium ratio being maintained at about 1:3. Of course it can be inferred from the constant osmolality of milk that the higher the lactose concentration the lower must be the ionic content.

The virtually constant osmolality also imposes constraints on milk composition. Thus the maximum possible lactose concentration is about 300 mmol/l and the maximum cationic or anionic strength approximately 150 mmol/l, to quote the two extremes.

PATHWAYS AND MECHANISMS OF ION TRANSPORT

In considering the basis for the determination of the ionic concentrations in milk it is necessary to distinguish between the two routes which have been proposed to account for the passage of ions and small molecules across the mammary secretory epithelium. The first is the *transcellular* route, that is across the secretory cell so that substances like sodium and potassium must cross the basal and apical (luminal) cell membranes. The second is the *paracellular* route, that is between the cells so that substances can to some extent equilibrate directly between plasma and milk without having to pass through the cells.

Transcellular Route

The scheme suggested by Linzell & Peaker (1971a,b) for the transport of the major ions between extracellular fluid, intracellular fluid and milk is shown in Fig. 4A. The evidence on which this scheme is based, derived from measurements of the concentration and electrical gradients between the three compartments and from permeability studies mainly in the guinea-pig in full lactation and in the goat, will not be considered in detail (see Linzell & Peaker, 1971a; Peaker, in press a) but the salient features are that the intracellular concentration of potassium is held high, and sodium low, by a typical sodium pump on the basolateral membrane, and that sodium and potassium are distributed passively across the apical membrane between intracellular fluid and milk according to the electrical potential difference across that membrane. Therefore with milk being electrically positive with respect to the inside of the cell, the concentrations of sodium and potassium are lower in milk than in intracellular fluid, but the ratio between these two ions is similar in both compartments at about 1:3. Therefore this scheme can account for the sodium:potassium ratio in milk but the question arises—what generates the electrical potential difference across the apical membrane, thereby keeping the milk sodium and potassium content low?

From the relationships between lactose, sodium and potassium concentrations (Fig. 3) and the scheme shown in Fig. 4A, it can be inferred

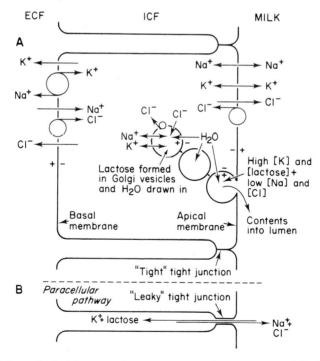

FIG. 4. A. Scheme for ion, lactose and water movements between extracellular fluid (ECF), intracellular fluid (ICF) and milk, suggested by Linzell & Peaker (1971a,b). The directions of electrical potential differences are indicated by plus and minus signs. B. The effect of a paracellular pathway through leaky tight junctions.

that in intraspecific and interspecific variations in milk composition, the higher lactose concentrations in milk should be associated with a higher electrical potential across the apical membrane (i.e. milk should be more positive with respect to the inside of the cell); this has recently been confirmed (Peaker, in press b, and unpublished observations). Studies on the mechanism by which the apical membrane potential is established have recently led to the suggestion that it is the secretion of lactose and consequent osmotic water movements which create the potential (Peaker, in press a,b).

As explained by other contributors to this Symposium (p. 57 and p. 78), there is excellent evidence that lactose is formed in the lumen of the Golgi apparatus and vesicles carry it along with casein to the apex of the cell where the vesicles discharge their contents into the lumen of the alveolus by exocytosis. Therefore, lactose is formed effectively in an extracellular site and, being a small molecule and unable to permeate cell membranes, draws water osmotically from the inside of the cell into the vesicles and into milk. In fact this is believed to be the main mechanism by which bulk water movements occur into milk since in the isolated per-fused goat mammary gland, when lactose secretion was arrested by omitting glucose from the perfusate, the secretion of water also fell in a similar manner (Hardwick, Linzell & Price, 1961; see Linzell & Peaker, 1971a). Incidentally it was found that in these circumstances, when the lactose concentration in milk as well as its rate of secretion decreased, the concentrations of sodium and potassium increased. Similarly, the trans-epithelial potential changed in a manner consistent with a fall in the apical membrane potential, and this is one line of evidence that the secretion of lactose and water keeps the sodium and potassium content of milk low by creating a potential difference across the apical membrane (Peaker, in press a,b).

The hypothesis to relate lactose secretion, water flow and ion move-ments is that the osmotic movement of water across the Golgi vesicle membrane and the apical membrane could establish the potential differ-ence between the inside of the cell and milk, there being two biophysical phenomena (both involving water flow across a membrane bearing fixed charges) which could cause charge separation and the creation of a potential difference of sufficient magnitude with the correct polarity; mobile ions, in this case the cations sodium and potassium, would then be distributed according to this potential, as in classical electrokinetic theory. Since in all probability the membranes of the secretory vesicles from the Golgi apparatus become part of the apical membrane following exocytosis, the ionic composition of milk could be determined in the vesicles and continue across the apical membrane; this hypothesis is also shown in Fig. 4A.

If the mechanism suggested does operate then one might envisage that the variation in the composition of the aqueous phase in different species could be explained by differences in such variables as the rate of lactose synthesis (for which Brew, 1970, provides an enzymological basis) and the properties of the membranes in terms of osmotic water permeability, density of fixed charges, etc. Therefore quantitative rather than qualitative differences in the basic mechanism could explain milk composition in the nine species shown in Fig. 3, but as will be explained below qualitative modifications are required to account for some physiological conditions and for the composition of the aqueous phase in some species.

Chloride movements into milk are still enigmatic. Although it has been suggested that active mechanisms could be involved (Linzell & Peaker, 1971a,b) this is not the only possible explanation since, for waterflow-induced potentials to be generated, the permeability of the apical membrane to anions must be low. Furthermore, the secretion and transport of other inorganic (phosphate) and organic (citrate) anions also has to be considered in relation to the preservation of electrical neutrality. Indeed the lack of a statistically significant correlation between lactose and chloride concentrations between species (Fig. 3) and in day-to-day variation in goats (Linzell & Peaker, 1971d, 1974) must be due to variation in the contribution to the total anionic strength of other anions. Therefore it is clear that this is one of the many problems on which more work must be done.

Paracellular Route

It is obvious that the scheme in Fig. 4A cannot solely apply in conditions or species in which the sodium:potassium ratio in milk is quite unlike that of intracellular fluid. In fact in some cases the concentration of sodium may exceed that of potassium.

In some of these conditions, which are characterized by increases in the concentrations of sodium and chloride in milk, and decreases in potassium and lactose, it has been established in the goat that, in contrast to the normal situation in lactation, disaccharides can permeate the secretory epithelium. The first of these changes in milk composition to be studied was the effect of exogenous oxytocin, and in view of the permeability to isotopically-labelled sucrose and lactose and the lack of effect of oxytocin on intracellular ionic composition, it was suggested that oxytocin in some way disrupts the mammary epithelium thus allowing ions and small molecules to pass between extracellular fluid and milk down concentration gradients established by transcellular mechanisms (Fig. 4B). Therefore, sodium, chloride and bicarbonate enter milk while lactose, potassium, citrate and phosphate pass in the reverse direction; milk pH also shifts in consequence (Linzell & Peaker, 1971d; Linzell & Peaker, 1975b; Linzell, Peaker & Taylor, 1975; Linzell, Mepham & Peaker, 1976). Since these changes were quickly reversed when treatment ceased, it was suggested that the effect could involve changes in the structure of the "tight junctions" or zonulae occludentes which connect neighbouring cells, by altering them to "leaky" tight junctions and allowing paracellular movements to occur.

Later studies on the change in composition of mammary secretions in the goat in late pregnancy and at parturition also led to the conclusion that a paracellular pathway exists in late pregnancy but that the epithelium becomes tight near term when milk of normal composition begins to be produced. This conclusion was also partly based on a change in permeability to labelled sucrose and lactose because in pregnancy when sodium and chloride concentrations in the secretion were high, and potassium and lactose low, these substances were found to pass from blood to milk and vice versa. Moreover, the entry of sodium and chloride (calculated from ^{24}Na and ^{36}Cl fluxes) into milk was higher than during established lactation, and further calculations indicated that the additonal entry could entirely be accounted for by paracellular movements. The passage of [^{14}C]-sucrose from blood into the secretion was found to be positively correlated with the sodium concentration of the secretion, normal milk composition being achieved when the sucrose entry was effectively zero. It was therefore proposed that the change to normal composition at about the time of parturition in the goat involves an alteration in the permeability of the tight junctions from leaky to truly tight (Linzell & Peaker, 1973, 1974). This proposal was made solely from physiological evidence, and it was gratifying that independent, morphological studies indicated that such a change occurs. Pitelka, Hamamoto, Duafala & Nemanic (1973) showed, using freeze-fracture techniques, that in the mouse mammary gland a change in the structure of the tight junctions from a diffuse network of relatively few ridges to a typical tight network of compact, abundant ridges between the lumen and the interstitial space, occurs between late pregnancy and the first few days post-partum.

Apart from the changes in milk composition and in epithelial permeability associated with the proposed paracellular pathway, other changes are evident. In normal day-to-day variation in milk composition in goats, the relationships between lactose, potassium, sodium and chloride are similar to those in variation between species (Fig. 3), i.e. lactose inversely correlated with sodium and potassium. Since with a paracellular pathway lactose and potassium move together out of milk, the correlation becomes positive, and with sodium and chloride moving into milk, the lactose concentration becomes inversely and significantly correlated with chloride as well as with sodium (Fig. 5) (Linzell & Peaker, 1971d, 1974). Furthermore the paracellular pathway short-circuits the two sides of the epithelium, and this can be detected as a change in the transepithelial potential difference, with a normally negative or, in some cases, positive potential falling towards zero.

With this background on some of the methods used to detect a paracellular pathway, it is clearly possible to consider species like the rabbit in which the milk has an "unusual" ionic composition.

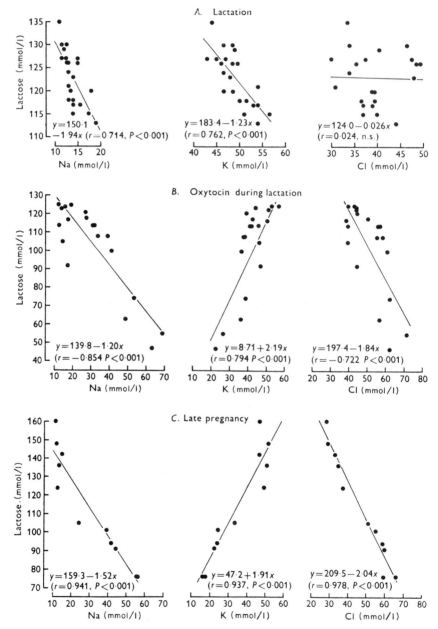

FIG. 5. Relationships between lactose and ion concentrations. A. In normal day-to-day variation in a lactating goat. B. The same goat treated with oxytocin. C. In a non-lactating goat during late pregnancy. In B and C a paracellular pathway is believed to be present. From Linzell & Peaker (1974) with permission.

Rabbit

The rabbit is interesting not only because the composition of the aqueous phase of milk differs from that of other eutherian mammals for which adequate data exist (Table I) but also because the composition changes markedly during the relatively short period of lactation (see Cowie, 1969; Gachev, 1965, 1971; Peaker & Taylor, 1975). The concentrations of

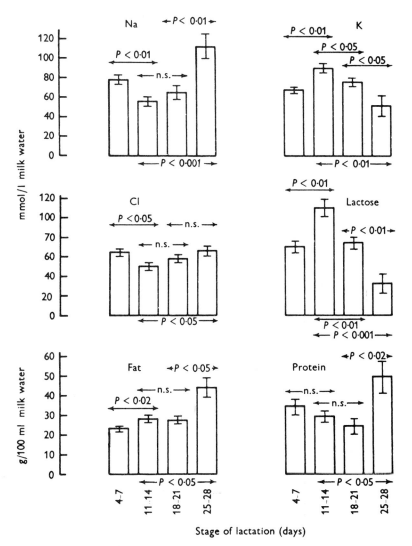

FIG. 6. Milk composition at different stages of lactation in the rabbit. Mean±s.e.mean, n.s. = not significant. From Peaker & Taylor (1975) with permission.

lactose and potassium are low, and sodium and chloride high, even at the height of lactation, while later, although the milk yield remains high, sodium and chloride increase still further while potassium and lactose decrease (Fig. 6).

Recent studies have shown that the mammary epithelium is permeable to sucrose and lactose throughout lactation (Fig. 7), and the correlation between the concentrations of lactose and potassium in milk is positive (i.e. as in non-lactating goats in late pregnancy or in lactating goats treated with oxytocin). Furthermore, a marked transepithelial potential differ-ence (milk negative) would be expected in this species with a low milk lactose concentration if only transcellular processes were involved (Peaker, in press a,b). However, it was found that the transepithelial potential is close to zero which would suggest that the epithelium is

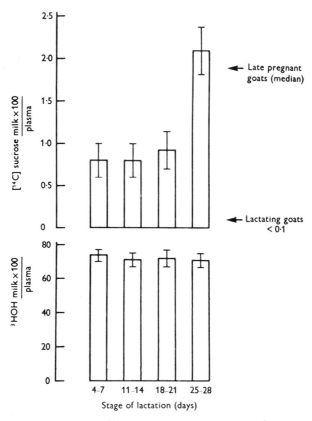

FIG. 7. Changes in the passage of $[^{14}C]$-sucrose and tritiated water (^{3}HOH) from blood to milk in the rabbit (mean ± s.e.mean). The concentrations (in milk water) are expressed as percentages of the concentrations in plasma 25 min after the start of an intravenous infusion. From Peaker & Taylor (1975). Comparable data for goats are shown on the right.

short-circuited by a shunt pathway. Thus it appears that in the rabbit a paracellular pathway persists throughout lactation (Peaker & Taylor, 1975).

In late lactation (25–28 days), when milk sodium and chloride are even higher, further experiments demonstrated that the epithelium is more permeable to [^{14}C]-sucrose in the blood to milk direction and to [^{14}C]-lactose passing from milk to blood (Fig. 7); the sucrose entry in milk was found to be positively correlated with milk sodium, and inversely correlated with potassium and lactose concentrations. Furthermore, the entry of ^{24}Na and ^{36}Cl into milk paralleled the changes in milk composition, and calculations indicated that this could be attributed to increased paracellular movements of these ions. Therefore it was concluded that the permeability of the paracellular route increases in late lactation.

These studies on the rabbit indicate that it is possible to modify the basic scheme proposed for ion transport by transcellular processes (Fig. 4A) in a manner similar to that suggested to explain similar changes in milk composition that occur, for example, during oxytocin treatment or in late pregnancy. This does not mean that the transcellular mechanisms are the same (and with a persistent paracellular pathway a direct study of transcellular processes in the rabbit appears not to be feasible) but that the results can be explained by a simple modification to the basic scheme, i.e. by including a paracellular route for ions and small molecules to move into or out of milk.

Another interesting facet to the work on the rabbit is that the administration of prolactin has been found to reverse or prevent the changes in milk composition that occur in late lactation (see Cowie, 1969; Gachev, 1971). Therefore Linzell, Peaker & Taylor (1975) studied the effects of this hormone on the permeability of the mammary epithelium and found that, as well as reversing the changes in composition, prolactin also reduced the entry of [^{14}C]-sucrose into milk to levels characteristic of earlier in lactation; calculated paracellular movements of sodium and chloride were similarly decreased. In other words, it appears that prolactin could affect the structure of the tight junctions of the secretory epithelium. Prolactin failed to affect milk composition or permeability earlier in lactation and so it would seem that the presence of a paracellular pathway throughout lactation is not simply due to a relative lack of prolactin.

Guinea-pig

The guinea-pig has a short lactation and the young may eat solid food shortly after birth. The milk sodium: potassium ratio of about 1: 3 is only retained for about seven days and after this time sodium and chloride rise while potassium and lactose fall; these changes are accompanied by a marked decrease in the rate of milk secretion (Fig. 8) (Mepham & Beck, 1973; Peaker, Jones, Goode & Linzell, 1975). Although permeability measurements have not yet been done after seven days it would seem

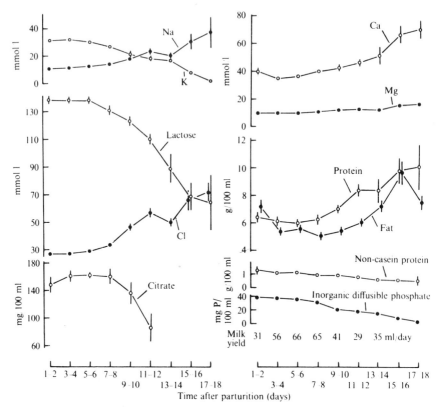

FIG. 8. Mean changes (± s.e.mean) in milk composition during lactation in the guinea-pig. Yields are from Mepham & Beck (1973). From Peaker et al. (1975) with permission.

probable, by analogy with the other situations in which similar changes in composition occur, that paracellular movements supervene. Exogenous prolactin failed to affect milk composition in late lactation in this species, in contrast to the rabbit (Peaker et al., 1975).

Other conditions

In addition to the physiological conditions described, the changes that occur during involution of the mammary gland, near oestrus in some goats and cows (Peaker & Linzell, 1974) and during milk accumulation, can all be explained at least in part by the presence of a paracellular pathway. In fact most of the problems associated with milk sampling (p. 115) are related to the loss of integrity of the mammary epithelium and the movement of substances between cells as a result of oxytocin treatment, excessive periods of milk accumulation, sampling during colostrum formation, involution, etc. Therefore it is only too easy to conclude that paracellular movements are a normal part of mammary function at all stages of lactation or in various species, when they may well be an artifact induced by the experimental procedure.

Possible Changes During Immunoglobulin Transport

It is possible that an alteration in ion transport by the secretory cell occurs when immunoglobulins are being transported into mammary secretions. If these proteins are carried across the cell by pinocytosis (see the chapter by Lascelles, this volume) then membrane could be carried from the basal side of the cell to the apical membrane and become part of it. This membrane material would contain sodium pumps and their orientation on the apical membrane would be such that they would exchange sodium in the cell for potassium in milk (Fig. 9). Therefore one might expect the

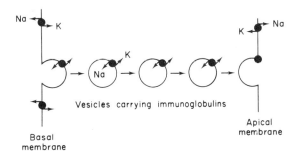

FIG. 9. A possible mechanism by which sodium pumps are carried to the apical membrane during secretion of immunoglobulins.

sodium concentration in milk to rise and that of potassium to fall, while leaving the chloride and lactose levels unchanged. In fact such an alteration in composition has been observed in lactating goats given oestrogens when immunoglobulin movement into milk was initiated (Peaker & Linzell, 1974). Although this process may occur in late pregnancy during colostrum formation, it may not be evident because the paracellular pathway would mask such changes. Incidentally, the effect could not simply be due to the transport of sodium and chloride into milk in pinocytotic vesicles because the lactose concentration would be expected to fall as a result of dilution with extracellular fluid; this did not occur.

It is possible that such a process of membrane transfer could occur normally in lactation in some species to an extent such that milk composition could be affected. Investigation of species like the viscacha (Table I) with apparently "normal" lactose concentrations but with a somewhat high sodium and low potassium in milk could therefore prove profitable.

COMPLEX IONS IN MILK

There is little comparative information on calcium, magnesium and phosphate in the aqueous phase of milk and a great deal of work remains to be done on the mechanism by which they are secreted. However, somewhat more is known of citrate in milk.

Citrate forms one of the main buffer systems of milk and from the pH it would be expected that half would be present as $HCit^{2-}$ and half as Cit^{3-}. The latter ion chelates calcium and magnesium to form soluble complexes which act as monovalent anions and it is therefore unlikely that Cit^{3-} is present in the free form. In species like the rat where the concentration of citrate in milk is low, ATP-citrate lyase is present in the cytosol fraction of mammary homogenates, whereas in ruminants which have a high milk concentration of citrates the levels of this enzyme are very low (see Bauman, Mellenberger & Derrig, 1973). This evidence would suggest that the citrate in milk is derived from that in the cytosol. Recent studies have shown that the apical membrane is impermeable to citrate and since the time-course of secretion of citrate from labelled precursors was similar to that of lactose and casein, Linzell, Mepham & Peaker (1976) have suggested that citrate, like lactose and casein, is secreted in Golgi vesicles and that the Golgi apparatus in some way concentrates citrate from the cytosol.

SOME UNRESOLVED QUESTIONS

Eared Seals

Having stressed the importance of lactose secretion in the movement of water into milk, it is rather embarassing to find a group of animals with no lactose nor indeed any carbohydrate in milk, and important questions on ion transport and water secretion remain. A number of investigators have established that in the fat- and protein-rich milk of several species of the family Otariidae, the eared seals (fur seals and sea-lions), virtually no carbohydrate can be detected (see, for example, Pilson & Kelly, 1962; Pilson, 1965; Ashworth, Ramaiah & Keyes, 1966; Bonner, 1968; Kerry & Messer, 1968; Schmidt, Walker & Ebner, 1971).

Through Dr M. R. Payne of the British Antarctic Survey we have obtained samples of milk from the fur seal *Arctocephalus tropicalis*; these were frozen immediately after their rapid removal from shot animals and only thawed immediately before analysis. As in other species the osmolality was similar to, or perhaps a little higher than, that of blood plasma at 330 and 332 mosmol/kg water. Preliminary data indicate that this is due to ionized substances in milk; extensive analysis failed to reveal carbohydrate in other than trace amounts. Moreover, the sodium: potassium ratio was not 1:3, and so no clues yet exist as to the mechanism of water movement or ion transport across the mammary epithelium in this group of interesting pinnipeds.

Marsupials and Monotremes

Although research on lactation in marsupials and monotremes has intensified, surprisingly little is known of the ionic composition of milk in these diverse forms. Bentley & Shield (1962) found that prior to about 140 days post-partum, milk sodium was higher than potassium in the macropod marsupial *Setonix brachyurus*; after this time and up to 200 days, when sampling ceased, there was a reversal with potassium higher than sodium (Fig. 10). Although the sodium: potassium ratio did not reach 1:3 in this

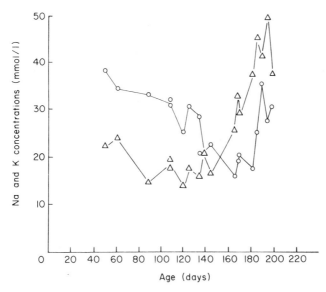

FIG. 10. Sodium (○) and potassium (△) concentrations in the milk of the marsupial *Setonix brachyurus* in relation to the age of the young. From Bentley & Shield (1962) with permission.

later stage, the young had been separated for 12–16 h so there is the possibility that paracellular movements may have supervened.

From time to time we have taken milk samples from the wallabies (*Macropus rufogriseus*) kept at the Institute of Animal Physiology, Babraham. Although this study is not yet complete, preliminary data indicate that a sodium: potassium ratio of 1:3 obtains in females with an advanced joey in the pouch. For example, in samples taken from one gland being sucked by a joey weighing 945 g, the sodium concentration was 12 mmol/l and the potassium 36 mmol/l. At this stage one might envisage a secretory mechanism the same as that shown in Fig. 4A. By contrast, the sodium: potassium ratio was less than 1 : 3 earlier in lactation. Therefore it might be suggested that at early stages a paracellular pathway is present across the mammary epithelium.

MILK IONIC COMPOSITION IN RELATION TO METABOLISM OF THE YOUNG

From what is known of the composition of the aqueous phase of milk in various mammalian species, it does appear, by extrapolation if not from actual experiments, that there is a common mechanism for ion transport across the mammary epithelium. Moreover a relatively simple modification to the basic scheme—the addition of a paracellular pathway—can be invoked to account for milk composition in some species or for changes during lactation. Therefore the composition of the aqueous phase can be varied rather widely but so far we have little knowledge of the evolutionary strategies employed by different species.

Although teleonomic arguments can sometimes be applied to explain differences in the ionic composition of milk, for example that the fall in the milk potassium concentration occurs when young rabbits begin to eat solid food rich in potassium, physiological studies of the young need to be done in relation to the yield and composition of the milk supplied by the mother. Another question is of course whether, in anything but gross terms, the ionic composition of milk *per se* has been subject to selection pressure or whether the concentrations of ions in various species have arisen merely as a consequence of other biochemical events in the mammary gland. In the latter case it must then be assumed that young mammals have sufficient flexibility to cope with milk of rather different ionic contents. Again, studies on the young in terms of physiological maturity of the organs involved in salt and water metabolism and rate of growth (potassium being required for the fluid inside new cells) should prove of value. There is perhaps slightly more support for the view that milk composition is related to the physiological state of the young in a particular species, and there is increasing evidence from studies on human infants that the higher ionic concentration of cow's milk may in some circumstances have deleterious short-term and long-term effects.

REFERENCES

Arman, P., Kay, R. N. B., Goodall, E. D. & Sharman, G. A. M. (1974). The composition and yield of milk from captive deer (*Cervus elaphus* L). *J. Reprod. Fertil.* **37**: 67–84.
Ashworth, U. S., Ramaiah, G. D. & Keyes, M. C. (1966). Species difference in the composition of milk with special references to the northern fur seal. *J. Dairy Sci.* **49**: 1206–1211.

Bauman, D. E., Mellenberger, R. W. & Derrig, R. G. (1973). Fatty acid synthesis in sheep mammary tissue. *J. Dairy Sci.* **56**: 1312–1318.

Bentley, P. J. & Shield, J. W. (1962). Metabolism and kidney function in the pouch young of the macropod marsupial, *Setonix brachyurus. J. Physiol., Lond.* **164**: 127–137.

Bonner, W. N. (1968). The fur seal of South Georgia. *Scient. Rep. Br. Antarctic. Surv.* No. 56: 1–81.

Brew, K. (1970). Lactose synthetase: evolutionary origins, structure and control. In *Essays in Biochemistry* **6**: 93–118. Campbell, P. N. & Dickens, F. (eds). London and New York: Academic Press.

Cowie, A. T. (1969). Variations in the yield and composition of the milk during lactation in the rabbit and the galactopoietic effect of prolactin. *J. Endocr.* **44**: 437–450.

Davies, D. T. & White, J. C. D. (1960). The use of ultrafiltration and dialysis in isolating the aqueous phase of milk and in determining the partition of milk constituents between the aqueous and disperse phases. *J. Dairy Res.* **27**: 171–190.

Gachev (Gatschew), E. P. (1965). The physiological and chemical characteristics of lactation in rabbits (Translated). *Biol. Zbl.* **84**: 447–460.

Gachev, E. P. (1971). Changes in osmotically active milk components during lactation. *C.r. Acad. bulg. Sci.* **24**: 543–546.

Hardwick, D. C., Linzell, J. L. & Price, S. M. (1961). The effect of glucose and acetate on milk secretion by the perfused goat udder. *Biochem. J.* **80**: 37–45.

Kerry, K. R. & Messer, M. (1968). Intestinal glycosidases of three species of seals. *Comp. Biochem. Physiol.* **25**: 437–446.

Linzell, J. L. (1967). The effect of very frequent milking and of oxytocin on the yield and composition of milk in fed and fasted goats. *J. Physiol., Lond.* **190**: 333–346.

Linzell, J. L., Mepham, T. B. & Peaker, M. (1976). The secretion of citrate into milk. *J. Physiol., Lond.* **260**: 739–750.

Linzell, J. L. & Peaker, M. (1971a). Mechanism of milk secretion. *Physiol. Rev.* **51**: 564–597.

Linzell, J. L. & Peaker, M. (1971b). Intracellular concentrations of sodium, potassium and chloride in the lactating mammary gland and their relation to the secretory mechanism. *J. Physiol., Lond.* **216**: 663–700.

Linzell, J. L. & Peaker, M. (1971c). The permeability of mammary ducts. *J. Physiol., Lond.* **216**: 701–716.

Linzell, J. L. & Peaker, M. (1971d). The effects of oxytocin and milk removal on milk secretion in the goat. *J. Physiol., Lond.* **216**: 717–734.

Linzell, J. L. & Peaker, M. (1973). Changes in mammary gland permeability at the onset of lactation in the goat: an effect on tight junctions? *J. Physiol., Lond.* **230**: 13–14P.

Linzell, J. L. & Peaker, M. (1974). Changes in colostrum composition and in the permeability of the mammary epithelium at about the time of parturition in the goat. *J. Physiol., Lond.* **243**: 129–151.

Linzell, J. L. & Peaker, M. (1975a). Efficacy of the measurement of electrical conductivity of milk for the detection of subclinical mastitis in cows: detection of infected cows at a single visit. *Br. vet. J.* **131**: 447–461.

Linzell, J. L. & Peaker, M. (1975b). The distribution and movements of carbon dioxide, carbonic acid and bicarbonate between blood and milk in the goat. *J. Physiol., Lond.* **244**: 771–782.

Linzell, J. L., Peaker, M. & Taylor, J. C. (1975). The effects of prolactin and oxytocin on milk secretion and on the permeability of the mammary epithelium in the rabbit. *J. Physiol., Lond.* **253**: 547–563.

Macy, I. C., Kelly, H. J. & Sloan, R. E. (1953). *The composition of milks (human, cow and goat).* Washington, D.C.: Natl. Acad. Sci., Natl. Res. Council Publ. 254.

Mepham, T. B. & Beck, N. F. G. (1973). Variation in the yield and composition of milk throughout lactation in the guinea-pig (*Cavia porcellus*). *Comp. Biochem. Physiol.* **45A**: 273–281.

Peaker, M. (in press a). Ion and water transport in the mammary gland. In *Lactation* **4**. Larson, B. L. (ed.). New York and London: Academic Press.

Peaker, M. (in press b). Milk composition in relation to potential difference across the mammary epithelium: a possible mechanism for the determination of the composition of the aqueous phase of milk. *J. Physiol., Lond.*

Peaker, M., Jones, C. D. R., Goode, J. A. & Linzell, J. L. (1975). Changes in milk composition during lactation in the guinea-pig and the effects of prolactin. *J. Endocr.* **67**: 307–308.

Peaker, M. & Linzell, J. L. (1974). The effects of oestrus and exogenous estrogens on milk secretion in the goat. *J. Endocr.* **61**: 231–240.

Peaker, M. & Taylor, J. C. (1975). Milk secretion in the rabbit: changes during lactation and the mechanism of ion transport. *J. Physiol., Lond.* **253**: 527–545.

Pilson, M. E. Q. (1965). Absence of lactose from the milk of Otarioidea, a superfamily of marine mammals. *Am. Zool.* **5**: 220.

Pilson, M. E. Q. & Kelly, A. L. (1962). Composition of the milk from *Zalophus californianus*, the California sea lion. *Science, N.Y.* **135**: 104–105.

Pitelka, D. R., Hamamoto, S. T., Duafala, J. G. & Nemanic, M. K. (1973). Cell contacts in the mouse mammary gland. I. Normal gland in postnatal development and the secretory cycle. *J. cell Biol.* **56**: 797–818.

Schmidt, D. V., Walker, L. E. & Ebner, K. E. (1971). Lactose synthetase activity in Northern fur seal milk. *Biochim. biophys. Acta* **252**: 439–442.

Symp. zool. Soc. Lond. (1977) No. 41, 135–163

Comparative Endocrinology of Mammary Growth and Lactation

ISABEL A. FORSYTH and T. J. HAYDEN

National Institute for Research in Dairying,
Shinfield, Reading, England

SYNOPSIS

The ability to secrete milk is a characteristic of all female mammals. Mammary development in monotremes and marsupials is similar in non-pregnant and pregnant females and further development depends on the stimulus of suckling by the young. Thus virgin marsupials are capable of suckling young transferred to their pouch at the appropriate time after oestrus. In the bitch, fox and ferret mammary development again follows a similar course after ovulation in pregnancy and pseudopregnancy, but in most eutherian mammals extensive lobulo-alveolar and secretory development of the mammary gland only occurs in pregnancy. This development to enable the onset of copious milk secretion at parturition, which is so characteristic of eutherian mammals, must have evolved together with mechanisms for prolonging the life of the corpus luteum. Unless adequate nutrition could be immediately provided for more mature young, there would have been no selective advantage in extending intra-uterine life.

Mammary function is regulated by hormones but only a few species of eutherian mammal have been studied in any detail. Classic experiments in ovariectomized, adrenalectomized, hypophysectomized rats and mice have shown that injection of oestrogen, progesterone, adrenal steroids, prolactin and growth hormone combined in an appropriate sequence will develop an atrophic duct system into a fully lactating gland. There is also strong evidence for the involvement in mammary development of hormones from the placenta, and in particular placental lactogen secreted by fetal trophoblast. Placental lactogens have now been identified, and in some cases isolated, from representative species in four orders of eutherian mammals (Primates, Rodentia, Lagomorpha and Artiodactyla) and may prove to be even more widely distributed. Lactation is maintained in all species through the action of the suckling or milking stimulus in releasing hormones and removing milk from the gland, but there are species variations in the minimal hormonal requirement which will initiate or sustain milk secretion. Recent work indicates that hormone action is mediated by specific receptors either in the cytoplasm (steroids) or on the cell surface (protein hormones). The induction of receptors must form a crucial step in cell differentiation and the ability of cells to recognize and respond to hormones.

INTRODUCTION

The ability to secrete milk for the nutrition of the young is an essential characteristic of female mammals. There is considerable controversy about the evolutionary relationships of the three groups of living mammals, the monotremes, marsupials and eutherian or placental mammals, and the extent to which their similar features are the result of common ancestry or independent parallel evolution from the mammal-like reptiles. The mammary glands of monotremes lack teats but in histological structure (Griffiths, Elliott, Leckie & Schoefl, 1973), in features of milk composition (Griffiths, Elliott *et al.*, 1973; Tyndale-Biscoe, 1973; Jenness, 1974) and in the milk ejection response to oxytocin (Griffiths, Elliott *et al.*, 1973) all three groups show remarkable similarity. It is possible, though by no means certain, that milk production developed in the therapsids, reptilian ancestors of the mammals (see Jenness, 1974; Morriss, 1975).

MAMMARY DEVELOPMENT IN MONOTREMES AND MARSUPIALS

Monotremes are probably monoestrous and in both the echidna (*Tachyglossus aculeatus*) and the platypus (*Ornithorhynchus anatinus*) the mammary glands show seasonal development. During the breeding season the gland differentiates from generally closed ducts to a branched tubular system, even in non-pregnant females. Incubation of an egg and suckling of the hatched young provides the stimulus for the formation of alveoli and full functional development (Griffiths, Elliott *et al.*, 1973).

Some dasyurid marsupials are monoestrous, but the majority, like most eutherian mammals, show a polyoestrous pattern. The duration of pregnancy is almost always shorter than one oestrous cycle and, if the newborn young are prevented from reaching the teat, oestrus recurs at the expected time. The mammary gland develops to a similar extent in both non-mated and pregnant females. Lactation can be initiated in unmated females by transferring newborn young to their pouches at the appropriate time after oestrus. Young which attach to a teat show normal growth rates and usually only the suckled mammary gland produces milk which changes in composition as lactation progresses, showing increases in fat and protein content. The red kangaroo (*Megaleia rufa*) has four mammary glands of which two may be regressing, quiescent or redifferentiating and two functional but at different stages of development, one supplying a pouch young attached to the teat and the other a joey at heel, the two milks being of quite different composition (see Sharman, 1962, 1970; Griffiths, McIntosh & Leckie, 1972; Tyndale-Biscoe, 1973).

These observations suggest that the hormones secreted after ovulation are sufficient to produce limited lobulo-alveolar growth and that suckling initiates milk secretion, with the amount and rate of milk withdrawal having important local effects. Involvement of the pituitary is shown by cessation of lactation within two days in hypophysectomized tamar wallabies, *Macropus eugenii* (Hearn, 1973). It is probable that prolactin is one of the hormones involved and this has been identified and partially purified from red kangaroo pituitaries (Farmer & Papkoff, 1974). The rate of regression of a previously suckled mammary gland is slower in lactating than non-lactating female bush possums, *Trichosurus vulpecula* (Sharman, 1962), which also indicates the importance of systemic hormonal effects. However, the way in which local stimuli interact with the same hormonal environment to produce the variety of developmental stages seen in the mammary glands of the red kangaroo is not understood (Sharman, 1970; Griffiths, McIntosh *et al.*, 1972).

LACTOGENESIS IN MONOTREMES, MARSUPIALS AND EUTHERIANS

The young of both monotremes and marsupials are very small, less than 1 g in weight, when suckling begins. Most of their growth and development takes place during a lengthy post-natal period. Although no complete lactation curves are apparently available, it is clear from growth curves, which indirectly reflect milk yields, that the amount of milk produced increases very slowly (see Fig. 1). There are also progressive

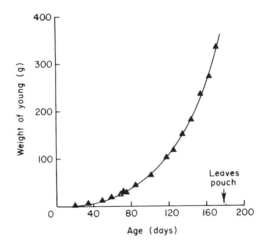

FIG. 1. Growth curve of the pouch young of *Setonix brachyrus*. Data from Shield (1966).

increases in fat and protein content of the milk, the secretion initially produced being a clear fluid (Griffiths, 1968; Tyndale-Biscoe, 1973). By contrast eutherian mammals are born at a much more advanced stage of development, which approximates to that of a marsupial ready to spend some time outside the pouch, and make accordingly much greater nutritional demands on the mother. During the first 70 days of life the single pouch young of the quokka, *Setonix brachyurus*, increases in weight from about 0·25 g to 30 g (Shield, 1966). During the same period post-partum, the Dutch rabbit, which has a similar adult body weight to *Setonix*, has completed its lactation, producing some 4000 g of milk, and reared a litter of about six to eight young with a total birth weight of some 400 g. During a 31-day period on an exclusively milk diet the weight of such a litter increases fourfold to 1600 g (Cowie, 1969).

In the monotremes and marsupials a slow, progressive mammary response to the suckling stimulus and milk removal, as a means of developing the gland to a fully lactating state, is clearly adequate to meet the demands of the young. Regular application of the suckling or milking stimulus induces mammary growth and milk secretion in non-pregnant women, rats and ruminants, an effect shown in goats to be abolished by pituitary stalk section and thus mediated by release of anterior pituitary hormones (Cowie, Knaggs, Tindal & Turvey, 1968). The changes of appearance of the secretion and the nature of the yield curves in ovariec-tomized virgin goats milked twice daily is very reminiscent of observations of the onset of milk secretion in marsupials.

In eutherians, the period which monotremes and marsupials spend attached to the teat is replaced by one of intra-uterine development, nourished by a chorio-allantoic placenta. Preparation of the mammary gland in eutherians, to produce milk in adequate volumes and of a relatively constant calorific value from the start of lactation, requires preparatory changes in the gland during pregnancy, and, as will be discussed below, there is evidence in several species that placental hor-mones are involved in producing these changes. Thus the onset of copious milk secretion at parturition, *lactogenesis stage II* in the very useful terminology proposed by Fleet, Goode, Hamon, Laurie, Linzell & Peaker (1975), is a feature of milk secretion in all eutherian mammals. Mammary growth during pregnancy produces lobulo-alveolar development with a massive increase in cell numbers, and mammary alveolar cells become terminally differentiated, acquiring the ability to make and secrete specific milk products, which in many species accumulate in the gland as pre-colostrum; this is termed *lactogenesis stage I* by Fleet *et al.* (1975).

Colostrum is replaced soon after parturition by normal milk in increasing quantities until a peak of lactation is reached, followed by declining yields, often coinciding with weaning of the young, and finally gland involution (see Cowie & Tindal, 1971). Lactation itself consists of milk secretion, involving synthesis of milk constituents from blood pre-cursors in alveolar cells and the passage of milk into the alveolar lumen, and milk removal in which stored milk is passively removed from the milk sinuses and is actively ejected from the lumen of the alveoli and from the smaller ducts by the action of the milk-ejection reflex. Other papers in this Symposium deal with triggering mechanisms for lactogenesis stage II and with milk removal (see the chapters by Kuhn and Cross, this volume). Hormonal control of lactogenesis stage I and milk secretion will be considered here.

EXPERIMENTAL INVESTIGATION OF HORMONAL DEPENDENCE IN THE EUTHERIAN MAMMARY GLAND

Rat and Mouse

Minimal hormonal requirements for mammogenesis and lactogenesis

As in many areas of reproductive physiology, studies are most extensive in the rat (Lyons, 1958; Cowie & Lyons, 1959) and the mouse (Nandi, 1959; Nandi & Bern, 1960) and it is possible to give a fairly complete account, based on experiments in endocrinectomized animals, of the sequential hormone action required to develop an atrophic duct system into a fully lactating gland. Lyons (1958) summarized the results which he and his colleagues obtained over a period of 25 years, using the Long-Evans strain of rat. During this period purified steroid and anterior pituitary hormones became available and techniques were perfected for the surgical removal of endocrine organs. Extensive study of the C3H strain of mouse was made by Nandi (1959) and the following account is based on these classic studies (see Cowie & Tindal, 1971, for additional references).

In adolescent rats and mice mammary ducts show a phase of allometric growth, i.e. they grow faster than the body. This growth phase begins shortly before the onset of oestrous cycles, is abolished by ovariectomy and restored in ovariectomized immature females with intact pituitaries by injection of oestrogen. However, in hypophysectomized, adrenalectomized, ovariectomized (triply-operated) immature female rats and mice with an atropic duct system, duct growth comparable to that occurring at puberty required injection of oestrogen + ox growth hormone + adrenal steroid. Lobulo-alveolar development similar to that seen in the second half of pregnancy was obtained by adding a high dose of progesterone and sheep prolactin. Finally, withdrawal of ovarian steroids and growth hormone, with the continued injection of prolactin and adrenal steroids, induced milk secretion.

The action of the various hormone complexes on the mammary gland is a direct one. This is shown by the induction of both lobulo-alveolar growth and milk secretion in rodent mammary tissues by exposure to hormones *in vitro*; the results are, generally, in good agreement with those obtained *in vivo* (see Forsyth, 1971). Mammary tissue from immature mice will only show lobulo-alveolar growth in culture if the animals are first given priming injections of oestrogen + progesterone (Ichinose & Nandi, 1964, 1966). The minimal hormonal requirement for lobulo-alveolar growth *in vitro* is then insulin + prolactin + aldosterone (Singh & Bern, 1969; Wood, Washburn, Mukherjee & Banerjee, 1975). In tissue from 3- to 4-week-old unprimed female rats insulin + oestradiol + progesterone + prolactin + aldosterone gave extensive lobulo-alveolar growth (Dilley & Nandi, 1968).

A predominant role for protein hormones in stimulating lobulo-alveolar development is indicated. Prolactin + growth hormone will induce lobulo-alveolar growth in triply-operated rats in the absence of steroids, if given in sufficient doses (Talwalker & Meites, 1961) while insulin + prolactin stimulate formation of small lobules of alveoli in immature rat mammary gland cultured in vitro (Dilley & Nandi, 1968). However, steroid hormones are involved in normal mammary gland growth, acting both directly on the gland and indirectly via the release of anterior pituitary hormones. The role of insulin in in vitro systems has been much discussed (Forsyth & Jones, 1976) but it now seems unlikely that, as originally proposed (see Topper & Oka, 1974), it has a specific mitogenic effect on the mammary gland.

As indicated above, prolactin + an adrenal corticoid (+insulin in in vitro systems) is the minimal hormonal requirement for the initiation or maintenance of secretion in the rat and mouse mammary gland (see also Cowie, 1957; Rivera & Bern, 1961). In some mouse strains, particularly the C3H, ox growth hormone can replace prolactin in producing lobulo-alveolar growth and milk secretion (Nandi & Bern, 1960; Rivera, 1964). The basis for the lactogenic activity of ruminant growth hormone in, for example, C3H (but not strain A) mice is not fully understood, but may relate to similarity in structure between prolactin and growth hormone (Wallis, 1975) and/or receptor specificity. The response of rodent mammary tissue to purified homologous anterior pituitary hormones has not so far been tested.

The onset of secretory activity occurs very late in pregnancy in the rat, lactose being detected in the gland only one to two days before parturition (Kuhn, 1969). In many species, including the mouse, secretory activity can be detected during the last third of pregnancy (see Denamur, 1971).

Mammary growth in pregnancy and the role of placental lactogen

Although experiments such as those outlined above establish minimal hormonal requirements for the various phases of mammary growth and function, they do not necessarily determine what is actually occurring in normal development. In 1958, Lyons emphasized the importance of placental hormones in mammary stimulation. Experiments during the 1930s and 1940s had already clearly established that the rat placenta secretes a luteotrophic and mammotrophic hormone (termed rat chorionic mammotrophin by Lyons) which will allow maintenance of the corpus luteum and mammary development in rats hypophysectomized after the 11th day of pregnancy (see Lyons, 1958, for earlier references). Rat placental lactogen has recently been isolated (Robertson & Friesen, 1975) as a protein with an estimated molecular weight of 22 000. Its potency was estimated in radioreceptor assays and compared with that of human placental lactogen. In a receptor assay based on rabbit mammary gland, which is specific for lactogenic hormones, its activity was about 40% that of sheep prolactin but about fourfold greater than that shown by human placental lactogen. In a growth hormone receptor assay both human and rat placental lactogens showed very little activity. Placental lactogen has also been identified but not yet isolated in the mouse (Kohmoto & Bern, 1970; Kohmoto, 1975).

The quantitative contribution made by the placenta towards stimulation of mammary development has been studied by several authors in both mice and rats using mammary weight and/or nucleic acid content as indices of growth. Nagasawa & Yanai (1971) showed a significant positive correlation between the degree of mammary development one day before term and the number and weight of placentae in primiparous mice in which the number of placentae and fetuses had been adjusted surgically to between one and 12 on day 8 of pregnancy. In rats, removal of fetuses plus fetal placentae on day 12 or 16 of pregnancy reduced the weight of the mammary glands to control, non-pregnant levels by day 21, but removal of fetuses alone on day 16 was without effect. Removal of fetuses on day 12 was somewhat inhibitory but significant mammary growth nevertheless occurred by day 21 (Desjardins,Paape & Tucker, 1968). Rats hypophysectomized on days 11, 12 or 15 of pregnancy and killed on day 20 had the same mammary development as control or sham-operated animals, unless they aborted (Anderson, 1975). A minimum of three fetoplacental units was needed to maintain normal mammary development in the absence of the pituitary.

A role for placental lactogen, synergizing with ovarian, adrenal and limited placental steroid secretion, in producing mammary development in pregnancy in the rat and mouse is also indicated by measurement of hormone concentrations. Using a radioreceptor assay, Shiu, Kelly & Friesen (1973) studied placental lactogen in rat serum throughout pregnancy. There was a steady rise between days 8 and 12 to peak levels of 1500 ng/ml, a marked fall to 180 ng/ml on days 14–16, and a further rise on days 17 to 21 to 800 ng/ml. Levels fell dramatically just before parturition. By contrast, prolactin levels were low in the second half of pregnancy, rising only just before term. The peak of hormonal activity on day 12 had been observed in previous studies using bioassay methods (see Forsyth, 1974; Matthies, 1974). A biphasic pattern of placental lactogen secretion is also reported for the mouse (Kelly, Tsushima, Shiu &Friesen, 1976).

In summary, complex synergisms are involved in all stages of mammary development and function in rats and mice, involving pituitary, ovarian, adrenal, placental and probably other metabolic hormones such as insulin and thyroxine. Onset of oestrogen secretion seems to be the initiator of duct growth at puberty. In early pregnancy mammary growth may be particularly dependent on prolactin. Up to about day 6 the extent of mammary growth is similar in pregnant and pseudopregnant animals (Brookreson & Turner, 1959; Anderson & Turner, 1968) and during this time two daily surges of prolactin secretion occur, initiated by the stimulus of mating and associated with initial maintenance of the corpus luteum (see Smith & Neill, 1976). After this stage it is probable that placental lactogen serves to maintain the corpus luteum and stimulate mammary growth, while prolactin is again involved in maintaining milk secretion. The increase in prolactin secretion in late pregnancy may be under ovarian control, since it is prevented by ovariectomy in late pregnancy (Bridges & Goldman, 1975). During lactation prolactin release is stimulated by suckling, with exteroceptive stimuli from the litter becoming increasingly important as lactation progresses (see Grosvenor & Mena, 1974).

Rabbit

Changes in the rabbit mammary gland in pseudopregnancy, pregnancy, and in post-partum and induced lactation, have been comprehensively studied using light and electron microscopy (Turner & Gardner, 1931; Bousquet, Fléchon & Denamur, 1969), measurement of weight and nucleic acid content (Denamur, 1961; Lu & Anderson, 1973) and synthesis of milk specific products (Strong & Dils, 1972; Mellenberger & Bauman, 1974a,b). Development of the mammary gland is rather similar in pregnant and pseudopregnant rabbits until day 16 after ovulation, and lobules of alveoli are formed by this time. Between days 18 and 22 the mammary gland of the pregnant rabbit shows ultrastructural evidence of differentiation and the synthesis of lactose and of $C_{8:0}$ and $C_{10:0}$ fatty acids which are characteristic of rabbit milk. A second marked increase in synthetic rates occurs about the time of parturition.

The minimal hormonal combination which will stimulate lobulo-alveolar growth in the rabbit has not been clearly established, but the mammary gland was found to be more responsive to ovarian steroids after hypophysectomy than in rodents or goats. Oestrogen + progesterone given together with insulin + thyroxine + cortisone gave limited lobulo-alveolar development (Norgren, 1968). In ovariectomized rabbits given oestradiol duct growth was stimulated, but lobulo-alveolar development was observed only adjacent to locally-implanted progesterone (Chatterton, 1971). Growth hormone apparently does not stimulate mammary growth (Norgren, 1968) but prolactin is clearly mammogenic since it stimulated growth as well as milk secretion when injected locally into mammary gland sectors of pseudopregnant rabbits (Lyons, 1942). Rabbits hypophysectomized on day 14 of pregnancy and maintained on progesterone did not show normal mammary development unless prolactin was also given (Denamur & Delouis, 1972).

Acting on already developed lobulo-alveolar tissue, prolactin is a sufficient stimulus in the rabbit to induce or sustain milk secretion. This has been conclusively demonstrated by experiments *in vivo* on adrenalectomized, adrenalectomized-ovariectomized, hypophysectomized and triply-operated rabbits (see Cowie, Hartmann & Turvey, 1969; Denamur, 1971) and confirmed *in vitro* when the response occurred even in the absence of insulin (Forsyth, Strong & Dils, 1972). The importance of prolactin for rabbit lactation is further emphasized by the drastic inhibition of milk yield by bromocryptine, an ergot drug which suppresses prolactin secretion, and the restoration of milk yield following prolactin injection (Taylor & Peaker, 1975).

In hypophysectomized lactating rabbits, ox growth hormone had little effect on milk yield, but did delay regression of the gland. Sheep prolactin restored milk yields to pre-operative levels and there was no evidence that adrenal corticoids improved the response (Cowie, Hartmann *et al.*, 1969; Hartmann, Cowie & Hosking, 1970). Nevertheless, injection of ACTH or adrenal corticoids into pseudopregnant rabbits will stimulate milk secretion and also increase the sensitivity to locally-injected prolactin (Chadwick, 1971) so that there may be circumstances in the intact animal when concentrations of adrenal steroids rather than prolactin limit milk production.

The nature of the lactogenic stimulus which first initiates the limited synthesis of milk products between days 18 and 22 of pregnancy in the rabbit is not known. Using co-culture or bioassay methods, Forsyth (1974) and Talamantes (1975) failed to find evidence for the secretion of placental lactogen. However, Bolander & Fellows (1976c) have recently isolated a hormone from the placentae of late-pregnant rabbits which appears to fulfil criteria for a placental lactogen, although having relatively low activity in both prolactin and growth hormone specific radioreceptor assays. A molecular weight of about 20 000, isoelectric point of 6·1 and similarities in amino acid composition to both rabbit growth hormone and rat prolactin are reported. It is tempting to suggest that this hormone may provide the first lactogenic stimulus during pregnancy in the rabbit. However, the study by Denamur & Delouis (1972) clearly shows that hypophysectomized, pregnant rabbits maintained on progesterone do not show normal mammary growth or synthesize lactose unless prolactin is injected. The relative contributions of prolactin and placental lactogen to mammary development in the rabbit must await the clear distinction of the two hormones and measurement of both through pregnancy.

Ruminants

Hormonal control of the udder in ruminants has recently been reviewed (see Convey, 1974; Forsyth & Hart, 1976). There is a phase of rapid lobulo-alveolar development in mid-pregnancy and the specific secretory abilities of the udder are established by the last third of gestation, as in the rabbit. There is strong presumptive evidence that placental lactogen is involved in producing these changes.

The effect of hormones on mammary growth after removal of the pituitary has been studied in hypophysectomized-ovariectomized goats (Cowie, Tindal & Yokoyama, 1966). Moderate lobulo-alveolar development was obtained in response to oestrogen + progesterone + prolactin + growth hormone + ACTH, but oestrogen + progesterone were without effect in the absence of anterior pituitary hormones. In sheep, hypophysectomy after day 50 does not affect the normal course of pregnancy and live young are born at term. The pregnancy is maintained by placental steroid secretion, and mammary development, although not as extensive as in intact controls, does occur and a transient lactation takes place (Denamur & Martinet, 1961), which suggests that the placenta can, at least in part, substitute for the absence of the pituitary in promoting udder growth.

Following identification of placental lactogens by bioassay methods in goats, sheep, cows and deer (Buttle, Forsyth & Knaggs, 1972; Forsyth, 1973; Buttle & Forsyth, 1976), their isolation has been reported from placentae of sheep (Handwerger, Maurer, Barrett, Hurley & Fellows, 1974; Martal & Djiane, 1975; Chan, Robertson & Friesen, 1976) and cows (Bolander & Fellows, 1976a). Some of the properties of the preparations are summarized in Table I. Notable is the high growth hormone-like activity reported for ovine placental lactogen especially by Chan et al. (1976) and the low specific activity of bovine placental lactogen.

TABLE I

Properties of ruminant placental lactogen preparations

| | Species of origin | | | |
| | Sheep | | | Cow |
Reference	1,2	3	4	2,5
Molecular weight	22 500	20 000	22 000	22 150
Isoelectric point	6·7	7·2	8·8	5·9
No. of residues	192	—	—	196 199
No. of tryptophans	2	—	—	2
No. of half-cystine	6	—	—	6
Biological activity*				
prolactin-like	100%	100%	100%	0·13%
growth hormone-like	20%	—	100%	0·06%

* Approximate activities by comparison with ruminant prolactin or growth hormone in radioreceptor or bioassays. See references for full details.
  1. Handwerger *et al.* (1974). 2. Fellows, Bolander, Hurley & Handwerger (1976). 3. Martal & Djiane (1975). 4. Chan, Robertson & Friesen (1976). 5. Bolander & Fellows (1976a).

Levels of total lactogenic activity have been measured in the circulation of pregnant goats by bioassay (Buttle *et al.*, 1972) and radioreceptor assay (C. R. Thomas, see Forsyth & Hart, 1976; Kelly, Tsushima, *et al.*, 1976) and in sheep by radioreceptor assay (Kelly, Robertson & Friesen, 1974; Djiane & Kann, 1975; Kelly, Tsushima *et al.*, 1976). Since prolactin levels are low in pregnant goats and sheep until shortly before parturition, total lactogenic activity in pregnancy is largely accounted for by placental lactogen (see Forsyth & Hart, 1976; Kelly, Tsushima *et al.*, 1976 for details and references). Levels of total lactogenic activity show considerable individual variation, but all these studies show a similar pattern of increase from about mid-pregnancy to reach peak values which may be in excess of 1 μg/ml between 100 and 140 days (length of pregnancy 148 days). In a small series of goats, multiple pregnancies were associated with higher levels of placental lactogen (see Forsyth & Hart, 1976). In the pregnant cow, bioassay (Buttle & Forsyth, 1976) and radioreceptor assay (Kelly, Tsushima *et al.*, 1976) have failed to give any reliable indication of the presence of placental lactogen in the circulation. However, using a radioimmunoassay Bolander, Ulberg & Fellows (1976) have detected changes in bovine placental lactogen which follow closely the pattern previously established in the goat and sheep and suggest that failure to measure it in previous assays is due to the low specific activity of the hormone (see Table I). Between day 200 and term (280 days) levels of bovine placental lactogen showed a plateau which was significantly higher in dairy cows (1103 ± 342 ng/ml) than in beef cows (650 ± 37 ng/ml).

In ruminants, therefore, the close association between rising levels of placental lactogen from about mid-pregnancy, now established from several studies, and the rapid phase of morphological and secretory mammary development in the second half of pregnancy, when both prolactin and growth hormone levels remain low, suggests an important role for placental lactogen in this development. It may be that in pregnancy the placenta substitutes for the pituitary in the production of both prolactin and growth hormone-like activity. Ovine placental lactogen is reported to have very significant growth-promoting activity by Chan *et al.* (1976) and the purified bovine hormone also has low but significant activity (Bolander & Fellows, 1976a,b).

The induction of udder growth and lactation in infertile or nonpregnant ruminants with intact pituitaries by injection of oestrogen + progesterone has been the subject of study for many years because of its potential economic importance (see Cowie & Tindal, 1971). Recently Smith and co-workers have introduced much shorter injection periods for treatment of cows (Smith & Schanbacher, 1973) and the ewe has also been studied (see Fulkerson, McDowell & Fell, 1975). The hormonal events in induced lactogenesis are still imperfectly understood and may be complicated by hormone release in response to early application of the milking stimulus. Fulkerson *et al.* (1975) have claimed that during the development of the udder in the ewe, induced by the injection of low doses of oestrogen plus high doses of progesterone, plasma prolactin concentrations are not usually raised. However, the episodic nature of prolactin secretion necessitates very careful study for such a claim to be upheld and in a recent experiment on goats Hart (1976) found that prolactin levels were progressively raised by a somewhat similar treatment. When bromocryptine was given in a dosage adequate to suppress the steroid- and milking-induced rises in prolactin, then there was no increase in udder size and almost no milk was secreted. It should also be emphasized that while events in induced lactation may mimic those in normal pregnancy, the mechanisms will not be the same, because of the absence of the placenta and its endocrine contribution.

Restoration of milk yield to pre-operative levels in hypophysectomized lactating goats and sheep requires injection of prolactin + growth hormone + adrenal steroids + thyroid hormone. Once lactation has been re-established in the goat, prolactin can be withdrawn without affecting milk yield for several weeks (see Cowie & Tindal, 1971; Denamur, 1971). The importance of prolactin for the initiation, and of growth hormone for the maintenance, of established lactation in ruminants is further emphasized by experiments using bromocryptine. In cows, suppression of the peak of prolactin secretion at parturition drastically reduced subsequent milk yield (Schams, Reinhardt & Karg, 1972). However, prolonged administration of bromocryptine during established lactation in cows (Karg, Schams & Reinhardt, 1972) and goats (Hart, 1973) abolished milking-induced rises in prolactin, reducing plasma concentrations to sub-basal levels without significantly reducing milk yields.

Other Eutherians

Information on other species including primates is very limited. Beck (1972) administered oestrogen + progesterone + human placental lactogen to female rhesus monkeys and obtained some lobulo-alveolar and secretory development but it was not as extensive as in mid-pregnant monkeys. A number of attempts have been made to stimulate human mammary tissue in culture (see Flaxman, 1974) but the results are somewhat conflicting. Ceriani, Contesso & Nataf (1972) increased lobulo-alveolar differentiation and obtained some secretion *in vitro* in response to insulin + sheep prolactin + ox growth hormone + aldosteron + oestradiol + progesterone, i.e. a combination known to be effective for rodent mammary tissue. Human growth hormone appears to play no essential role in human mammary development and lactation, since these events occur normally in dwarfs with a specific growth hormone deficiency (Rimoin, Holzman, Merimee, Rabinowitz, Barnes, Tyson & McKusick, 1968). In women, levels of both prolactin and placental lactogen rise steadily in pregnancy, an unusual situation since in most species, including monkeys, prolactin levels are low until term (Friesen, Shome, Belanger, Hwang, Guyda & Myers, 1972). The use of human hormones *in vitro* may help further to elucidate their role (Dilley & Kister, 1975).

The importance of prolactin for the maintenance of lactation in several species has been demonstrated by the administration of bromocryptine; this suppresses the initiation of lactation and established milk secretion in women and also suppresses lactation in pigs and dogs (del Pozo & Flückiger, 1973).

There is a large literature on the induction of mammary growth and lactation in animals with intact pituitaries (see Folley, 1952, 1956; Cowie & Tindal, 1971). This has been considered here only selectively, since the interpretation of such studies is often difficult. Administered hormones interact with, and may also release, endogenous hormones, levels of which have rarely been measured, especially in the older studies when appropriate methods were not available.

HORMONE RECEPTORS IN MAMMARY GLAND

Interaction between a hormone and its target tissue is now known to be mediated by the presence of specific receptors. Steroid hormones are bound by high-affinity binding proteins in the cytoplasm and a modified receptor steroid complex undergoes translocation to the nucleus (see O'Malley & Schrader, 1976). Protein hormones are also bound by receptors, which current opinion suggests are situated on the cell membrane and transmit their signal by some type of second messenger system. Intracellular binding sites for insulin and prolactin (Goldfine & Smith, 1976; Posner & Bergeron, 1976) have also been detected, but it is not yet known whether these are functional.

The investigation of hormone receptors in the mammary gland is still at a relatively early stage, but it is already clear that an understanding of how the various receptors are induced may provide a key to the synergistic interaction of the many hormones to which the mammary gland is directly responsive. Measurement of cytosol receptors for steroid hormones in human breast cancers is proving of great value in selecting patients for appropriate treatment (McGuire, 1975).

Oestrogen Receptors

Several authors have identified specific receptors for 17β-oestradiol in the cytoplasmic fraction of mammary glands from lactating rats (Gardner & Wittliff, 1973a; Hseuh, Peck & Clark, 1973; Leung, Jack & Reiney, 1976) and mice (Shyamala & Nandi, 1972; Richards, Shyamala & Nandi, 1974). The sedimentation and binding properties of the receptor have been reported. It is a protein and may contain a bound metal ion (Shyamala & Yeh, 1975). The cytoplasmic oestrogen receptor of mammary gland has been shown to undergo translocation to the nucleus (Shyamala & Nandi, 1972; Hseuh et al., 1973; Leung et al., 1976). Although Auricchio, Rotondi & Bresciani (1976) found no change in the amount of 17β-oestradiol bound per unit weight of cytosol protein between pregnant, lactating and ovariectomized virgin female rats, other authors have detected alterations in the level of oestrogen receptors in different physiological states. For example, the levels of oestradiol receptors have been found to be much lower in virgin and pregnant compared with lactating female rats (Gardner & Wittliff, 1973a). Hseuh *et al.* (1973) found a marked increase in cytoplasmic oestrogen receptor between days 1 and 10 of lactation, reaching a peak on day 21. The increase was unaffected by ovariectomy and was therefore apparently not dependent on ovarian oestrogen. Recently, Leung *et al.* (1976) have reported changes in cytoplasmic oestradiol receptor levels during pregnancy, lactation and involution. Levels were very variable in the later half of pregnancy, rose through lactation and fell dramatically after weaning. Similar findings have recently been obtained in our laboratory (Fig. 2, T. J. Hayden,

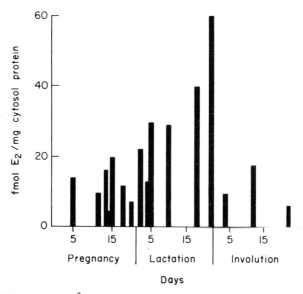

FIG. 2. Specific binding of ^{3}H-oestradiol-17β(E$_2$) by cytosol from rat mammary gland during pregnancy and lactation. 200 μl of 100 000 g supernatant from 1 : 5 (w/v) homogenates of rat mammary gland were incubated in the presence of (a) ^{3}H-oestradiol-17β (0·2–4·0 pmol) or (b) ^{3}H-oestradiol plus 4 nmol unlabelled steroid, for 18 h at 4°C. Specific binding is the difference between (a) and (b) after treatment with 500 μl dextran-coated charcoal (0·5% charcoal: 0·0025% dextran) to remove unbound steroid.

unpublished results). During lactation, the endogenous nuclear receptor-oestrogen complex was, however, found to change very little (Hseuh *et al.*, 1973) owing, apparently, to low blood oestrogen levels and not to a failure of translocation.

The physiological significance of the changes in cytoplasmic oestrogen receptor levels is not yet known. It does, however, appear likely that they are under hormonal control. Prolactin has been shown to increase oestrogen receptor concentrations in mammary gland and uterus (Leung & Saski, 1973), in mammary tumours (Saski & Leung, 1974; Vignon &

Rochefort, 1976) and in liver (Chamness, Costlow & McGuire, 1975), while progesterone (Leung & Saski, 1973) and prostaglandin (Jacobson, 1974) are inhibitory.

Progesterone Receptors

There is much less information available about progesterone receptors in mammary gland. This has in part been due to technical problems, since progesterone will bind to glucocorticoid receptors in cytoplasm, to corticosterone-binding globulin and to high-capacity, low-affinity non-specific binding sites. Use of a highly active synthetic progestin, 17,21-dimethyl-19-nor-pregna-4,9-dione-3,20-dione (R5020, Roussel-Uclaf) has recently enabled detection of a progesterone receptor in the immature uterus of mice, rats, rabbits and guinea-pigs (Philibert & Raynaud, 1973, 1974). It has also been used to detect a progesterone receptor in mammary tumours of women (Horwitz & McGuire, 1975) and rats (Labrie, Kelly, Philibert & Raynaud, 1976). Using this same compound, recent experiments (T. J. Hayden, unpublished) have shown high levels of a progesterone receptor in rat mammary gland in early pregnancy which fall progressively to late pregnancy and remain at low levels in the gland through lactation and involution (Fig. 3).

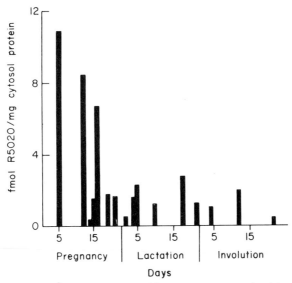

FIG. 3. Specific binding of ^{3}H-R5020 by cytosol from rat mammary gland during pregnancy and lactation. 200 μl of 100 000 g supernatant from 1 : 5 (w/v) homogenates of rat mammary gland were incubated in the presence of (a) ^{3}H-R5020 (0·2–4·0 pmol) or (b) ^{3}H-R5020 plus 4 nmol unlabelled steroid for 18 hours at 4°C. Specific binding is the difference between (a) and (b) after treatment with 500 μl dextran-coated charcoal (0·5% charcoal, 0·0025% dextran) to remove unbound steroid.

Glucocorticoid Receptors

Specific receptors for glucocorticoids have been detected in the cytosol of lactating rat (Gardner & Wittliff, 1973b) and mouse (Shyamala, 1973) mammary glands and also in mammary tumours in both species (Shyamala, 1974; Goral & Wittliff, 1975). To avoid problems associated with binding to the corticosterone-binding globulin, these studies have used either triamcinolone acetonide or dexamethasone. Sedimentation, binding and specificity properties of the binding protein have been investigated and a possible role of sulphydryl groups in the binding reaction demonstrated (Gardner & Wittliff, 1973b; Shyamala, 1973). Nuclear translocation has been shown to occur *in vivo* (Goral & Wittliff, 1975). Using *in vitro* incubation of GR strain mouse mammary tumour, Shyamala (1975) showed that the ability of various corticoids to block nuclear localization of dexamethasone correlated well with their known potency as glucocorticoids. On the other hand, oestradiol and progesterone competed for binding sites in the cytosol if present at high enough concentrations, but were not translocated to specific glucocorticoid receptors in the nucleus.

Using cytosol prepared similarly on a wet weight basis from mammary tissue of virgin, late-pregnant and lactating rats, Gardner & Wittliff (1973b) found more specific binding in pregnant than in virgin rats but lactating rats showed by far the greatest binding. Correcting for protein content of the cytosol, Chomczynski & Zwierzchowski (1976) have recently reported that in mice, receptor activity rises from day 10 to peak levels on days 16 and 17 of pregnancy, decreasing rapidly before parturition to relatively constant levels on days 1–5 of lactation.

Macromolecules binding glucocorticoids have also been identified in lactating mammary glands of voles (Turnell, Beers & Wittliff, 1974) and cows (see Gorewit & Tucker, 1976).

Prolactin Receptors

Although there is overwhelming evidence of the effectiveness of prolactin in stimulating growth and secretory activity in rodent mammary gland both *in vivo* and *in vitro*, considerable difficulty has been experienced in measuring prolactin binding in normal rat mammary tissue. This is in marked contrast with the ease with which prolactin receptors can be measured in mammary tissue from late-pregnant and lactating rabbits (Shiu & Friesen, 1974a) and in other tissues such as adrenals, kidneys and liver from a variety of species including the rat (Posner, Kelly, Shiu & Friesen, 1974), although the physiological effect of prolactin on many of these latter tissues is still uncertain. It appears that the binding capacity for prolactin of mammary epithelial cells in rats and mice is relatively small. In preliminary experiments ruminant mammary tissue also showed low percentage binding of prolactin (Forsyth & Hart, 1976).

Nevertheless, prolactin binding has been demonstrated to membrane-rich fractions of mouse mammary gland (Frantz, MacIndoe & Turkington, 1974) and, using incubation techniques, to normal mouse (Sakai, Kohmoto & Johke, 1975) and rat (Costlow, Buschow & McGuire, 1974) mammary tissue. Binding has also been demonstrated to mammary tumours of rats and mice, but its predictive value in relation to the hormone responsiveness of the tumours has so far proved limited (Costlow et al., 1974; Frantz et al., 1974; Kelly, Bradley, Shiu, Meites & Friesen, 1974; Turkington, 1974; DeSombre, Kledzik, Marshall & Meites, 1976; Holdaway & Friesen, 1976).

Frantz et al. (1974) found that prolactin binding was greater in lactating than in pregnant mouse mammary tissue. A more extensive survey of prolactin binding during pregnancy and lactation in the rat has been carried out by Holcomb, Costlow, Buschow & McGuire (1976), using a slice incubation technique. They found an increase between days 8 and 11 of pregnancy and a further dramatic seven- to eightfold increase shortly after parturition to levels which were maintained through lactation. Levels of binding were also increased following removal of the ovaries and uterus in the second half of pregnancy and it was suggested that high levels of placental lactogen mask prolactin receptors present in the mammary gland at this time, while the more modest levels of circulating prolactin released in response to suckling post-partum apparently do not have this effect. The slice technique has the serious disadvantage that no account was taken of mammary gland growth and the increase in cell numbers occurring during pregnancy. Nevertheless, by studying prolactin binding to membrane-rich fractions from rat mammary gland, a similar pattern of results was obtained, although the changes, particularly at parturition, were of smaller magnitude (Fig. 4; T. J. Hayden, unpublished results).

The prolactin receptor of rabbit mammary gland has been studied extensively by Shiu & Friesen (1974a,b; 1976); they have succeeded in solubilizing, and partially purifying, the receptor and preparing an antiserum to it. The antiserum inhibited binding of iodinated prolactin to the membrane receptor in rabbit mammary tissue and also blocked the physiological action of prolactin on rabbit mammary gland explants without interfering with insulin stimulated events.

The specificity of prolactin binding to mammary tissue has been investigated, and iodinated prolactin can be displaced from binding to rodent and rabbit mammary tissue only by hormones, such as prolactin itself from various species, placental lactogen and primate growth hormone, all of which are known to have lactogenic activity in suitable assay systems (see Shiu et al., 1973). In dd strain mice Sakai et al. (1975) were unable to detect any competition for binding to mammary tissue between iodinated sheep prolactin and unlabelled ox growth hormone, although this strain does show a lactogenic response to ox growth hormone in vitro. They concluded that the mammary gland has separate binding sites for

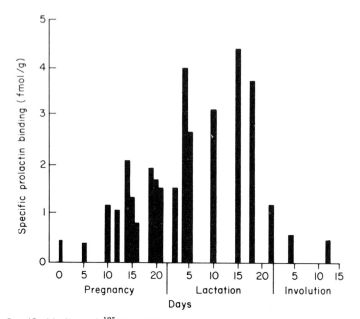

FIG. 4. Specific binding of ^{125}I-labelled ovine prolactin to rat mammary gland during pregnancy and lactation. Membrane-rich fractions prepared according to the method of Shiu, Kelly & Friesen (1973) were incubated for 18 h at 4°C. Specific binding is the difference between radioactivity bound in the presence of ^{125}I-prolactin only and that bound in the presence of 2000 ng of unlabelled prolactin.

prolactin and growth hormone. Sheth, Ranadive & Sheth (1974) showed binding of human placental lactogen to mouse mammary tissue. Ovine placental lactogen binds to rabbit mammary tissue and it is generally assumed that placental lactogen and prolactin bind to the same receptor sites on the basis of similarity of specific binding and of displacement curves (see Bolander, Hurley, Handwerger & Fellows, 1976).

Experiments on rat liver indicate that prolactin receptors may be induced by prolactin itself (Posner, Kelly & Friesen, 1975a), by thyroxine and by oestrogen (Gelato, Marshall, Boudreau, Bruni, Campbell & Meites, 1975; Posner, Kelly & Friesen, 1975b). The effectiveness of oestrogen appears to depend on an intact pituitary and may be mediated via stimulation of prolactin secretion (Posner, Kelly & Friesen, 1975a; Kelly, Posner & Friesen, 1975). Preliminary experiments on Sprague-Dawley rats suggest that the prolactin receptor of the mammary gland is similarly controlled (T. J. Hayden, unpublished).

Other Hormones

O'Keefe & Cuatrecasas (1974) detected specific binding of insulin to isolated mouse mammary cells. The binding was similar in virgin and pregnant animals but increased three- to fourfold in lactation, expressed on a DNA basis.

In recent experiments (T. J. Hayden, unpublished) binding of tri-iodothyronine (T_3) to nuclei of rat mammary cells has been detected. Preliminary characterization of the binding shows considerable similarity to the properties reported for T_3 binding to rat liver nuclei (see Oppenheimer & Surks, 1975). Levels of binding are increased threefold during pregnancy, and show a further slight increase to mid-lactation, followed by a decline which continues during gland involution following weaning.

GENERAL CONCLUSIONS: THE EVOLUTION OF ENDOCRINE SUPPORT FOR THE MAMMARY GLAND

The hormonal control of mammary growth and lactation has been studied in so few eutherian mammals that only limited general conclusions can be drawn. Mammary growth in pregnancy can be best assessed by combining histological examination with measurement of DNA content which reflects cell numbers. By the end of pregnancy 60–100% of mammary growth has been completed, growth continuing into early lactation in some species such as the rat, mouse, hamster and rabbit (see Lu & Anderson, 1973). In rats and mice, experiments in endocrinectomized animals have shown that mammary development equivalent to that seen in late pregnancy can be stimulated by the injection of oestrogen + progesterone + adrenal steroids + growth hormone + prolactin; metabolic hormones such as insulin and thyroxine may also be involved. Specific receptors have been demonstrated in mammary tissue for all these hormones with the exception of growth hormone. Placental lactogen, secreted by fetal trophoblast, is implicated as a contributory factor in mammary development in rats, mice and ruminants, since concentrations of this hormone in plasma are elevated from mid-pregnancy while those of prolactin and growth hormone remain low.

Placental lactogens have now been identified, and in some cases isolated, from representative species in four orders of eutherian mammals: the rodents, lagomorphs, primates and ruminant artiodactyls. So far they have largely been identified on the basis of their prolactin-like activity and they clearly belong to the growth hormone–prolactin family (Wallis, 1975). However, recent findings by Fellows and co-workers (see Fellows et al., 1976 and Table I) suggest that ruminant placental lactogens resemble prolactin in features of structure, while the primate placental lactogens very closely resemble primate growth hormone (Wallis, 1975). In addition, Kelly, Tsushima et al. (1976) have observed that growth hormone-like and prolactin-like activity, measured by receptor assays in the same sera, show different patterns of change in both pregnant sheep and in women. It is possible that two separate placental hormones exist, related respectively to prolactin or growth hormone. Alternatively, this type of placental hormone may have arisen more than once. The evolutionary relationships and distribution of such hormones in other species and orders is clearly a matter of great interest.

The young of eutherians, retained within the female genital tract and nourished by a chorio-allantoic placenta, are born at a relatively advanced stage and immediately make considerable nutritional demands on the mother. In guinea-pigs, the young are so mature at birth that they may be able to survive independent of maternal care, but this is unusual and most species are totally dependent on milk for some period after birth. As already indicated, the mammary gland must therefore be developed during pregnancy to exhibit the onset of copious milk secretion at parturition which is so typical of eutherian mammals. Indeed, there would have been no selective advantage in prolonging the period of intra-uterine life of the fetus unless the mammary gland was capable of producing a high milk yield within days of parturition. It follows that mechanisms for prolonging the luteal phase of the oestrous cycle and suppressing ovulation, and for developing the mammary gland in pregnancy must have been evolved together and that it may be instructive to consider their hormonal control in this light.

Mechanisms for the recognition of pregnancy vary widely between species (see Heap, Perry & Challis, 1973). In some carnivores, such as ferrets and dogs, pregnancy and pseudopregnancy are the same length and maternal recognition of pregnancy may not be involved at all. The corpora lutea of pregnancy and pseudopregnancy are similar in size and duration of function, and it is notable that mammary development is also indistinguishable in pregnant and pseudopregnant ferrets (Hammond & Marshall, 1930), dogs (Turner, 1939) and probably foxes (Rowlands & Parkes, 1935).

In the rat, mouse and rabbit the stimulus of an infertile mating leads to pseudopregnancy, but its duration is about half that of pregnancy and mammary development proceeds only to about a mid-pregnancy stage. In other species, such as domestic ruminants, pseudopregnancy does not occur and an infertile mating is followed by a return to oestrus at the expected time with no substantial change in the mammary gland.

Thus it seems that the beginning of mammary development in pregnancy is associated with, and possibly dependent on, rescue of the corpus luteum. Both mammary growth and corpus luteum are influenced by hormone complexes, and in many species prolactin, or particularly in the second half of pregnancy placental lactogen, is an important component of the complex regulating both structures. Luteinizing hormone is also important for maintenance of the corpus luteum, acting both directly and, notably in the rabbit, via oestrogen secretion. Oestrogen also acts on the developing mammary gland together with progesterone. The latter is secreted by the corpus luteum and also the placenta in some species and is essential for the maintenance of pregnancy. While stimulating mammary growth and lobulo-alveolar formation, progesterone apparently restrains the development of full secretory activity until term, although it is not inhibitory in established lactation. The mechanisms by which progesterone produces these effects are, however, still poorly understood.

It is well recognized that the young maintain the mother's lactation, via the suckling stimulus, by bringing about the release from her anterior pituitary of the necessary hormone complex. In some species such as the rat, exteroceptive stimuli from the litter are also involved. There are marked species differences in the minimal hormonal combination which will initiate or sustain lactation. Prolactin alone is effective in rabbits, prolactin + adrenal corticoids in rats and mice, prolactin + growth hormone + thyroid hormone + adrenal corticoids in sheep and goats. In ruminants, growth hormone is apparently more important than prolactin in maintaining an established lactation. The suckling stimulus also releases oxytocin which facilitates, and in some species is essential for, efficient milk removal (see Cowie & Tindal, 1971; Grosvenor & Mena, 1974; Tindal, 1974a,b). Milk allowed to accumulate rather rapidly causes a decline in the biosynthetic capacity of the mammary gland (Jones, 1968).

Thus functional mammary development, on which successful reproduction and the raising of young depend, can be regarded as a response to stimuli largely provided by the young. In monotremes, marsupials and a few eutherians, like the bitch and ferret, the initial development of the mammary gland is probably dependent on ovarian, adrenal and pituitary secretions which are produced following ovulation and independently of the presence of an embryo in the uterus. In most eutherians, however, extensive mammary growth normally only occurs during pregnancy, as a response to the presence of the fetus and with the involvement of the hormones secreted by fetal trophoblast. In all species hormones released by the suckling stimulus, together with milk removal, are then responsible for any further development and the function of the lactating gland. In a number of groups such as monotremes, marsupials, myomorph rodents and ruminants, there seems to be an inverse relationship between the proportion of mammary growth which occurs in lactation and the degree of maturity of the young at birth.

ACKNOWLEDGEMENT

We are grateful to the Cancer Research Campaign who provided financial support for T. J. Hayden.

REFERENCES

Anderson, R. R. (1975). Mammary gland growth in the hypophysectomized pregnant rat. *Proc. Soc. exp. Biol. Med.* **148**: 283–287.

Anderson, R. R. & Turner, C. W. (1968). Mammary gland growth during pseudopregnancy and pregnancy in the rat. *Proc. Soc. exp. Biol. Med.* **128**: 210–214.

Auricchio, F., Rotondi, A. & Bresciani, F. (1976). Oestrogen receptor in mammary gland cytosol of virgin, pregnant and lactating mice. *Molec. cellular Endocr.* **4**: 55–60.

Beck, P. (1972). Lactogenic activity of human chorionic somatomammotropin in rhesus monkeys. *Proc. Soc. exp. Biol. Med.* **140**: 183–187.

Bolander, F. F. & Fellows, R. E. (1976a). Purification and characterization of bovine placental lactogen. *J. biol. Chem.* **251**: 2703–2708.

Bolander F. F. & Fellows, R. E. (1976b). Growth-promoting and lactogenic activities of bovine placental lactogen. *J. Endocr.* **71**: 173–174.

Bolander, F. F. & Fellows, R. E. (1976c). The purification and characterization of rabbit placental lactogen. *Biochem. J.* **159**: 775–782.

Bolander, F. F., Hurley, T. W., Handwerger, S. & Fellows, R. E. (1976). Localization and specificity of binding of subprimate placental lactogen in rabbit tissues. *Proc. natn. Acad. Sci., U.S.A.* **73**: 2932–2935.

Bolander, F. F., Ulberg, L. C. & Fellows, R. E. (1976). Circulating placental lactogen levels in dairy and beef cattle. *Endocrinology* **99**: 1273–1278.

Bousquet, M., Fléchon, J. E. & Denamur, R. (1969), Aspects ultrastructuraux de la glande mammaire de lapine pendant la lactogénèse. *Z. Zellforsch. mikrosk. Anat.* **96**: 418–436.

Bridges, R. S. & Goldman, B. D. (1975). Ovarian control of prolactin secretion during late pregnancy in the rat. *Endocrinology* **97**: 496–498.

Brookreson, A. D. & Turner, C. W. (1959). Normal growth of mammary gland in pregnant and lactating mice. *Proc. Soc. exp. Biol. Med.* **102**: 744–745.

Buttle, H. L. & Forsyth, I. A. (1976). Placental lactogen in the cow. *J. Endocr.* **68**: 141–146.

Buttle, H. L., Forsyth, I. A. & Knaggs, G. S. (1972). Plasma prolactin measured by radioimmunoassay and bioassay in pregnant and lactating goats and the occurrence of a placental lactogen. *J. Endocr.* **53**: 483–491.

Ceriani, R. L., Contesso, G. P. & Nataf, B. M. (1972). Hormone requirement for growth and differentiation of the human mammary gland in organ culture. *Cancer Res.* **32**: 2190–2196.

Chadwick, A. (1971). Lactogenesis in pseudopregnant rabbits treated with adrenocorticotrophin and adrenal steroids. *J. Endocr.* **49**: 1–8.

Chamness, G. C., Costlow, M. E. & McGuire, W. L. (1975). Estrogen receptor in rat liver and its dependence on prolactin. *Steroids* **26**: 363–371.

Chan, J. S., Robertson, H. A. & Friesen, H. G. (1976). The purification and characterization of ovine placental lactogen. *Endocrinology* **98**: 65–76.

Chatterton, R. T. (1971). Progesterone and mammary gland development. In *The sex steroids. Molecular mechanisms*: 345–382. McKerns, K. W. (ed.). New York: Meredith Corporation.

Chomczynski, P. & Zwierzchowski, L. (1976). Mammary glucorticoid receptor of mice in pregnancy and in lactation. *Biochem. J.* **158**: 481–483.

Convey, E. M. (1974). Serum hormone concentrations in ruminants during mammary growth, lactogenesis, and lactation. A review. *J. Dairy Sci.* **57**: 905–917.

Costlow, M. E., Buschow, R. A. & McGuire, W. L. (1974). Prolactin receptors in an estrogen receptor-deficient mammary carcinoma. *Science, N.Y.* **184**: 85–86.

Cowie, A. T. (1957). The maintenance of lactation in the rat after hypophysectomy. *J. Endocr.* **16**: 135–147.

Cowie, A. T. (1969). Variations in the yield and composition of the milk during lactation in the rabbit and the galactopoietic effect of prolactin. *J. Endocr.* **44**: 437–450.

Cowie, A. T., Hartmann, P. E. & Turvey, A. (1969). The maintenance of lactation in the rabbit after hypophysectomy. *J. Endocr.* **43**: 651–662.

Cowie, A. T., Knaggs, G. S., Tindal, J. S. & Turvey, A. (1968). The milking stimulus and mammary growth in the goat. *J. Endocr.* **40**: 243–252.

Cowie, A. T. & Lyons, W. R. (1959). Mammogenesis and lactogenesis in hypophysectomized, ovariectomized, adrenalectomized rats. *J. Endocr.* **19**: 29–32.

Cowie, A. T. & Tindal, J. S. (1971). *The physiology of lactation.* London: Edward Arnold.

Cowie, A. T., Tindal, J. S. & Yokoyama, A. (1966). The induction of mammary growth in the hypophysectomized goat. *J. Endocr.* **34**: 185–195.

Denamur, R. (1961). Étude de la glande mammaire de la lapine pendant la gestation et la lactation. *Annls Endocr.* **22**: 768–776.

Denamur, R. (1971). Hormonal control of lactogenesis. *J. Dairy Res.* **38**: 237–264.

Denamur, R. & Delouis, C. (1972). Effects of progesterone and prolactin on the secretory activity and the nucleic acid content of the mammary gland of pregnant rabbits. *Acta endocr., Copenh.* **70**: 603–618.

Denamur, R. & Martinet, J. (1961). Effets de l'hypophysectomie et de la section de la tige pituitaire sur la gestation de la brebis. *Annls Endocr.* **22**: 755–759.

Desjardins, C., Paape, M. J. & Tucker, H. A. (1968). Contribution of pregnancy, fetuses, fetal placentas and deciduomas to mammary gland and uterine development. *Endocrinology* **83**: 907–910.

DeSombre, E. R., Kledzik, G., Marshall, S. & Meites, J. (1976). Estrogen and prolactin receptor concentrations in rat mammary tumors and response to endocrine ablation. *Cancer Res.* **36**: 354–358.

Dilley, W. G. & Kister, S. J. (1975). *In vitro* stimulation of human breast tissue by human prolactin. *J. natn. Cancer Inst.* **55**: 35–36.

Dilley, W. G. & Nandi, S. (1968). Rat mammary gland differentiation *in vitro* in the absence of steroids. *Science, N.Y.* **161**: 59–60.

Djiane, J. & Kann, G. (1975). Mise en évidence de l'activité lactogène et mesure dans le sérum de l'activité prolactinique du placenta chez la brebis au cours de la gestation. *C.r. hebd. Séanc. Acad. Sci., Paris.* **280**: 2785–2788.

Farmer, S. W. & Papkoff, H. (1974). Studies on the anterior pituitary hormones of the kangaroo. *Proc. Soc. exp. Biol. Med.* **145**: 1031–1036.

Fellows, R. E., Bolander, F. F., Hurley, T. W. & Handwerger, S. (1976). Isolation and characterization of bovine and ovine placental lactogen. In *Growth hormone and related peptides*: 315–326. Pecile, A. & Müller, E. E. (eds). Amsterdam: Excerpta Medica.

Flaxman, A. (1974). *In vitro* studies of the normal human mammary gland. *J. invest. Derm.* **63**: 48–57.

Fleet, I. R., Goode, J. A., Hamon, M. H., Laurie, M. S., Linzell, J. L. & Peaker, M. (1975). Secretory activity of goat mammary glands during pregnancy and the onset of lactation. *J. Physiol., Lond.* **251**: 763–773.

Folley, S. J. (1952). Lactation. In *Marshall's physiology of reproduction* **2**: 525–647. 3rd edn. Parkes, A. S. (ed.). London: Longmans Green.

Folley, S. J. (1956). *The physiology and biochemistry of lactation*. Edinburgh: Oliver & Boyd.

Forsyth, I. A. (1971). Organ culture techniques and the study of hormone effects on the mammary gland. *J. Dairy. Res.* **38**: 419–444.

Forsyth, I. A. (1973). Secretion of a prolactin-like hormone by the placenta in ruminants. In *Le corps jaune*: 239–255. Denamur R. & Netter, A. (eds). Paris: Masson et Cie.

Forsyth, I. A. (1974). The comparative study of placental lactogenic hormones. A review. In *Lactogenic hormones, fetal nutrition and lactation*: 49–67. Josimovich, J. B., Reynolds, M. & Cobo, E. (eds). New York: John Wiley.

Forsyth, I. A. & Hart, I. C. (1976). Mammotrophic hormones in ruminants. In *Growth hormone and related peptides*: 422–432. Pecile, A. & Müller, E. E. (eds). Amsterdam: Excerpta Medica.

Forsyth, I. A. & Jones, E. A. (1976). Organ culture of mammary gland and placenta in the study of hormone action and placental lactogen secretion. In *Organ culture in biomedical research*: 201–221. Balls, M. & Monnickendam, M. A. (eds). Cambridge: Cambridge University Press.

Forsyth, I. A., Strong, C. R. & Dils, R. (1972). Interactions of insulin, cortico-sterone and prolactin in promoting milk-fat synthesis by mammary explants from pregnant rabbits. *Biochem. J.* **129**: 929–935.

Frantz, W. L., MacIndoe, J. H. & Turkington, R. W. (1974). Prolactin receptors: characterstics of the particulate fraction binding activity. *J. Endocr.* **60**: 485–497.

Friesen, H., Shome, B., Belanger, C., Hwang, P., Guyda, H. & Myers, R. (1972). The synthesis and secretion of human and monkey placental lactogen (HPL and MPL) and pituitary prolactin (HPr and MPr). In *Growth and growth hormone*: 224–238. Pecile, A. & Müller, E. E. (eds). Amsterdam: Excerpta Medica.

Fulkerson, W. J., McDowell, G. H. & Fell, L. R. (1975). Artificial induction of lactation in ewes: the role of prolactin. *Aust. J. biol. Sci.* **28**: 525–530.

Gardner, D. G. & Wittliff, J. L. (1973a). Specific estrogen receptors in the lactating mammary gland of the rat. *Biochemistry* **12**: 3090–3096.

Gardner, D. G. & Wittliff, J. L. (1973b). Characterization of a distinct glucocorticoid-binding protein in the lactating mammary gland of the rat. *Biochim. biophys. Acta* **320**: 617–627.

Gelato, M., Marshall, S., Boudreau, M., Bruni, J., Campbell, G. A. & Meites, J. (1975). Effects of thyroid and ovaries on prolactin binding activity in rat liver. *Endocrinology* **96**: 1292–1296.

Goldfine, I. D. & Smith, G. J. (1976). Binding of insulin to isolated nuclei. *Proc. natn. Acad. Sci., U.S.A.* **73**: 1427–1431.

Goral, J. E. & Wittliff, J. L. (1975). Comparison of glucocorticoid-binding proteins in normal and neoplastic mammary tissues of the rat. *Biochemistry* **14**: 2944–2952.

Gorewit, R. C. & Tucker, H. A. (1976). Corticoid binding in mammary tissue slices from lactating cows. *J. Dairy Sci.* **59**: 232–240.

Griffiths, M. (1968). *Echidnas*. Oxford: Pergamon Press.

Griffiths, M., Elliott, M. A., Leckie, R. M. C. & Schoefl, G. I. (1973). Observations on the comparative anatomy and ultrastructure of mammary glands and on the fatty acids of the triglycerides in platypus and echidna milk fats. *J. Zool., Lond.* **169**: 255–279.

Griffiths, M., McIntosh, D. L. & Leckie, R. M. C. (1972). The mammary glands of the red kangaroo with observations on the fatty acid components of the milk triglycerides. *J. Zool., Lond.* **166**: 265–272.

Grosvenor, C. E. & Mena, F. (1974). Neural and hormonal control of milk secretion and milk ejection. In *Lactation* 1: 227–276. Larson, B. L. & Smith, V. R. (eds). New York and London: Academic Press.

Hammond, J. & Marshall, F. H. A. (1930). Oestrus and pseudo-pregnancy in the ferret. *Proc. R. Soc.* **105**: 607–630.

Handwerger, S., Maurer, W., Barrett, J., Hurley, T. & Fellows, R. E. (1974). Evidence for homology between ovine and human placental lactogens. *Endocrine Res. Commun.* **1**: 403–413.

Hart, I. C. (1973). Effect of 2-bromo-α-ergocryptine on milk yield and the level of prolactin and growth hormone in the blood of the goat at milking. *J. Endocr.* **57**: 179–180.

Hart, I. C. (1976). Prolactin, growth hormone, insulin and thyroxine: their possible roles in steroid induced mammary growth and lactation in the goat. *J. Endocr.* **71**: 41–42P.

Hartmann, P. E., Cowie, A. T. & Hosking, Z. D. (1970). Changes in enzymic activity, chemical composition and histology of the mammary glands and blood metabolites of lactating rabbits after hypophysectomy and replacement therapy with sheep prolactin, human growth hormone or bovine growth hormone. *J. Endocr.* **48**: 433–448.

Heap, R. B., Perry, J. S. & Challis, J. R. G. (1973). The hormonal maintenance of pregnancy. In *Handbook of physiology*, Section 7, Endocrinology, **2** (2): 217–260. Greep, R. O. (ed.). Washington, D.C.: American Physiological Society.

Hearn, J. P. (1973). Pituitary inhibition of pregnancy. *Nature, Lond.* **241**: 207–208.

Holcomb, H., Costlow, M., Buschow, R. & McGuire, W. L. (1976). Prolactin binding in rat mammary gland during pregnancy and lactation. *Biochim. biophys. Acta* **428**: 104–112.

Holdaway, I. M. & Friesen, H. G. (1976). Correlation between hormone binding and growth response of rat mammary tumor. *Cancer Res.* **36**: 1562–1567.

Horwitz, K. B. & McGuire, W. L. (1975). Specific progesterone receptors in human breast cancer. *Steroids* **25**: 497–505.

Hseuh, A. J. W., Peck, E. J. & Clark, J. H. (1973). Oestrogen receptors in the mammary gland of the lactating rat. *J. Endocr.* **58**: 503–511.

Ichinose, R. R. & Nandi, S. (1964). Lobuloalveolar differentiation in mouse mammary tissues *in vitro*. *Science, N.Y.* **145**: 496–497.

Ichinose, R. R. & Nandi, S. (1966). Influence of hormones on lobulo-alveolar differentiation of mouse mammary glands *in vitro*. *J. Endocr.* **35**: 331–340.

Jacobson, H. I. (1974). Oncolytic action of prostaglandins. *Cancer Chemother.* **58**: 503–511.

Jenness, R. (1974). The composition of milk. In *Lactation* 3: 3–107. Larson, B. L. & Smith, V. R. (eds). New York and London: Academic Press.

Jones, E. A. (1968). The relationship between milk accumulation and enzyme activities in the involuting rat mammary gland. *Biochim. biophys. Acta* **177**: 158–160.

Karg, H., Schams, D. & Reinhardt, V. (1972). Effects of 2-Br-α-ergocryptine on plasma prolactin level and milk yields in cows. *Experientia* **28**: 574–576.

Kelly, P. A., Bradley, C., Shiu, R. P. C., Meites, J. & Friesen, H. G. (1974). Prolactin binding to rat mammary tumor tissue. *Proc. Soc. exp. Biol. Med.* **146**: 816–819.

Kelly, P. A., Posner, B. I. & Friesen, H. G. (1975). Effects of hypophysectomy, ovariectomy, and cycloheximide on specific binding sites for lactogenic hormones in rat liver. *Endocrinology* **97**: 1408–1415.

Kelly, P. A., Robertson, H. A. & Friesen, H. G. (1974). Temporal pattern of placental lactogen and progesterone secretion in sheep. *Nature, Lond.* **248**: 435–437.

Kelly, P. A., Tsushima, T., Shiu, R. P. C. & Friesen, H. G. (1976). Lactogenic and growth hormone-like activities in pregnancy determined by radioreceptor assays. *Endocrinology* **99**: 765–774.

Kohmoto, K. (1975). Synthesis of two lactogenic proteins by the mouse placenta *in vitro. Endocr. jap.* **22**: 275–278.

Kohmoto, K. & Bern, H. A. (1970). Demonstration of mammotrophic activity of the mouse placenta in organ culture and by transplantation. *J. Endocr.* **48**: 99–107.

Kuhn, N. J. (1969). Progesterone withdrawal as the lactogenic trigger in the rat. *J. Endocr.* **44**: 39–54.

Labrie, F., Kelly, P. A., Philibert, D. & Raynaud, J. P. (1976). Specific progesterone receptors in dimethylbenzanthracene (DMBA)-induced mammary tumors. *Steroids* **27**: 395–404.

Leung, B. S., Jack, W. M. & Reiney, C. G. (1976). Estrogen receptor in mammary glands and uterus of rats during pregnancy, lactation and involution. *Biochemistry* **7**: 89–95.

Leung, B. S. & Saski, G. H. (1973). Prolactin and progesterone effect on specific estradiol binding in uterine and mammary tissues *in vitro. Biochem. biophys. Res. Commun.* **55**: 1180–1187.

Lu, M.-H. & Anderson, R. R. (1973). Growth of the mammary gland during pregnancy and lactation in the rabbit. *Biol Reprod.* **9**: 538–543.

Lyons, W. R. (1942). The direct mammotrophic action of lactogenic hormone. *Proc. Soc. exp. Biol. Med.* **51**: 308–311.

Lyons, W. R. (1958). Hormonal synergism in mammary growth. *Proc. R. Soc. (B)* **149**: 303–325.

McGuire, W. L. (1975). Current status of estrogen receptors in human breast cancer. *Cancer* **36**: suppl. 638–644.

Martal, J. & Djiane, J. (1975). Purification of a lactogenic hormone in sheep placenta. *Biochem. biophys. Res. Commun.* **65**: 770–778.

Matthies, D. L. (1974). Placental peptide hormones affecting fetal nutrition and lactation. Effects of rodent chorionic mammotrophin. In *Lactogenic hormones, fetal nutrition and lactation*: 297–334. Josimovich, J. B., Reynolds, M. & Cobo, E. (eds). New York: John Wiley.

Mellenberger, R. W. & Bauman, D. E. (1974a). Metabolic adaptations during lactogenesis. Fatty acid synthesis in rabbit mammary tissue during pregnancy and lactation. *Biochem. J.* **138**: 373–379.

Mellenberger, R. W. & Bauman, D. E. (1974b). Metabolic adaptations during lactogenesis. Lactose synthesis in rabbit mammary tissue during pregnancy and lactation. *Biochem. J.* **142**: 659–665.

Morriss, G. (1975). Placental evolution and embryonic nutrition. In *Comparative placentation. Essays in structure and function*: 87–107. Steven, D. H. (ed.). London and New York: Academic Press.

Nagasawa, H. & Yanai, R. (1971). Quantitative participation of placental mammotropic hormones in mammary development during pregnancy of mice. *Endocr. jap.* **18**: 507–510.

Nandi, S. (1959). Hormonal control of mammogenesis and lactogenesis in the C3H/He Crgl mouse. *Univ. California Publs Zool.* **65**: 1–128.

Nandi, S. & Bern, H. A. (1960). Relation between mammary-gland responses to lactogenic hormone combinations and tumor susceptibility in various strains of mice. *J. natn. Cancer Inst.* **24**: 907–933.

Norgren, A. (1968). Modifications of mammary development of rabbits injected with ovarian hormones. *Acta Univ. Lund.* Section II, **4**: 41.

O'Keefe, E. & Cuatrecasas, P. (1974). Insulin receptors in murine mammary cells: comparison in pregnant and nonpregnant animals. *Biochim. biophys. Acta* **343**: 64–77.

O'Malley, B. W. & Schrader, W. T. (1976). The receptors of steroid hormones. *Scient. Am.* **234**: 32–43.

Oppenheimer, J. H. & Surks, M. I. (1975). Biochemical basis of thyroid hormone action. In *Biochemical actions of hormones* **3**: 119–157. Litwack, G. (ed.). New York and London: Academic Press.

Philibert, D. & Raynaud, J. P. (1973). Progesterone binding in the immature mouse and rat uterus. *Steroids* **22**: 89–98.

Philibert, D. & Raynaud, J. P. (1974). Progesterone binding in the immature rabbit and guinea-pig uterus. *Endocrinology* **94**: 627–632.

Posner, B. I. & Bergeron, J. J. M. (1976). Intracellular polypeptide hormone receptors. *58th Annual Meeting of the Endocrine Society: Program and Abstracts*: 165.

Posner, B. I., Kelly, P. A. & Friesen, H. G. (1975a). Prolactin receptors in rat liver: possible induction by prolactin. *Science, N.Y.* **187**: 57–59.

Posner, B. I., Kelly, P. A. & Friesen, H. G. (1975b). Induction of a lactogenic receptor in rat liver: influence of estrogen and the pituitary. *Proc. natn. Acad. Sci., U.S.A.* **71**: 2407–2410.

Posner, B. I., Kelly, P. A., Shiu, R. P. C. & Friesen, H. G. (1974). Studies of insulin, growth hormone and prolactin binding: tissue distribution, species variation and characterization. *Endocrinology* **95**: 521–531.

del Pozo, E. & Flückiger, E. (1973). Prolactin inhibition: experimental and clinical studies. In *Human prolactin*: 291–301. Pasteels, J. L., Robyn, C. & Ebling, F. J. G. (eds). Amsterdam: Excerpta Medica.

Richards, J. E., Shyamala, G. & Nandi, S. (1974). Estrogen receptor in normal and neoplastic mouse mammary tissues. *Cancer Res.* **34**: 2764–2772.

Rimoin, D. L., Holzman, G. B., Merimee, T. J., Rabinowitz, D., Barnes, A. C., Tyson, J. E. A. & McKusick, V. A. (1968). Lactation in the absence of human growth hormone. *J. clin. Endocr. Metab.* **28**: 1183–1188.

Rivera, E. M. (1964). Differential responsiveness to hormones of C3H and A-mouse mammary tissues in organ culture. *Endocrinology* **74**: 853–864.

Rivera, E. M. & Bern, H. A. (1961). Influence of insulin on maintenance and secretory stimulation of mouse mammary tissues by hormones in organ-culture. *Endocrinology* **69**: 340–353.

Robertson, M. C. & Friesen, H. G. (1975). The purification and characterization of rat placental lactogen. *Endocrinology* **97**: 621–629.

Rowlands, I. W. & Parkes, A. S. (1935). The reproductive processes of certain mammals. VIII. Reproduction in foxes. *Proc. zool. Soc. Lond.* **1935**: 823–841.

Sakai, S., Kohmoto, K. & Johke, T. (1975). A receptor site for prolactin in lactating mouse mammary tissues. *Endocr. jap.* **22**: 379–387.

Saski, G. H. & Leung, B. S. (1974). Prolactin stimulation of estrogen receptor *in vitro* in 7,12,dimethyl-benz(A) anthracene-induced mammary tumors. *Res. Comm. Chem. Path. Pharmacol.* **8**: 409–412.

Schams, D., Reinhardt, V. & Karg, H. (1972). Effects of 2-Br-α-ergokryptine on plasma prolactin level during parturition and onset of lactation in cows. *Experientia* **28**: 697–699.

Sharman, G. B. (1962). The initiation and maintenance of lactation in the marsupial, *Trichosurus vulpecula*. *J. Endocr.* **25**: 375–385.

Sharman, G. B. (1970). Reproductive history of marsupials. *Science, N.Y.* **167**: 1221–1228.

Sheth, N. A., Ranadive, K. J. & Sheth, A. R. (1974). *In vitro* binding of radioiodinated human placental lactogen to murine mammary gland. *Eur. J. Cancer* **10**: 653–660.

Shield, J. (1966). Oxygen consumption during pouch development of the macropod marsupial *Setonix brachyurus*. *J. Physiol., Lond.* **187**: 257–270.

Shiu, R. P. C. & Friesen, H. G. (1974a). Properties of a prolactin receptor from the rabbit mammary gland. *Biochem. J.* **140**: 301–311.

Shiu, R. P. C. & Friesen, H. G. (1974b). Solubilization and purification of a prolactin receptor from the rabbit mammary gland. *J. biol. Chem.* **249**: 7902–7911.

Shiu, R. P. C. & Friesen, H. G. (1976). Blockade of prolactin action by an antiserum to its receptors. *Science, N. Y.* **192**: 259–261.

Shiu, R. P. C., Kelly, P. A. & Friesen, H. G. (1973). Radioreceptor assay for prolactin and other lactogenic hormones. *Science, N.Y.* **180**: 968–971.

Shyamala, G. (1973). Specific cytoplasmic glucocorticoid hormone receptors in lactating mammary glands. *Biochemistry* **12**: 3085–3089.

Shyamala, G. (1974). Glucocorticoid receptors in mouse mammary tumors. Specific binding of glucocorticoids in the cytoplasm. *J. biol. Chem.* **249**: 2160–2163.

Shyamala, G. (1975). Glucocorticoid receptors in mouse mammary tumours: specific binding to nuclear components. *Biochemistry* **14**: 437–444.

Shyamala, G. & Nandi, S. (1972). Interactions of 6,7-^{3}H-17β-estradiol with the mouse lactating mammary tissue *in vivo* and *in vitro*. *Endocrinology* **91**: 861–867.

Shyamala, G. & Yeh, Y.-F. (1975). Is the estrogen receptor of mammary glands a metallo-protein? *Biochem. biophys. Res. Commun.* **64**: 408–415.

Singh, D. V. & Bern, H. A. (1969). Interaction between prolactin and thyroxine in mouse mammary gland lobulo-alveolar development *in vitro*. *J. Endocr.* **45**: 579–583.

Smith, M. S. & Neill, J. D. (1976). Termination at midpregnancy of the two daily surges of plasma prolactin initiated by mating in the rat. *Endocrinology* **98**: 696–701.

Smith, R. L. & Schanbacher, F. L. (1973). Hormone-induced lactation in the bovine. 1. Lactational performance following injections of 17β-oestradiol and progesterone. *J. Dairy Sci.* **56**: 738–743.

Strong, C. R. & Dils, R. (1972). Fatty acid biosynthesis in rabbit mammary gland during pregnancy and early lactation. *Biochem. J.* **128**: 1303–1309.

Talamantes, F. (1975). Comparative study of the occurrence of placental prolactin among mammals. *Gen. comp. Endocr.* **27**: 115–121.

Talwalker, P. K. & Meites, J. (1961). Mammary lobulo-alveolar growth induced by anterior pituitary hormones in adreno-ovariectomized and adreno-ovariectomized-hypophysectomized rats. *Proc. Soc. exp. Biol. Med.* **117**: 121–124.

Taylor, J. C. & Peaker, M. (1975). Effects of bromocryptine on milk secretion in the rabbit. *J. Endocr.* **67**: 313–314.

Tindal, J. S. (1974a). Hypothalamic control of secretion and release of prolactin. *J. Reprod. Fert.* **39**: 437–461.

Tindal, J. S. (1974b). Stimuli that cause the release of oxytocin. In *Handbook of physiology*, Section 7, Endocrinology, 4(1): 257–267. Greep, R. O. & Astwood, E. B. (eds). Washington, D.C: American Physiological Society.

Topper, Y. J. & Oka, T. (1974). Some aspects of mammary gland development in the mature mouse. In *Lactation* 1: 327–348. Larson, B. L. & Smith, V. R. (eds). New York and London: Academic Press.

Turkington, R. W. (1974). Prolactin receptors in mammary carcinoma cells. *Cancer Res.* **34**: 758–763.

Turnell, R. W., Beers, P. C. & Wittliff, J. L. (1974). Glucocorticoid-binding macromolecules in the lactating mammary gland of the vole. *Endocrinology* **95**: 1770–1773.

Turner, C. W. (1939). *The comparative anatomy of the mammary glands.* Columbia, Missouri: University Cooperative Store.

Turner, C. W. & Gardner, W. U. (1931). The relation of the anterior pituitary hormones to the development and secretion of the mammary gland. *Univ. Mo. agr. exp. Sta. Res. Bull.* No. 158.

Tyndale-Biscoe, H. (1973). *Life of marsupials.* London: Edward Arnold.

Vignon, F. & Rochefort, H. (1976). Regulation of estrogen receptors in ovarian-dependent rat mammary tumors. 1. Effects of castration and prolactin. *Endocrinology* **98**: 722–729.

Wallis, M. (1975). The molecular evolution of pituitary hormones. *Biol. Rev.* **50**: 35–98.

Wood, B. G., Washburn, L. L., Mukherjee, A. S. & Banerjee, M. R. (1975). Hormonal regulation of lobulo-alveolar growth, functional differentiation and regression of whole mouse mammary gland in organ culture. *J. Endocr.* **65**: 1–6.

Symp. zool. Soc. Lond. (1977) No. 41, 165–192

Lactogenesis: the Search for Trigger Mechanisms in Different Species

N. J. KUHN

Department of Biochemistry, University of Birmingham, Birmingham, England

SYNOPSIS

This paper deals with the question of identifying the particular hormonal change, or changes, which serve to trigger the onset of copious lactation near the end of pregnancy. Attention is drawn to the two concepts of lactogenic control, (a) as a "release" of the mammary gland from inhibition by steroid hormones, especially progesterone, and (b) as a "push" by prolactational hormones, especially glucocorticoid and prolactin. Results of experiments in which normal lactogenesis is advanced or modified by presentation or withdrawal of these hormones *in vivo* or *in vitro* are summarized. It is shown that induction of lactogenesis in pregnancy by prolactin occurs probably only in the rabbit, while most other procedures leading to lactogenesis are explicable in terms of progesterone withdrawal. Data on the timing of hormonal changes are listed, and the problems of relating them accurately to the timing of lactogenesis are pointed out. The unsuitability of many commonly used criteria of lactogenesis is discussed, and the desirability of measuring rates of milk solid synthesis is stressed.

While the collection of these results in different species is emphasized, it is concluded that, with certain tentative exceptions, there is no reason to suppose that all species do not use the same hormonal trigger mechanism, and that at present progesterone withdrawal constitutes the most promising candidate as lactogenic trigger.

INTRODUCTION

Three-quarters of a century have now elapsed since serious attempts were first made to account for the onset of copious maternal lactation that normally accompanies birth of the young. Although it now seems that the biochemical and morphological changes underlying the mammary reponse will prove to be very complex, there is a widespread hope—if not always belief—that the response is initiated and co-ordinated by a relatively simple change in the hormonal status of the mother. One feels instinctively that such an important physiological event as lactogenesis must be triggered by a signal that is simple, loud and clear. Recent years have seen a great surge of interest in mammalian endocrinology, aided by the development of techniques for the assay of very low hormone concentrations in the blood. Despite this, no simple unambiguous hormonal mechanism has yet been accepted for the control of lactogenesis in any one species.

There are in principle only two experimental approaches to the problem. The one approach is to disturb the mammary gland by known procedures that either prevent normal lactogenesis from occurring, or else precipitate it at times when it would not otherwise occur. Table I lists

TABLE I

Procedures or administered substances that have been used to initiate lactation in vivo

Ovariectomy, or luteal extirpation	Choriogonadotrophin
Hysterectomy, or abortion	Prostaglandin $F_{2\alpha}$
Stress	Aminoglutethimide
Milking	Adrenaline
Prolactin	Perphenazine
Glucocorticoid	Chlorpromazine
Oestrogen	

the many different procedures that have been used for this purpose. The other approach is simply to map the changing concentration, or flux, of each substance believed to be centrally involved in normal lactogenesis. Thus where the first approach perturbs the system, the second one analyses it; by combining the information from these two sources it should be possible to arrive at a clear account of lactogenesis. In practice both approaches have been followed, but incompletely; so in the meantime one is obliged to weigh the theories that have been proposed against this imperfect evidence.

Underlying all the hypotheses that have been put forward are essentially two conceptual mechanisms, as was recognized by Nelson as early as 1934. Lactogenesis might be due to a "release" of mammary function from an existing restraint, or else it might be due to a "push" from factors that favour lactation. Such factors are usually taken as being hormonal and have frequently been described as "lactogenic". I shall rather use the more neutral expression "prolactational". Although such hormonal factors will be a major subject of this review, it would be as well to remain aware of other prolactational factors such as blood supply to the mammary gland, and milk removal by the young.

The many publications dealing with hormonal aspects of lactogenesis have been the subject of several sensitive and comprehensive reviews (Cowie, 1969; Denamur, 1971; Tucker, 1974), so that this paper need not attempt a complete coverage of the literature. Rather, it will try to summarize and critically assess the nature and the quality of the available evidence, and to interpret it in the light of the possible mechanisms mentioned above.

PERTURBATIONS OF LACTOGENESIS

The Concept of Release from Inhibition

It was probably the Viennese doctor Halban who, in 1905, first gave clear expression to the view that during pregnancy the mammary gland was under an ovarian or placental influence that promoted its growth while restraining its function (Halban, 1905). He drew attention to the parallel that exists in this respect between the mammary gland and the uterus, basing his views on his observations of women undergoing normal, aborted or prolonged pregnancy, or experiencing placental retention after parturition. He did not use the term "hormone" which, indeed, was coined only that year by Starling (1905). The concept was later extended by demonstrations of lactation being precipitated by the removal of the corpora lutea from pregnant goats (Drummond-Robinson & Asdell, 1926), and of the ovaries, conceptus, or both from pregnant rats (Selye, Collip & Thomson, 1933; Collip, Selye & Thomson, 1933; Shinde, Ôta & Yokoyama, 1965; Liu & Davis, 1967; Deis, 1968; Kuhn, 1969a), mice (Bradbury, 1932), guinea-pigs (Nelson, 1934), rabbits (Denamur & Delouis, 1972), ewes (Hartmann, Trevethan & Shelton, 1973) and cows (Shirley, Emery, Convey & Oxender, 1973). Administered progesterone can largely prevent such induced, or normal, lactogenesis in rats (Deis, 1968; Yokoyama, Shinde & Ôta, 1969; Kuhn, 1969a; Davis, Wikman-Coffelt & Eddington, 1972; Murphy, Ariyanayagam & Kuhn, 1973) where the effect appears to be specific to this steroid (Kuhn, 1969b), and in mice (Turkington & Hill, 1969), sheep (Hartmann et al., 1973) and rabbits (Denamur & Delouis, 1972), although the measured criterion of lactogenesis was not always the same. Progesterone can also inhibit the lactogenesis induced by the provision of prolactational hormones to pseudopregnant rabbits and to mammary explants from mice and rabbits (Turkington & Hill, 1969; Denamur & Delouis, 1972; Assairi, Delouis, Gaye, Houdebine, Ollivier-Bousquet & Denamur, 1974; Delouis, 1975).

In the ovarohysterectomized guinea-pig oestrogen alone could completely prevent the onset of lactation (Nelson, 1934), but this hormone appears not to affect the mammary tissue of mice (Turkington & Hill, 1969) or rats (Kuhn, 1969a; Davis et al., 1972) in this way. In experiments with different species the inhibitory action of progesterone has been variously enhanced (Meites & Sgouris, 1953; Denamur, Delouis & Gaye, 1970), reversed (Hartmann et al., 1973) or unaffected (Kuhn, 1972a) by oestrogen.

In summary, therefore, the removal of the conceptus and/or ovaries during pregnancy precipitates lactation in those species so far examined. This is most generally reversed by progesterone, although few species have been tested for this. The role of oestrogen is still obscure; its anti-lactogenic effect in the guinea-pig deserves reinvestigation, especially since in this species the uterus seems to be refractory to progesterone (Zarrow, Anderson & Callantine, 1963; Porter, 1969; Donovan & Peddie, 1974).

The administration of certain chemicals or hormones can also cause "withdrawal" of progesterone in pregnant animals. Aminoglutethimide inhibits progesterone synthesis at the stage of cholesterol side-chain cleavage (Gower, 1974), whereas human choriogonadotrophin and prostaglandin $F_{2\alpha}$ both induce luteal 20α-hydroxysteroid dehydrogenase, the action of which reduces the progesterone secretion of the corpus luteum in pregnant rats (Kuhn, 1972b; Rodway & Kuhn, 1975; Deis, 1971). In the case of the first two agents, progesterone has been shown to inhibit the lactogenesis thus induced.

The use of progesterone on explants of mammary tissue maintained *in vitro* has revealed that its inhibition can be exerted directly upon the tissue itself (Turkington & Hill, 1969; Delouis, 1975), a view supported by experiments on hypophysectomized or prolactin-treated rabbits (Meites & Sgouris, 1953; Assairi *et al.*, 1974).

From an historical standpoint it is interesting to speculate why, since the discovery of progesterone in the early 1930s, it took almost 40 years before its direct inhibition of lactogenesis was demonstrated. Walker & Mathews (1949) apparently just missed the effect because they chose too small a dose. Others were apparently put off by the report that even massive doses do not inhibit established lactation (Folley, 1942), a finding that still requires explanation.

The Concept of a Push by Prolactational Hormones

This idea was introduced, quite unexpectedly, when Stricker & Grueter (1928) observed lactation in pseudopregnant rabbits that had been injected with crude extracts of anterior pituitary. The effect was independently confirmed by Corner (1930), and similar extracts were shown to improve or re-establish lactation in the bitch, the cow and the sow (Grueter & Stricker, 1929). With the benefit of hindsight one can probably ascribe this action of pituitary extract to its content of prolactin and corticotrophin.

Two experimental approaches subsequently established the minimal hormonal requirements for the onset of lactation and its maintenance. In the one case animals with suitably developed mammary glands were subjected to hypophysectomy, adrenalectomy and ovariectomy and injected with various combinations of pure hormones to induce lactation (Lyons, Li & Johnson, 1958; Cowie & Lyons, 1959; Nandi & Bern, 1961). In the other case explants of mammary tissue were removed from pregnant or pseudopregnant animals and incubated in a physiological medium suitably reinforced with combinations of pure hormones (for review, see Forsyth, 1971). The two approaches have shown an absolute requirement for prolactin and glucocorticoid for the rat and mouse, although prolactin may be replaced by choriomammotrophin (Turkington & Topper, 1966; Turkington, 1968) or by human somatotrophin (Rivera, Forsyth & Folley, 1967), and, in the case of certain mouse strains, by somatotrophin of other species (Nandi & Bern, 1960; Rivera *et al.*, 1967). Glucocorticoid may also be replaced *in vitro* by a mineralocorticoid (Turkington, Juergens & Topper, 1967). Insulin is absolutely necessary for lactogenesis to occur in tissues maintained *in vitro*, as many people have found, and is also required for the maintenance of normal lactation *in vivo* (Walters & McLean, 1968; Martin & Baldwin, 1971), especially in the late stage of lactation of the rat (Kumaresan & Turner, 1965). As yet, however, it has not been possible to show a dependence on this hormone of lactogenesis *in vivo* (Kyriakou & Kuhn, 1973).

In the intact rabbit both the onset and the maintenance of lactation are satisfied by prolactin alone (insulin was also present), since neither adrenalectomy nor hypophysectomy (with prolactin replacement) appeared to impair lactational performance in the short term (Cowie & Watson, 1966; Denamur, 1969; Cowie, Hartmann & Turvey, 1969). Human choriomammotrophin can replace prolactin (Friesen, 1966). Cortisol can also be dispensed with for the synthesis of milk fat, lactose and galactosyltransferase (EC 2.4.1.22) by explants of rabbit mammary tissue (Strong, Forsyth & Dils, 1972; Delouis, 1975).

Studies on the hypophysectomized goat have established the importance of thyrotrophin and somatotrophin in addition to prolactin and glucocorticoid for the full maintenance of lactation (Cowie, 1969). The situation in the cow seems to be similar, but with the interesting exception that prolactin apparently plays some role in lactogenesis but little or none in established lactation (Karg, Schams & Reinhardt, 1972; Schams, Reinhardt & Karg, 1972; Kaprowski & Tucker, 1973).

Nobody has apparently tried to induce lactation in pregnant animals by injecting insulin, but it is possible to do so with prolactin or glucocorticoid. Injected prolactin readily induces lactation in the pregnant rabbit (see Denamur, 1971) but apparently in other species it can only initiate or maintain lactation in the non-pregnant or lactating state. Perphenazine, (Ben-David, Danon, Benveniste, Weller & Sulman, 1971), adrenaline (Meites, 1959; Desclin, 1960) and chlorpromazine (Talwalker, Meites, Nicoll & Hopkins, 1960) can each induce lactation in non-pregnant animals, apparently by virtue of their ability to stimulate the secretion of prolactin from the pituitary. Oestrogen has also been used to induce lactation, especially in farm animals (Turner & Meites, 1941), and this can perhaps also be ascribed to its well documented stimulation of prolactin secretion (Niswender, Chen, Midgeley, Meites & Ellis, 1969; Chen & Meites, 1970). An alternative, or additional, explanation may lie in the stimulation by oestrogen of prostaglandin $F_{2\alpha}$ release from the uterus (Challis, Harrison, Heap, Horton & Poyser, 1972), since this substance is known to be luteolytic in several species and to cause lactogenesis in the rat (Deis, 1971).

Large doses of natural or synthetic glucocorticoid have successfully been used to induce lactation in pregnant rabbits, rats and cows (Talwalker, Nicoll & Meites, 1961; Meites, Hopkins & Talwalker, 1963; Tucker & Meites, 1965; Karg & Schams, 1971). More physiological doses are ineffective in the rat (Davis & Liu, 1969).

Complications in the Interpretations of "Release" and "Push" Phenomena

Care is required in the interpretation of many of the effects described above and, indeed, it is now clear that some of the early interpretations must be revised. In early work with virgin animals it was not always realized that a suitable lactogenic response requires not merely an appropriate stimulus but also a well developed mammary gland. Such development can be achieved by therapy with oestrogen and progesterone. However in cases where such treatment was stopped in exchange for treatment with the prolactational hormones under investigation, the ensuing lactational response has been frequently considered only in terms of its dependence on these latter hormones, so that the importance of the "withdrawal" of the steroid hormones has been overlooked (see, for example, Lyons et al., 1958; Fulkerson, McDowell & Fell, 1975). Similarly the emphasis on the hormonal requirements for secretion by mammary explants taken from pregnant animals has tended to overshadow the important fact of their simultaneous removal from the progesterone and oestrogen environment of the donor animals.

In cases where a pregnant animal has been "pushed" into lactation by the administration of glucocorticoid it has generally been implied that plasma levels of this hormone were being made up to values that would support lactation, where before they had been too low to do so (Talwalker, Nicoll et al., 1961; Nandi & Bern, 1961; Gala & Westphal, 1967). Yet this explanation is unconvincing in the case of the rabbit where lactation can proceed even in the absence of adrenal support (see Cowie & Tindal, 1971:147). Moreover the doses that have been employed are very great, especially where the highly potent synthetic glucocorticoids are concerned, so that one cannot rule out some pharmacological reaction— perhaps producing a stress that in turn stimulates the release of prolactin (see Meites & Clemens, 1972). Most germane to the argument, however, is the long known action of glucocorticoid in impairing the maintenance of normal gestation (Burdick & Konanz, 1941; Seifter, Christian & Ehrich, 1951). In cows, sheep and rabbits this hormone has been successfully used to induce premature delivery, which is preceded also by lactogenesis and by a premature fall in the concentration of progesterone (Tucker & Meites, 1965; Adams & Wagner, 1969; Schams, Reinhardt et al., 1972; Kendall & Liggins, 1972). Therefore until further documentation is available it seems probable that the lactogenic effect of glucocorticoid in pregnant animals is secondary to an induced withdrawal of progesterone. The mechanism by which glucocorticoid achieves this is unclear.

Some comment is required on prolactin, the hormone most frequently taken to be "lactogenic". The striking feature to note is that despite its ability to induce lactation in the pregnant rabbit—which it does most readily—all demonstrations of its lactogenic properties in other species appear to be possible only in the absence of progesterone. It has been unfortunate that the expression "lactogenic" carries the double meaning of "favouring lactation" and of "initiating lactation".

At the same time one must consider whether experiments designed to achieve a release of the mammary gland from inhibition do not simultaneously bring about an increase in the plasma concentration of prolactational hormones. Thus, an elevation of plasma prolactin levels has been noted in pregnant rats receiving prostaglandin $F_{2\alpha}$ (Vermouth & Deis, 1972) and in cows subjected to early Caesarean delivery (Shirley et al., 1973), but has not been observed after ovariectomy of the pregnant rat (Simpson, Simpson & Kulkarni, 1973). But it seems unlikely that such increases in prolactin can account for the ensuing lactogenesis, in view of the doubtful role of prolactin in the cow (see references on p. 169 above) and of the ineffectiveness of prolactin given to the pregnant rat (Talwalker, Nicoll et al., 1961; Kuhn, 1969a).

Whether or not the withdrawal of progesterone is always followed by a rise in plasma prolactin levels—and at normal parturition it certainly is—the occurrence of lactogenesis in hypophysectomized, parturient rats (Selye, Collip & Thomson, 1934; Nelson, 1936; Bintarningsih, Lyons, Johnson & Li, 1958; Abraham, Cady & Chaikoff, 1960), in mice (Newton & Richardson, 1940), in guinea-pigs (Pencharz & Lyons, 1934) and apparently also in the human (Kaplan, 1961) certainly implies that no such rise is necessary for at least the onset of lactogenesis. Nevertheless some may wish to remain cautious on this point since it is established that at least a minimum of some prolactin-like hormone is essential for lactogenesis, and this is almost certainly satisfied in the hypophysectomized animal by choriomammotrophin or even by fetal pituitary prolactin. Experiments have not yet ruled out a surge of these hormones in the hypophysectomized, parturient animal, although the possibility does not seem strong at present. In this context it may be noted that fetal glucocorticoid can sometimes enter the maternal circulation, although its impact on normal lactogenesis is unknown (Dupouy, Coffigny & Magne, 1975).

TIME ANALYSIS OF EVENTS ASSOCIATED WITH LACTOGENESIS

The Timing of Hormonal Changes at Normal Lactogenesis

Concentrations of plasma progesterone during pregnancy are now known for a large number of species (Table II). A fairly uniform pattern emerges in which the relatively high concentration that characterizes

TABLE II

Determinations of plasma progesterone during pregnancy and parturition

Animal used	Reference
Mouse	Murr, Stabenfeldt, Bradford & Geschwind (1974)
	Virgo & Bellward (1974)
Hamster	Lukaszewska & Greenwald (1970)
Rat	Grota & Eik-Nes (1967)
	Wiest *et al.* (1968)
	Kuhn (1969a)
	Morishige, Pepe & Rothchild (1973)
	Labhsetwar & Watson (1974)
	Pepe & Rothchild (1974)
Ferret	Heap & Hammond (1974)
Guinea-pig	Challis, Heap & Illingworth (1971)
Rabbit	Challis, Davies & Ryan (1973)
	Baldwin & Stabenfeldt (1974)
Bitch	Jones, Boyns, Cameron, Bell, Christie & Parkes (1973)
Ewe	Bassett, Oxborrow, Smith & Thorburn (1969)
	Stabenfeldt, Drost & Franti (1972)
	Chamley, Buckmaster, Cerini, Cumming, Goding, Obst, Williams & Winfield (1973)
	Hartmann *et al.* (1973)
	Kelly, Robertson & Friesen (1974)
Sow	Masuda, Anderson, Henricks & Melampy (1967)
	Robertson & King (1974)
Goat	Heap & Linzell (1966)
Rhesus monkey	Neill, Johansson & Knobil (1969)
Human	Short (1961)
	Wiest (1967)
Cow	Short (1958)
	Donaldson *et al.* (1970)
	Hunter, Erb, Randel, Garverick, Callahan & Harrington (1970)
	Stabenfeldt, Osburn & Ewing (1970)
	Henricks, Dickey, Hill & Johnston (1972)
	Schams, Hoffmann, Fischer, Marz & Karg (1972)
	Smith *et al.* (1973)

most of pregnancy falls during the last few days to reach a very low value at parturition itself. Careful inspection of the published data reveals, in very many cases, that the fall is apparently biphasic. A sharp drop during the one or two days immediately before parturition is preceded by a more dignified descent initiated several days—30 to 40 days in the bitch— earlier (see, for example, Kuhn, 1969a; Donaldson, Bassett & Thorburn, 1970; Smith, Edgerton, Hafs & Convey, 1973). This biphasic pattern is all the more interesting because studies in the rat have indicated a dual mechanism for the withdrawal of progesterone at the end of pregnancy. The early, gradual, withdrawal reflects a declining net rate of Δ^4-3-keto steroid synthesis, while the final rapid withdrawal coincides with a shift in the nature of the secreted steroids from progesterone to its metabolite 20α-hydroxypregn-4-en-3-one (Hashimoto, Henricks, Anderson & Melampy, 1968; Wiest, Kidwell & Balogh, 1968; Kuhn, 1969a; Kuhn & Briley, 1970). It is important to note that the latter hormone does not block lactogenesis in the rat (Kuhn, 1969b). 17α, 20α-Dihydroxypregn-4-en-3-one may play a similar role in the placenta of the ewe (Anderson, Flint & Turnbull, 1975). In the guinea-pig a prepartum fall of progesterone has not been clearly demonstrated, perhaps because the total hormone that is measured mainly reflects that bound to the unusually high level of cortisol binding globulin in this species.

Primates appear to be exceptions to the pattern described above, since analysis of plasma progesterone in the human and the rhesus monkey indicates a fall only after loss of the placenta at parturition. It is therefore of particular interest that in these species lactogenesis occurs only one or two days after delivery (Geschickter, 1945; Speert, 1948; Del Pozo,

TABLE III

Determinations of plasma prolactin and choriomammotrophin during pregnancy and parturition

Animal used	Reference
	PROLACTIN
Mouse	Murr, Bradford & Geschwind (1974)
	Sinha *et al.* (1974)
Rat	Amenomori, Chen & Meites (1970)
	Linkie & Niswender (1972)
	Morishige *et al.* (1973)
	Simpson, Simpson, Sinha & Schmidt (1973)
	Labhsetwar & Watson (1974)
Ewe	Arai & Lee (1967)
	Chamley *et al.* (1973)
	Kelly, Robertson *et al.* (1974)
Goat	Hart (1972)
Human	L'Hermite & Robyn (1972)
	Del Pozo *et al.* (1976)
Cow	Schams & Karg (1970)
	CHORIOMAMMOTROPHIN
Rat	Shiu, Kelly & Friesen (1973)
Human	Kaplan & Grumbach (1965a)
Ewe	Kelly, Robertson *et al.* (1974)

Flückiger & Lancranjan, 1976), and may be further delayed by placental retention (Halban, 1905).

Fewer species have yet been analysed for plasma prolactin (Table III), but in most cases a parturient surge appears to be initiated during the ultimate or penultimate day of pregnancy. Two groups of authors have reported a double peak of prolactin at parturition (Labhsetwar & Watson, 1974; Sinha, Selby & Vanderlaan, 1974). In women, however, prolactin levels rise progressively during the last two-thirds of pregnancy (L'Hermite & Robyn, 1972; Del Pozo et al., 1976). In most species the prolactin is sustained at an elevated level during most of lactation, but in the cow it falls to a low value soon after parturition. The level rises greatly, and temporarily, after the onset of each suckling.

Choriomammotrophin (placental lactogen) has now been identified or characterized in a number of species including the human, rat, mouse, goat, sheep, monkey and cow (Kohmoto & Bern, 1970; Buttle, Forsyth & Knaggs, 1972; Friesen, 1973; Kelly, Robertson et al., 1974; Kelly, Shiu, Robertson & Friesen, 1975; Buttle & Forsyth, 1976; for review, see Forsyth, 1974). The potency of the human hormone is very similar to that of ovine prolactin when tested for its stimulation of casein and DNA synthesis in explants of mouse mammary gland (Turkington & Topper,

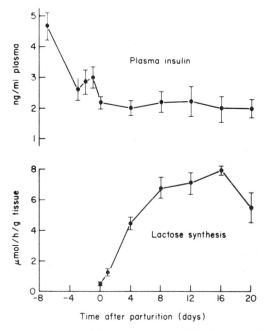

FIG. 1. Changes in the plasma insulin concentration and in the rate of mammary lactose synthesis in intact parturient or lactating rats (C. J. Wilde & N. J. Kuhn, unpub. obs.).

1966). In women the venous concentration rises during the last two-thirds of pregnancy to reach a value of about 6 μg/ml at term (Kaplan & Grumbach, 1965a,b). The serum concentration in rats peaks at 12–14 days and again at 17–21 days of pregnancy, exceeding 1 μg/ml at these times; it then falls just before parturition to a value of about 60 ng/ml, this coinciding with the early rise of prolactin concentration (Shiu *et al.*, 1973). Extremely high levels are also seen in ewes up to parturition (Kelly, Robertson *et al.*, 1974).

I am not aware of many studies of insulin levels at the time of lactogenesis. In the rat, however, values are high during pregnancy and fall to lower values at the time of lactogenesis, or just before (Sutter-Dub, Leclercq, Sutter & Jacquot, 1974; this paper, Fig. 1).

The changes in plasma glucocorticoid that have been reported to accompany lactogenesis (Table IV) are less easily summarized than those for the hormones already mentioned and will consequently be described in some detail.

TABLE IV

Determinations of plasma glucocorticoid during pregnancy and parturition

Animal used	Reference
Mouse	Gala & Westphal (1967)
Hamster	Brinck-Johnsen, Kilham & Margolis (1973)
Rat	Kuhn (1969a)
	Voogt, Sar & Meites (1969)
	Simpson, Simpson, Sinha *et al.* (1973)
	Ôta, Ôta & Yokoyama (1974)
	Dupouy *et al.* (1975)
Guinea-pig	Gala & Westphal (1967)
Rabbit	Gala & Westphal (1967)
	Baldwin & Stabenfeldt (1974)
Ewe	Chamley *et al.* (1973)
Sow	Baldwin & Stabenfeldt (1973)
Cow	Brush (1958)
	Adams & Wagner (1970)
	Smith *et al.* (1973)

In the pregnant rabbit cortisol and corticosterone concentrations are both low (about 10 ng/ml) until day 24. Thereafter cortisol, but not corticosterone, rises to reach a peak (about 90 ng/ml) at day 30 and then drops rapidly to a low value again by day 32 (Baldwin & Stabenfeldt, 1974). Gala & Westphal (1967) reported an increase in cortisol concentration between day 27–28 of pregnancy and day 1–2 of lactation, the intervening period not being examined.

Brush (1958) recorded large changes in plasma concentrations of 17-hydroxy corticosteroids of individual cows around the time of parturition, but it is difficult to make out a common trend in these data except that concentrations generally subsided after 4–5 days of lactation. Mean corticosteroid levels rose about twofold during the four days before parturition in the study of Adams & Wagner (1970), and three- to fourfold during the two days before parturition according to Smith et al. (1973), returning to lower values again after a few days.

In ewes, the rise in corticosteroid has been reported to occur mainly post-partum (Chamley et al., 1973).

The picture in the rat is confusing, owing partly to the infrequent sampling by some authors and perhaps also to the marked circadian variations that occur. From some reports there appears to be an increase in maternal corticosterone during days 15–21 of pregnancy, possibly due to the increasing activity of the fetal adrenals at this time (Simpson, Simpson, Sinha & Schmidt, 1973; Dupouy et al., 1975). In other reports a roughly similar increase has been seen in morning samples but not in afternoon samples (Ôta, Ôta et al., 1974; Kuhn, 1969a). Voogt et al. (1969) found levels to be elevated on the day of parturition, relative to the previous sample at day 20, but Kuhn (1969a) observed no elevation even in rats sampled during delivery.

In mice and guinea-pigs the concentration of corticosterone and cortisol respectively is reported to be high near the end of pregnancy and to fall over the time of parturition, but very few sample times were employed (Gala & Westphal, 1967). Cortisol concentrations in the pregnant hamster apparently rise greatly up to the end of pregnancy, especially during the last week (Brinck-Johnsen et al., 1973). In the sow a rise of corticosteroid seems to be associated particularly with farrowing (Baldwin & Stabenfeldt, 1973).

Any conclusions that may be drawn from such data may need to take into account the extensive binding of plasma corticosteroids to circulating "corticosteroid-binding globulin" (CBG) and to serum albumen. Moreover the concentration of CBG can itself change, particularly through the influence of oestrogen. Taking this factor into account, Gala & Westphal (1967) calculated that the concentration of unbound corticosteroid between late pregnancy and early lactation rises about fourfold, twofold and 1·5-fold in the rabbit, rat and mouse respectively, and falls by about two-thirds in the guinea-pig.

Viewing this evidence as a whole it appears that although the transition from pregnancy to lactation is often associated with increased concentrations of total or free glucocorticoid in the plasma, it is difficult to see any consistent relationship with the onset of lactation. Rather, the changes seem to reflect several different influences including (a) increased fetal adrenal activity, (b) the stress of parturition, especially in large animals, and (c) the effect of the young in stimulating the maternal production of glucocorticoid (Zarrow, Schlein, Denenberg & Cohen, 1972; Ôta, Harai, Unno, Sakaguchi, Tomogane & Yokoyama, 1974).

Problems in the Interpretation of Timing of Hormonal Change

Certain problems are encountered in the interpretation of these hormonal changes and their significance for lactogenesis. It is extremely difficult to pinpoint the exact onset of change in a hormone concentration because such change is already superimposed on shorter term changes such as a circadian rhythm and the pulsatile nature of hormone release. The onset of change may also be obscured by insufficient sampling times, by the effect of stress on the hormone level itself, or by the use of data averaged from the results of several animals. Nevertheless, the fall in concentration of progesterone seems to coincide with, or precede, the early rise in prolactin concentration, a relationship that is consistent with the inhibition of prolactin secretion that progesterone can exert (Chen & Meites, 1970). Alternatively, or additionally, the surge of prolactin may reflect the surge of oestrogen that occurs about this time (Shaikh, 1971; Neill, Freeman & Tillson, 1971; see also Bedford, Challis, Harrison & Heap, 1972: Caligaris, Astrada & Taleisnik, 1974). Against this must be set at least one case (sheep) where the oestrogen concentration clearly rose after the increase in prolactin (Chamley et al., 1973).

The significance of the prolactin surge as a possible lactogenic trigger is hard to assess accurately in view of the uncertainty of its exact timing in relation to that of lactogenesis (see below). This is further complicated if one recognizes that lactogenesis would follow such a trigger only after a lag period (see Chadwick, 1962). But it is the presence of choriomammotrophin during pregnancy, often in very high concentrations, which raises severe doubt as to whether the net concentration of prolactin-like hormone rises at all around the time of lactogenesis, or whether it may not even temporarily fall.

The Timing of Lactogenesis

It is at first sight surprising that the timing of lactogenesis itself is less well documented than that of the hormonal changes associated with it, but in fact considerable problems are involved. Ideally the changing rate of synthesis of each major milk solid (lactose, casein and, where they are prominent, shorter chain fatty acids) would be continuously monitored in single animals and compared with changing hormone levels measured in frequent samples of plasma from the same animal. Unfortunately no such technique is yet available. Among the few workers who have determined rates, Bauman and his colleagues have measured lactose and fat synthesis in mammary slices from the cow and rabbit (Mellenberger, Bauman & Nelson, 1973; Mellenberger & Bauman, 1974a,b). However only a few sampling times were chosen, and the actual rate increases that they observed were surprisingly small, so that no particular time emerged as defining the onset of lactation. In mammary explants taken from rabbits at different stages of pregnancy Strong & Dils (1972) observed a threefold increase in the rate of fatty acid synthesis on day 21, followed by a much larger increase initiated between day 27 and 30 of pregnancy; medium-chain fatty acids, characteristic of rabbit milk fat, first appeared on day 19. Large changes, well defined in time, were derived by Hartmann (1973) from the volume and composition of daily mammary secretion of cows approaching and entering parturition. Nearly all other workers have observed the prevailing levels of milk components in the tissue. The scoring of microscopically visible secretory granules or fat globules which appear progressively toward the end of pregnancy, or of microscopically visible milk, has gradually given way to the specific and quantitative measurement of lactose (Chadwick, 1962; Shinde et al., 1965; Kuhn & Lowenstein, 1967), casein (Yokoyama et al., 1969; Kuhn, 1972a) or citrate (Peaker & Linzell, 1975). Unfortunately these specific criteria have as yet been applied to only a few species. Lactose appears on the last day of pregnancy in the rat (Shinde et al., 1965; Wren, DeLauder & Bitman, 1965; Kuhn & Lowenstein, 1967), but about eight days before parturition in the rabbit (Denamur, 1969) and goat (Fleet, Goode, Hamon, Laurie, Linzell & Peaker, 1975). Casein clearly precedes lactose in the rat (Yokoyama et al., 1969; Kuhn, 1972a), but accompanies it in the rabbit (Bousquet, Fléchon & Denamur, 1969).

Two major problems attend the interpretation of such measurements on the lactogenic response. First, where animals must be killed before parturition in the course of tissue sampling, one loses the information as to how long before parturition each sample was taken. This becomes most critical, and least certain, when changes are sought at the very end of pregnancy. For the student of lactogenesis, of course, parturition is not a strictly essential reference time, but in fact most of our information on hormone changes currently relates to it. In such cases the mean time of parturition can be gauged from a parallel group of animals that are allowed to proceed to parturition. However such time-averaging can considerably blur the sharpness of an event as it may occur in each individual. We can speculate as to the sort of information that might be thus lost when we examine some recently determined values for the rate of lactose synthesis in intact, parturient and post-partum rats (see Fig. 1; Wilde & Kuhn, unpublished data obtained with a radioisotopic method). When compared with previous data on the timing and rate of accumulation of tissue lactose (Kuhn, 1969a; Murphy et al., 1973), the rate seen at parturition itself appears to be too small to account for the tissue lactose found at this time. Is it possible that the first few hours of lactogenesis see a large surge in the rate of lactose synthesis which, once the level of tissue lactose reaches a suitable value, subsides again until such time as the onset of suckling removes the milk and requires its rapid replacement? Evidence of such a biphasic response can, perhaps, be seen in recently published studies on the cow (Hartmann, 1973). Although this suggestion is of course speculative, it serves to illustrate the sort of important detail that could be obscured by our current ignorance of the exact timing during the critical period just before parturition.

The second problem lies in the interpretation of tissue levels of milk substances. Normally such tissue levels give no information on the rate of milk formation, which is a pity since it is the increase in the *rate* which especially characterizes lactogenesis. There may be one or two fortuitous exceptions to this statement. Thus in the rat build-up of tissue lactose occurs so rapidly, so close to parturition, and coincides so exactly with the increase in the two components of the rate-limiting enzyme lactose synthase—galactosyltransferase and α-lactalbumin—(Kuhn, 1969a; Murphy *et al.*, 1973) that there can be little doubt of the general relationship between tissue level and rate of synthesis. Yet the same cannot be said of casein in the rat, where the appreciable level that is seen two days before parturition probably reflects a slow accumulation over several days. It must be recognized that the amount of a milk substance that is found in the tissue at any one time reflects not only its rate of synthesis and time of accumulation, but also its destruction in the tissue, conversion to some other substance, or resorption into the blood stream. Even when these rates are very low, an imbalance maintained over a few days can markedly affect the net tissue content of the substance in question. In such a situation it is not to be expected that all substances will necessarily make their appearance at the same time. Thus lactose appears sometime before citrate in the secretion of goat udders before parturition, yet both are bona fide components of goat milk (Fleet *et al.*, 1975). Probably the most striking case of pre-partum "lactation" is the rabbit, where the mammary glands are filled with abundant milk several days before parturition, and where lactose and medium-chain fatty acids appear during days 19–22 of pregnancy, accompanied by an increased RNA/DNA ratio (Denamur, 1963, 1969; Strong & Dils, 1972). Clearly the appearance of these milk solids, before the further boost in their synthesis at parturition, is the result of some definite signal. But is one really to suppose that the *rate* of lactation is high at a time well before any young are there to benefit? And if so, then what happens to the milk? The phenomenon of pre-partum secretion in ruminants is sometimes instanced as an example of lactation occurring well before parturition. However, Hartmann's (1973) study of cows and heifers shows clearly that when attention is directed to yields, rather than just concentrations, of milk substances, the rate of secretion is indeed seen to rise rapidly very close to parturition.

Although the timing of lactogenesis should ideally be determined from rates of milk solid synthesis, there is naturally a great interest in the underlying biochemical changes of the tissue. Tissue levels of enzymes and of RNA have been of especial value in this respect, but they are not all of equal value for pinpointing the timing of events. Very early increases can be most reliably detected with those enzymes or factors which occur initially at a low concentration and where a given increase therefore represents a substantial percentage increase. α-Lactalbumin (McKenzie, Fitzgerald & Ebner, 1971) and lipoprotein lipase (McBride & Korn, 1963; Robinson, 1963) are examples. Enzymes which present a high tissue activity already before lactation begins generally display a small percentage increase at early lactation, although substantial increases may later become evident. Phosphoglucomutase (EC 2.7.5.1), nucleoside diphosphokinase (EC 2.7.4.6) and glycerol-3-phosphate dehydrogenase (EC 1.1.1.8) are examples of these (Baldwin & Milligan, 1966; Kuhn & Lowenstein, 1967; Murphy *et al.*, 1973). Since many enzymes fall into the latter category, or nearly so, the several studies that have been made of enzyme changes at lactogenesis have often not managed to pinpoint accurately the onset of increase in their tissue activities. The evidence suggests, however, that the time of increase may not be the same for all enzymes. Galactosyl transferase appears to increase slightly before α-lactalbumin in the rat (Murphy *et al.*, 1973) and in the mouse (according to Turkington, Brew, Vanaman & Hill, 1968; but not according to Palmiter, 1969). Again in the rat lactose synthase (probably in fact α-lactalbumin) increases about 12 h before glucose-6-phosphate dehydrogenase (EC 1.1.1.49; Kuhn, 1969a). It is possible that future studies will reveal a definite sequence of biochemical responses, representing a "fine structure" of the lactogenic response.

Where several mammary responses occur with different timing one is entitled to ask whether they may not actually be responding to different triggers. There is little information to help us on this score, but three particular cases are worth noting. Firstly, there is distinct evidence from studies of mouse and rabbit mammary explants that higher concentrations of progesterone are required to suppress the induction of galactosyl transferase than of α-lactalbumin (Turkington & Hill, 1969; Delouis, 1975). This implies that as progesterone concentrations progressively fall in the plasma of late pregnant animals, galactosyl transferase and α-lactalbumin will be sequentially induced; this in fact appears to be the case, as pointed out above. The second case refers to glucose-6-phosphate dehydrogenase, the induction of which in mouse mammary explants has been carefully studied by Green, Skarda & Barry (1971). It appears that the induction may owe more to an increased flux of glucose into the tissue than to direct hormone action. Unlike other mammary enzymes, glucose-6-phosphate dehydrogenase requires only insulin for its induction, and even this can be supplanted by a high concentration of glucose in the surrounding medium. This raises the possibility that lactogenesis is characterized by primary and secondary responses, where primary ones are due directly to the impact of the hormonal signal, and where secondary ones recognize intracellular changes brought about by the primary ones. Thirdly, there is evidence that although progesterone can prevent an increase of RNA in lactogenic rabbit mammary tissue, it does not prevent the increase in DNA (Assairi *et al.*, 1974). Although increased DNA synthesis does not seem to be essential for lactogenesis (Green *et al.*, 1971; Owens, Vonderhaar & Topper, 1973) it does appear normally to accompany or quickly follow this event (Greenbaum & Slater, 1957; Traurig, 1967; Stellwagen & Cole, 1969). It is therefore possible that this is a response to elevated concentrations of prolactin and not to decreased concentrations of progesterone *per se*.

DISCUSSION

In the previous sections I have tried to outline the different sorts of information that are available to us in our consideration of the hormonal mechanism of lactogenesis. I have also tried to indicate where our lack of detailed information prevents us from confidently describing the hormonal control in any one species. At present, therefore, there are no strong grounds for assuming that different species employ basically different mechanisms. Even for the guinea-pig, where the role of progesterone in events associated with parturition is in some doubt, we simply lack the information to comment usefully.

The sort of uncertainty which arises from our lack of detailed information is illustrated when one asks whether the surge of prolactin in the parturient rat comes early enough to influence the appearance of mammary lactose. From the prolactin data of Linkie & Niswender (1972) and the lactose data of Kuhn & Lowenstein (1967) it appears that both rise simultaneously. But if prolactin induces the formation of lactose only after a lag period, as Chadwick's (1962) observations on the rabbit show, then perhaps the prolactin surge does after all come too late. However, as long as the data have to be "collected" from different laboratories, and even different species, in this way any such conclusion must remain tentative and therefore unsatisfactory.

Nevertheless, for the design of future experiments we ought to try to consider the possibilities that exist. While the participation of numerous hormones in mammary function is recognized, experiments have repeatedly emphasized prolactin, glucocorticoid and progesterone as those most closely involved with the onset of copious lactation, and it is in terms of these (together with choriomammotrophin) that one can envisage three possible mechanisms for the triggering of lactogenesis.

(1) The plasma concentrations of glucocorticoid, prolactin and choriomammotrophin prevailing in pregnancy suffice for a reasonable rate of milk synthesis, but are overcome by the inhibitory action of progesterone. Falling concentrations of progesterone release the mammary gland from this inhibition and the rate of milk synthesis rises to a value that becomes limited by new factors—perhaps the concentrations of prolactin and glucocorticoids but maybe also by negative feedback from accumulating milk products. As suckling begins, the rate of milk synthesis

then rises further in response to the new elevated plasma concentrations of prolactin and glucocorticoid and to the continued removal of milk.

(2) The plasma concentrations of glucocorticoid and prolactin rise at the end of pregnancy to a point where they breach the threshold of inhibition by progesterone. Their continued rise to values characteristic of full lactation elicits a similar rise in the rate of milk synthesis, while the progesterone inhibition drops unobtrusively away.

(3) The rise in concentration of prolactational hormones and the fall in that of progesterone overlap so closely that no precedence can be ascribed to either. In effect there is a concerted shift from a hormonal milieu that is unfavourable to one that is favourable for lactation.

It has become clear that a very much finer mapping of the magnitude and timing of events associated with normal lactogenesis is needed before one can point confidently to one or other of these possibilities. Such experiments may be aided by others in which several hormonal concentrations are manipulated experimentally in a far more controlled manner than hitherto attempted. At the same time we need deeper understanding at the molecular level of what each hormone actually does in the target cell, and of what sort of antagonism really exists between progesterone and the prolactational hormones.

Yet until such information is available, one is allowed to test the direction of the wind and to draw one's tentative conclusions. In my view the wind is now blowing strongly in favour of mechanism (1) above. In its favour is the close correlation between progesterone withdrawal and the onset of lactation that is found almost wherever it has been sought, in different species and in both natural and experimental situations. Even in the pregnant rabbit the very early appearance of milk, and the later onset of copious lactation associated with parturition, may correlate with changes in the concentration of plasma progesterone. From the wider standpoint of reproductive endocrinology, too, progesterone withdrawal seems well suited to regulate the timing of lactogenesis; heralding, as it does, the end of pregnancy; and controlled, as it appears to be, by the maturing conceptus (Anderson *et al.*, 1975; Hickman-Smith & Kuhn, 1976). The alternative idea, that prolactational hormones "push" the mammary gland into lactation, suffers from the fact—not widely appreciated—that the rabbit is really the only species in which lactation is readily induced by injection of prolactin during pregnancy. Practically all the other "lactogenic" effects which have been readily demonstrated with this hormone are achieved only in the absence of progesterone and are therefore of doubtful relevance to the pregnant state. Moreover, when the high concentration of circulating choriomammotrophin is taken into account, there may actually be no increase at all in the net concentrations of prolactin-like material at the time of lactogenesis. The experimental induction of lactation by glucocorticoid is, as explained earlier, probably a pharmacological effect and actually secondary to an induced withdrawal of progesterone.

It is far more likely that prolactin and glucocorticoid, together with insulin, act to adjust the intensity of lactation rather than its timing, serving to inform the mammary gland of the demands of the young, the health of the mother and her state of nutrition. It is along these lines, I suggest, that the function of these "prolactational" hormones will be most profitably explored.

REFERENCES

Abraham, S., Cady, P. & Chaikoff, J. L. (1960). Glucose and acetate metabolism and lipogenesis in mammary explants of hypophysectomized rats in which lactation was hormonally induced. *Endocrinology* **66**: 280–288.

Adams, W. M. & Wagner, W. C. (1969). The elective induction of parturition in cattle, sheep and rabbits. *J. Am. Vet. Med. Ass.* **154**: 1396–1397.

Adams, W. M. & Wagner, W. C. (1970). The role of corticoids in parturition. *Biol. Reprod.* **3**: 223–228.

Amenomori, Y., Chen, C. L. & Meites, J. (1970). Serum prolactin levels in rats during different reproductive states. *Endocrinology* **86**: 506–510.

Anderson, A. B. M., Flint, A. P. F. & Turnbull, A. C. (1975). Mechanism of action of glucocorticoids in induction of ovine parturition: effect on placental steroid metabolism. *J. Endocr.* **66**: 61–70.

Arai, Y. & Lee, T. H. (1967). A double-antibody radioimmunoassay procedure for ovine pituitary prolactin. *Endocrinology* **81**: 1041–1046.

Assairi, L., Delouis, C., Gaye, P., Houdebine, L., Ollivier-Bousquet, M. & Denamur, R. (1974). Inhibition by progesterone of the lactogenic effect of prolactin in the pseudopregnant rabbit. *Biochem. J.* **144**: 245–252.

Baldwin, R. L. & Milligan, L. P. (1966). Enzymatic changes associated with the initiation and maintenance of lactation in the rat. *J. biol. Chem.* **241**: 2058–2066.

Baldwin, D. M. & Stabenfeldt, G. H. (1973). Endocrine changes in the sow: preparturient, parturient and postparturient. *Fedn Proc. Fedn Am. Socs exp. Biol.* **32**: 267.

Baldwin, D. M. & Stabenfeldt, G. H. (1974). Plasma levels of progesterone, cortisol and corticosterone in the pregnant rabbit. *Biol. Reprod.* **10**: 495–501.

Bassett, J. M., Oxborrow, T. J., Smith, I. D. & Thorburn, G. D. (1969). The concentration of progesterone in the peripheral plasma of the pregnant ewe. *J. Endocr.* **45**: 449–457.

Bedford, C. A., Challis, J. R. G., Harrison, F. A. & Heap, R. B. (1972). The role of oestrogens and progesterone in the onset of parturition in various species. *J. Reprod. Fert.* Suppl. **16**: 1–23.

Ben-David, M., Danon, A., Benveniste, R., Weller, C. P. & Sulman, F. G. (1971). Results of radioimmunoassays of rat pituitary and serum prolactin after adrenalectomy and perphenazine treatment in rats. *J. Endocr.* **50**: 599–606.

Bintarningsih, Lyons, W. M., Johnson, R. E. & Li, C. H. (1958). Hormonally-induced lactation in hypophysectomized rats. *Endocrinology* **63**: 540–548.

Bousquet, M., Fléchon, J. E. & Denamur, R. (1969). Aspects ultrastructuraux de la glande mammaire de lapine pendant la lactogénèse. *Z. Zellforsch. mikrosk. Anat.* **96**: 418–436.

Bradbury, J. T. (1932). Study of endocrine factors influencing mammary development and secretion in the mouse. *Proc. Soc. exp. Biol. Med.* **30**: 212–213.

Brinck-Johnsen, T., Kilham, L. & Margolis, G. (1973). Increased cortisol levels in the pregnant hamster. *Fedn Proc. Fedn Am. Socs exp. Biol.* **32**: 244.

Brush, M. G. (1958). Adrenocortical activity in bovine pregnancy and parturition. *J. Endocr.* **17**: 381–386.

Burdick, H. O. & Konanz, E. J. (1941). The effect of deoxycorticosterone acetate on early pregnancy. *Endocrinology* **28**: 555–560.

Buttle, H. L. & Forsyth, I. A. (1976). Placental lactogen in the cow. *J. Endocr.* **68**: 141–146.

Buttle, H. L., Forsyth, I. A. & Knaggs, G. S. (1972). Plasma prolactin measured by radioimmunoassay and bioassay in pregnant and lactating goats and the occurrence of a placental lactogen. *J. Endocr.* **53**: 483–491.

Caligaris, L., Astrada, J. J. & Taleisnik, S. (1974). Oestrogen and progesterone influence on the release of prolactin in ovariectomized rats. *J. Endocr.* **60**: 205–215.

Chadwick, A. (1962). The onset of lactose synthesis after injection of prolactin. *Biochem. J.* **85**: 554–558.

Challis, J. R. G., Davies, I. I. & Ryan, K. J. (1973). The concentrations of progesterone, estrone and estradiol-17β in the plasma of pregnant rabbits. *Endocrinology* **93**: 971–976.

Challis, J. R. G., Harrison, F. A., Heap, R. B., Horton, E. W. & Poyser, N. L. (1972). A possible role of oestrogens in the stimulation of prostaglandin $F_{2\alpha}$ output at the time of parturition in a sheep. *J. Reprod. Fertil.* **30**: 485–488.

Challis, J. R. G., Heap, R. B. & Illingworth, D. V. (1971). Concentrations of oestrogen and progesterone in the plasma of non-pregnant, pregnant and lactating guinea-pigs. *J. Endocr.* **51**: 333-345.

Chamley, W. A., Buckmaster, J. M., Cerini, M. E., Cumming, I. A., Goding, J. R., Obst, J. M., Williams, A. & Winfield, C. (1973). Changes in the levels of progesterone, corticosteroids, estrone, estradiol-17β, luteinizing hormone, and prolactin in the peripheral plasma of the ewe during late pregnancy and at parturition. *Biol. Reprod.* **9**: 30–35.

Chen, C. L. & Meites, J. (1970). Effects of estrogen and progesterone on serum and pituitary prolactin levels in ovariectomized rats. *Endocrinology* **86**: 503–505.

Collip, J. B., Selye, H. & Thomson, D. L. (1933). Further observations on the effect of hypophysectomy on lactation. *Proc. Soc. exp. Biol. Med.* **30**: 913.

Corner, G. W. (1930). The hormonal control of lactation. I. Non-effect of the corpus luteum. II. Positive action of extracts of the hypophysis. *Am. J. Physiol.* **95**: 43–55.

Cowie, A. T. (1969). General hormonal factors involved in lactogenesis. In *Lactogenesis: the initiation of milk secretion at parturition*: 157–169. Reynolds, M. & Folley, S. J. (eds). Philadelphia: University of Philadelphia Press.

Cowie, A. T., Hartmann, P. E. & Turvey, A. (1969). The maintenance of lactation in the rabbit after hypophysectomy. *J. Endocr.* **43**: 651–662.

Cowie, A. T. & Lyons, W. R. (1959). Mammogenesis and lactogenesis in hypophysectomized, ovariectomized, adrenalectomized rats. *J. Endocr.* **19**: 29–32.

Cowie, A. T. & Tindal, J. S. (1971). *The physiology of lactation.* London: Edward Arnold.

Cowie, A. T. & Watson, S. C. (1966). The adrenal cortex and lactogenesis in the rabbit. *J. Endocr.* **35**: 213–214.

Davis, J. W. & Liu, T. M. Y. (1969). The adrenal gland and lactogenesis. *Endocrinology* **85**: 155–160.

Davis, J. W., Wikman-Coffelt, J. & Eddington, C. L. (1972). The effect of progesterone on biosynthetic pathways in mammary tissue. *Endocrinology* **91**: 1011–1019.

Deis, R. P. (1968). Oxytocin test to demonstrate the initiation and end of lactation in rats. *J. Endocr.* **40**: 133–134.

Deis, R. P. (1971). Induction of lactogenesis and abortion by prostaglandin $F_{2\alpha}$ in pregnant rats. *Nature, Lond.* **229**: 568.

Delouis, C. (1975). Milk protein synthesis *in vitro*. Hormonal regulation. In *Modern problems in paediatrics* **15**: 16–30. Falkner, F. & Kretchmer, N. (eds). Basle: Karger.

Denamur, R. (1963). Les acides nucléiques de la glande mammaire pendant la gestation et la lactation chez la lapine. *C.r. hebd. Séanc. Acad. Sci., Paris* **256**: 4748–4750.

Denamur, R. (1969). Changes in the ribonucleic acids of mammary cells at lactogenesis. In *Lactogenesis: the initiation of milk secretion at parturition*: 53–64. Reynolds, M. & Folley, S. J. (eds). Philadelphia: University of Philadelphia Press.

Denamur, R. (1971). Hormonal control of lactogenesis. *J. Dairy Res.* **38**: 237–264.

Denamur, R. & Delouis, C. (1972). Effects of progesterone and prolactin on the secretory activity and the nucleic acid content of the mammary gland of pregnant rabbits. *Acta Endocr.* **70**: 603–618.

Denamur, R., Delouis, C. &. Gaye, P. (1970). Modifications par les stéroides ovariens des effets de la prolactine sur les acides nucléiques et nucléotides libres de la glande mammaire de lapine. *Arch. int. Pharmacodyn.* **186**: 182–184.

Desclin, L. (1960). Influence of reserpine, oxytocin and adrenaline on the structure, secretory activity and involution of mammary gland in virgin and postpartum rats. *Anat. Rec.* **136**: 182.

Donaldson, L. E., Bassett, J. M. & Thorburn, G. D. (1970). Peripheral plasma progesterone concentration of cows during puberty, oestrous cycles, pregnancy and lactation, and the effects of under-nutrition or exogenous oxytocin on progesterone concentration. *J. Endocr.* **48**: 599–614.

Donovan, B. T. & Peddie, M. J. (1974). Adrenal function, oestrogen and the control of parturition in the guinea pig. *J. Endocr.* **61**: lxxi.

Drummond-Robinson, G. & Asdell, S. A. (1926). The relation between the corpus luteum and the mammary gland. *J. Physiol., Lond.* **61**: 608–614.

Dupouy, J. P., Coffigny, H. & Magne, S. (1975). Maternal and foetal corticosterone levels during late pregnancy in rats. *J. Endocr.* **65**: 347–352.

Fleet, I. R., Goode, J. A., Hamon, M. H., Laurie, M. S., Linzell, J. L. & Peaker, M. (1975). Secretory activity of goat mammary glands during pregnancy and the onset of lactation. *J. Physiol., Lond.* **251**: 763–773.

Folley, S. J. (1942). Non-effect of massive doses of progesterone and deoxycorticosterone on lactation. *Nature, Lond.* **150**: 266.

Forsyth, I. A. (1971). Organ culture techniques and the study of hormone effects on the mammary gland. *J. Dairy Res.* **38**: 419–444.

Forsyth, I. A. (1974). The comparative study of placental lactogenic hormones: a review. In *Lactogenic hormones, fetal nutrition, and lactation*: 49–67. Josimovich, J. B. (ed.). New York: Wiley & Sons.

Friesen, H. G. (1966). Lactation induced by human placental lactogen and cortisone acetate in rabbits. *Endocrinology* **79**: 212–215.

Friesen, H. G. (1973). Placental protein and polypeptide hormones. In *Handbook of physiology* **2** (2): 295–309. Greep, R. O. & Astwood, E. B. (eds). Washington, D.C.: American Physiological Society

Fulkerson, W. J., McDowell, G. H. & Fell, L. R. (1975). Artificial induction of lactation in ewes: the role of prolactin. *Aust. J. biol. Sci.* **28**: 525–530.

Gala, R. R. & Westphal, U. (1967). Corticosteroid-binding activity in serum of mouse, rabbit and guinea pig during pregnancy and lactation: possible involvement in the initiation of lactation. *Acta Endocr.* **55**: 47–61.

Geschickter, C. F. (1945). *Diseases of the breast: diagnosis, pathology, treatment.* 2nd edn. Philadelphia: Lippencott.

Gower, D. B. (1974). Modifiers of steroid hormone metabolism: a review of their chemistry, biochemistry and clinical applications. *J. Steroid Biochem.* **5**: 501–523.

Green, C. D., Skarda, J. & Barry, J. M. (1971). Regulation of glucose 6-phosphate dehydrogenase formation in mammary organ culture. *Biochim. biophys. Acta* **244**: 377–387.

Greenbaum, A. L. & Slater, T. F. (1957). Studies on the particulate components of rat mammary gland. 2. Changes in the levels of the nucleic acids of the mammary glands of rats during pregnancy, lactation and mammary involution. *Biochem. J.* **66**: 155–161.

Grota, L. J. & Eik-Nes, K. B. (1967). Plasma progesterone concentrations during pregnancy and lactation in the rat. *J. Reprod. Fertil.* **13**: 83–91.

Grueter, F. & Stricker, P. (1929). Uber die Wirkung eines Hypophysenvorderlappenhormons auf die Auslösung der Milchsecretion. *Klin. Wochenschr.* **8**: 2322–2323.

Halban, J. (1905). Die innere Secretion von Ovarium und Placenta und ihre Bedeutung für die Function der Milchdrüse. *Arch. Gynaek.* **75**: 353–441.

Hart, I. C. (1972). A solid phase radioimmunoassay for ovine and caprine prolactin using sepharose 6B: its application to the measurement of circulating levels of prolactin before and during parturition in the goat. *J. Endocr.* **55**: 51–62.

Hartmann, P. E. (1973). Changes in the composition and yield of the mammary secretion of cows during the initiation of lactation. *J. Endocr.* **59**: 231–247.

Hartmann, P. E., Trevethan, P. & Shelton, J. N. (1973). Progesterone and oestrogen and the initiation of lactation in ewes. *J. Endocr.* **59**: 249–259.

Hashimoto, I., Henricks, D. M., Anderson, L. L. & Melampy, R. M. (1968). Progesterone and pregn-4-en-20α-ol-3-one in ovarian venous blood during various reproductive states in the rat. *Endocrinology* **82**: 333–341.

Heap, R. B. & Hammond, J. (1974). Plasma progesterone levels in pregnant and pseudopregnant ferrets. *J. Reprod. Fertil.* **39**: 149–152.

Heap, R. B. & Linzell, J. L. (1966). Arterial concentration, ovarian secretion and mammary uptake of progesterone in goats during the reproductive cycle. *J. Endocr.* **36**: 389–399.

Henricks, D. M., Dickey, J. F., Hill, J. R. & Johnston, W. E. (1972). Plasma estrogen and progesterone levels after mating, and during late pregnancy and postpartum in cows. *Endocrinology* **90**: 1336–1342.

L'Hermite, M. & Robyn, C. (1972). Prolactine hypophysaire humaine: détection radio-immunologique et taux au cours de la grossesse. *Annls Endocr.* **33**: 357–360.

Hickman-Smith, D. & Kuhn, N. J. (1976). A proposed sequence of hormones controlling the induction of luteal 20α-hydroxysteroid dehydrogenase and progesterone withdrawal in the late-pregnant rat. *Biochem. J.* **160**: 663–670.

Hunter, D. L., Erb, R. E., Randel, R. D., Garverick, H. A., Callahan, C. J. & Harrington, R. B. (1970). Reproductive steroids in the bovine. I. Relationships during late gestation. *J. Anim. Sci.* **30**: 47–59.

Jones, G. E., Boyns, A. R., Cameron, E. H. D., Bell, E. T., Christie, D. W. & Parkes, M. F. (1973). Plasma oestradiol, luteinizing hormone and progesterone during pregnancy in the beagle bitch. *J. Reprod. Fertil.* **35**: 187–189.

Kaplan, N. M. (1961). Successful pregnancy following hypophysectomy during the twelfth week of gestation. *J. clin. Endocr. Metab.* **21**: 1139–1145.

Kaplan, S. L. & Grumbach, M. M. (1965a). Serum chorionic "growth hormone-prolactin" and serum growth hormone in mother and fetus at term. *J. clin. Endocr. Metab.* **25**: 1370–1374.

Kaplan, S. L. & Grumbach, M. M. (1965b). Immunoassay for human chorionic "growth hormone-prolactin" in serum and urine. *Science, N.Y.* **147**: 751–753.

Kaprowski, J. A. & Tucker, H. A. (1973). Serum prolactin during various physiological states and its relationship to milk production in the bovine. *Endocrinology* **92**: 1480–1487.

Karg, H. & Schams, D. (1971). Discussion on the prolactin levels in bovine blood under different physiological conditions. In *Lactation*: 141–143. Falconer, I. R. (ed.). London: Butterworths.

Karg, H., Schams, D. & Reinhardt, V. (1972). Effects of 2-Br-α-ergocryptine on plasma prolactin level and milk yield in cows. *Experientia* **28**: 574–576.

Kelly, P. A., Robertson, M. C. & Friesen, H. G. (1974). Temporal pattern of placental lactogen and progesterone secretion in sheep. *Nature, Lond.* **248**: 435–437.

Kelly, P. A., Shiu, R. P. C., Robertson, M. C. & Friesen, H. G. (1975). Characterization of rat chorionic mammotropin. *Endocrinology* **96**: 1187–1195.

Kendall, J. Z. & Liggins, G. C. (1972). The effect of dexamethasone on pregnancy in the rabbit. *J. Reprod. Fertil.* **29**: 409–413.

Kohmoto, K. & Bern, H. A. (1970). Demonstration of mammotrophic activity of the mouse placenta in organ culture and by transplantation. *J. Endocr.* **48**: 99–107.

Kuhn, N. J. (1969a). Progesterone withdrawal as the lactogenic trigger in the rat. *J. Endocr.* **44**: 39–54.

Kuhn, N. J. (1969b). Specificity of progesterone inhibition of lactogenesis. *J. Endocr.* **45**: 615–616.

Kuhn, N. J. (1972a). Changes in the protein phosphorus content of rat mammary tissue during lactogenesis. *J. Endocr.* **55**: 219–220.

Kuhn, N. J. (1972b). The lactogenic action of human chorionic gonadotrophin in the rat. *Biochem. J.* **129**: 495–496.

Kuhn, N. J. & Briley, M. S. (1970). The roles of pregn-5-ene-3β,20α-diol and 20α-hydroxy steroid dehydrogenase in the control of progesterone synthesis preceding parturition and lactogenesis in the rat. *Biochem. J.* **117**: 193–201.

Kuhn, N. J. & Lowenstein, J. M. (1967). Lactogenesis in the rat. Changes in metabolic parameters at parturition. *Biochem. J.* **105**: 995–1002.

Kumaresan, P. & Turner, C. W. (1965). Effect of graded levels of insulin on lactation in the rat. *Proc. Soc. exp. Biol. Med.* **119**: 415–416.

Kyriakou, S. Y. & Kuhn, N. J. (1973). Lactogenesis in the diabetic rat. *J. Endocr.* **59**: 199–200.

Labhsetwar, A. P. & Watson, D. J. (1974). Temporal relationship between secretory patterns of gonadotropins, estrogens, progestins, and prostaglandin-F in periparturient rats. *Biol. Reprod.* **10**: 103–110.

Linkie, D. M. & Niswender, G. D. (1972). Serum levels of prolactin, luteinizing hormone and follicle stimulating hormone during pregnancy in the rat. *Endocrinology* **90**: 632–637.

Liu, T. M. Y. & Davis, J. W. (1967). Induction of lactation by ovariectomy of pregnant rats. *Endocrinology* **80**: 1043–1050.

Lukaszewska, J. H. & Greenwald, G. S. (1970). Progesterone levels in the cyclic and pregnant hamster. *Endocrinology* **86**: 1–9.

Lyons, W. R., Li, C. H. & Johnson, R. E. (1958). The hormonal control of mammary growth and lactation. *Recent Prog. Horm. Res.* **14**: 219–248.

McBride, O. W. & Korn, E. D. (1963). The lipoprotein lipase of mammary gland and the correlation of its activity to lactation. *J. Lipid Res.* **4**: 17–20.

McKenzie, L., Fitzgerald, D. K. & Ebner, K. E. (1971). Lactose synthetase activities in rat and mouse mammary glands. *Biochim. biophys. Acta* **230**: 526–530.

Martin, R. J. & Baldwin, R. L. (1971). Effects of alloxan diabetes on lactational performance and mammary tissue metabolism in the rat. *Endocrinology* **88**: 863–867.

Masuda, H., Anderson, L. L., Henricks, D. M. & Melampy, R. M. (1967). Progesterone in ovarian venous plasma and corpora lutea of the pig. *Endocrinology* **80**: 240–246.

Meites, J. (1959). Induction and maintenance of mammary growth and lactation in rats with acetylcholine or epinephrine. *Proc. Soc. exp. Biol. Med.* **100**: 750–754.

Meites, J. & Clemens, J. A. (1972). Hypothalamic control of prolactin secretion. *Vitamins & Hormones* **30**: 165–221.

Meites, J., Hopkins, T. F. & Talwalker, P. K. (1963). Induction of lactation in pregnant rabbits with prolactin, cortisol acetate or both. *Endocrinology* **73**: 261–264.

Meites, J. & Sgouris, J. T. (1953). Can the ovarian hormones inhibit the mammary response to prolactin? *Endocrinology* **53**: 17–23.

Mellenberger, R. W. & Bauman, D. E. (1974a). Metabolic adaptations during lactogenesis. Fatty acid synthesis in rabbit mammary tissue during pregnancy and lactation. *Biochem. J.* **138**: 373–379.

Mellenberger, R. W. & Bauman, D. E. (1974b). Metabolic adaptations during lactogenesis. Lactose synthesis in rabbit mammary tissue during pregnancy and lactation. *Biochem. J.* **142**: 659–665.

Mellenberger, R. W., Bauman, D. E. & Nelson, D. R. (1973). Metabolic adaptations during lactogenesis. Fatty acid and lactose synthesis in cow mammary tissue. *Biochem. J.* **136**: 741–748.

Morishige, W. K., Pepe, G. J. & Rothchild, I. (1973). Serum luteinizing hormone, prolactin and progesterone levels during pregnancy in the rat. *Endocrinology* **92**: 1527–1535.

Murphy, G., Ariyanayagam, A. D. & Kuhn, N. J. (1973). Progesterone and the metabolic control of the lactose biosynthetic pathway during lactogenesis in the rat. *Biochem. J.* **136**: 1105–1116.

Murr, S. M., Bradford, G. E. & Geschwind, I. I. (1974). Plasma luteinizing hormone, follicle stimulating hormone and prolactin during pregnancy in the mouse. *Endocrinology* **94**: 112–116.

Murr, S. M., Stabenfeldt, G. H., Bradford, G. E. & Geschwind, I. I. (1974). Plasma progesterone during pregnancy in the mouse. *Endocrinology* **94**: 1209–1211.

Nandi, S. & Bern, H. A. (1960). Relation between mammary gland responses to lactogenic hormone combinations and tumour susceptibility in various strains of mice. *J. natn. Cancer Inst.* **24**: 907–924.

Nandi, S. & Bern, H. A. (1961). The hormones responsible for lactogenesis in BALB/cCrgl mice. *Gen. comp. Endocr.* **1**: 195–210.

Neill, J. D., Freeman, M. E. & Tillson, S. A. (1971). Control of the proestrus surge of prolactin and luteinizing hormone secretion by estrogens in the rat. *Endocrinology* **89**: 1448–1453.

Neill, J. D., Johansson, E. D. B. & Knobil, E. (1969). Patterns of circulating progesterone concentrations during the fertile menstrual cycle and the remainder of gestation in the rhesus monkey. *Endocrinology* **84**: 45–48.

Nelson, W. O. (1934). Studies on the physiology of lactation. III. The reciprocal hypophyseal-ovarian relationship as a factor in the control of lactation. *Endocrinology* **18**: 33–46.

Nelson, W. O. (1936). Endocrine control of the mammary gland. *Physiol. Rev.* **16**: 488–526.

Newton, W. H. & Richardson, K. C. (1940). The secretion of milk in hypophysectomized pregnant mice. *J. Endocr.* **2**: 322–328.

Niswender, G. D., Chen, C. L., Midgeley, A. R., Meites, J. & Ellis, S. (1969). Radioimmunoassay for rat prolactin. *Proc. Soc. exp. Biol. Med.* **130**: 793–797.

Ôta, K., Harai, Y., Unno, H., Sakaguchi, S., Tomogane, H. & Yokoyama, A. (1974). Corticosterone secretion in response to suckling at various stages of normal and prolonged lactation in rats. *J. Endocr.* **62**: 679–680.

Ôta, K., Ôta, T. & Yokoyama, A. (1974). Plasma corticosterone concentrations and pituitary prolactin content in late pregnancy and their within-day fluctuations in the rat. *J. Endocr.* **61**: 21–28.

Owens, I. S., Vonderhaar, B. K. & Topper, Y. J. (1973). Concerning the necessary coupling of development to proliferation of mouse mammary epithelial cells. *J. biol. Chem.* **248**; 472–477.

Palmiter, R. D. (1969). Hormonal induction and regulation of lactose synthetase in mouse mammary gland. *Biochem. J.* **113**: 409–417.

Peaker, M. & Linzell, J. L. (1975). Citrate in milk: a harbinger of lactogenesis. *Nature, Lond.* **253**: 464.

Pencharz, R. I. & Lyons, W. R. (1934). Hypophysectomy in the pregnant guinea pig. *Proc. Soc. exp. Biol. Med.* **31**: 1131-1132.

Pepe, G. J. & Rothchild, I. (1974). A comparative study of serum progesterone levels in pregnancy and in various types of pseudopregnancy in the rat. *Endocrinology* **95**: 275–279.

Porter, D. G. (1969). Progesterone and the guinea-pig myometrium. In *Progesterone: its regulatory effect on the myometrium*: 79–86. (Ciba Foundation Study Group No. 34) Wolstenholme, G. F. W. & Knight, J. (eds). London: Churchill.

Del Pozo, E., Flückiger, E. & Lancranjan, I. (1976). Endogenous control of prolactin release. In *Basic applications and clinical uses of hypothalamic hormones*: 137–150. Salgado, A. L. C., Fernandez-Durango, R. & Del Campo, J. G. L. (eds). Amsterdam: Excerpta Medica.

Rivera, E. M., Forsyth, I. A. & Folley, S. J. (1967). Lactogenic activity of mammalian growth hormones *in vitro. Proc. Soc. exp. Biol. Med.* **124**: 859–865.

Robertson, H. A. & King, G. J. (1974). Plasma concentrations of progesterone, oestrone, oestradiol-17β and of oestrone sulphate in the pig at implantation, during pregnancy and at parturition. *J. Reprod. Fertil.* **40**: 133–141.

Robinson, D. S. (1963). Changes in the lipolytic activity of the guinea pig mammary gland at parturition. *J. Lipid Res.* **4**: 21–23.

Rodway, R. G. & Kuhn, N. J. (1975). Hormonal control of luteal 20α-hydroxysteroid dehydrogenase and Δ^5-3β-hydroxysteroid dehydrogenase during luteolysis in the pregnant rat. *Biochem. J.* **152**: 433–443.

Schams, D., Hoffmann, B., Fischer, S., Marz, E. & Karg, H. (1972). Simultaneous determination of LH and progesterone in peripheral bovine blood during pregnancy, normal and corticoid-induced parturition and the postpartum period. *J. Reprod. Fertil.* **29**: 37–48.

Schams, D. & Karg, H. (1970). Untersuchungen über Prolaktin im Rinderblut mit einer radioimmunologischen Bestimmungsmethode. *Zbl. Vet. Med.* A, **17**: 193–212.

Schams, D., Reinhardt, V. & Karg, H. (1972). Effects of 2-Br-α-ergokryptine on plasma prolactin level during parturition and onset of lactation in cows. *Experientia* **28**: 697–699.

Seifter, J., Christian, J. J. & Ehrich, W. E. (1951). Effect of cortisone and other steroids on the hibernating gland of the pregnant white rat. *Fedn Proc. Fedn Am. Socs exp. Biol.* **10**: 334.

Sclye, H., Collip, J. B. & Thomson, D. L. (1933). Anterior pituitary and lactation. *Proc. Soc. exp. Biol. Med.* **30**: 588–589.

Selye, H., Collip, J. B. & Thomson, D. L. (1934). Nervous and hormonal factors in lactation. *Endocrinology* **18**: 237–248.

Shaikh, A. A. (1971). Estrone and estradiol levels in the ovarian venous blood from rats during the estrous cycle and pregnancy. *Biol. Reprod.* **5**: 297–307.

Shinde, Y., Ôta, K. & Yokoyama, A. (1965). Lactose content of mammary glands of pregnant rats near term: effect of removal of ovary, placenta and foetus. *J. Endocr.* **31**: 105–114.

Shirley, J. E., Emery, R. S., Convey, E. M. & Oxender, W. D. (1973). Enzymic changes in bovine adipose and mammary tissue, serum and mammary tissue hormonal changes with initiation of lactation. *J. Dairy Sci.* **56**: 569–574.

Shiu, R. P. C., Kelly, P. A. & Friesen, H. G. (1973). Radioreceptor assay for prolactin and other lactogenic hormones. *Science, N.Y.* **180**: 968–971.

Short, R. V. (1958). Progesterone in the peripheral blood of pregnant cows. *J. Endocr.* **16**: 426–428.

Short, R. V. (1961). Progesterone. In *Hormones in blood*: 379–437. Gray, C. H. & Bacharach, A. L. (eds). London and New York: Academic Press.

Simpson, A. A., Simpson, M. H. W. & Kulkarni, P. N. (1973). Prolactin production and lactogenesis in rats after ovariectomy in late pregnancy. *J. Endocr.* **57**: 425–429.

Simpson, A. A., Simpson, M. H. W., Sinha, Y. N. & Schmidt, G. H. (1973). Changes in concentration of prolactin and adrenal corticosteroids in rat plasma during pregnancy and lactation. *J. Endocr.* **58**: 675–676.

Sinha, Y. N., Selby, F. W. & Vanderlaan, W. P. (1974). Relationship of prolactin and growth hormone to mammary function during pregnancy and lactation in the C3H/ST mouse. *J. Endocr.* **61**: 219–229.

Smith, V. G., Edgerton, L. A., Hafs, H. D. & Convey, E. M. (1973). Bovine serum estrogens, progestins and glucocorticoids during late pregnancy, parturition and early lactation. *J. Anim. Sci.* **36**: 391–396.

Speert, H. (1948). The normal and experimental development of the mammary gland of the rhesus monkey, with some pathological correlations. *Contr. Embryol.* **32**: 9–65.

Stabenfeldt, G. H., Drost, M. & Franti, C. E. (1972). Peripheral plasma progesterone levels in the ewe during pregnancy and parturition. *Endocrinology* **90**: 144–150.

Stabenfeldt, G. H., Osburn, B. I. & Ewing, L. L. (1970). Peripheral plasma progesterone levels in the cow during pregnancy and parturition. *Am. J. Physiol.* **218**: 571–575.

Starling, E. H. (1905). The chemical control of the functions of the body. *Lancet* **1905 (ii)**: 339–341.

Stellwagen, R. H. & Cole, R. D. (1969). Histone biosynthesis in the mammary gland during development and lactation. *J. biol. Chem.* **244**: 4878–4887.

Stricker, P. & Grueter, F. (1928). Action du lobe antérieur de l'hypophyse sur la montée laiteuse. *C.r. hebd. Séanc. Soc. Biol.* **99**: 1978–1980.

Strong, C. R. & Dils, R. (1972). Fatty acid biosynthesis in rabbit mammary gland during pregnancy and early lactation. *Biochem. J.* **128**: 1303–1308.

Strong, C. R., Forsyth, I. A. & Dils, R. (1972). The effects of hormones on milk-fat synthesis in mammary explants from pseudopregnant rabbits. *Biochem. J.* **128**: 509–519.

Sutter-Dub, M. T. Leclercq, R., Sutter, B. C. J. & Jacquot, R. (1974). Plasma glucose, progesterone and immunoreactive insulin levels in the lactating rat. *Horm. Metab. Res.* **6**: 297–300.

Talwalker, P. K., Meites, J., Nicoll, C. S. & Hopkins, T. F. (1960). Effects of chlorpromazine on mammary glands of rats. *Am. J. Physiol.* **199**: 1073–1076.

Talwalker, P. K., Nicoll, C. S. & Meites, J. (1961). Induction of mammary secretion in pregnant rats and rabbits by hydrocortisone acetate. *Endocrinology* **69**: 802–808.

Traurig, H. H. (1967). Cell proliferation in the mammary gland during late pregnancy and lactation. *Anat. Rec.* **157**: 489–504.

Tucker, H. A. (1974). General endocrinological control of lactation. In *Lactation* **1**: 277–326. Larson, B. L. & Smith, V. R. (eds). New York and London: Academic Press.

Tucker, H. A. & Meites, J. (1965). Induction of lactation in pregnant heifers with 9-fluoroprednisolone acetate. *J. Dairy Sci.* **48**: 403–405.

Turkington, R. W. (1968). Induction of milk-protein synthesis by placental lactogen and prolactin *in vitro*. *Endocrinology* **82**: 575–583.

Turkington, R. W., Brew, K., Vanaman, T. C. & Hill, R. L. (1968). The hormonal control of lactose synthetase in the developing mouse mammary gland. *J. biol. Chem.* **243**: 3382–3387.

Turkington, R. W. & Hill, R. L. (1969). Lactose synthetase: progesterone inhibition of the induction of α lactalbumin. *Science, N.Y.* **163**: 1458–1460.

Turkington, R. W., Juergens, W. G. & Topper, Y. J. (1967). Steroid structural requirements for mammary gland differentiation *in vitro*. *Endocrinology* **80**: 1139–1142.

Turkington, R. W. & Topper, Y. J. (1966). Stimulation of casein synthesis and histological development of mammary gland by human placental lactogen *in vitro*. *Endocrinology* **79**: 175–181.

Turner, C. W. & Meites, J. (1941). Does pregnancy suppress the lactogenic hormone of the pituitary? *Endocrinology* **29**: 165–171.

Vermouth, N. T. & Deis, R. P. (1972). Prolactin release induced by prostaglandin $F_{2\alpha}$ in pregnant rats. *Nature, Lond.* **238**: 248–250.

Virgo, B. B. & Bellward, G. D. (1974). Serum progesterone levels in the pregnant and postpartum laboratory mouse. *Endocrinology* **95**: 1486–1490.

Voogt, J. L., Sar, M. & Meites, J. (1969). Influence of cycling, pregnancy, labor and suckling on corticosterone-ACTH levels. *Am. J. Physiol.* **216**: 655–658.

Walker, S. M. & Mathews, J. I. (1949). Observations on the effects of prepartal and postpartal estrogen and progesterone treatment on lactation in the rat. *Endocrinology* **44**: 8–17.

Walters, E. & McLean, P. (1968). Effect of Alloxan-diabetes and treatment with anti-insulin serum on pathways of glucose metabolism in lactating rat mammary gland. *Biochem. J.* **109**: 407–417.

Wiest, W. G. (1967). Estimation of progesterone in biological tissues and fluids from pregnant women by double isotope derivative assay. *Steroids* **10**: 279–290.

Wiest, W. G., Kidwell, W. R. & Balogh, K. (1968). Progesterone catabolism in the rat ovary: a regulatory mechanism for progestational potency during pregnancy. *Endocrinology* **82**: 844–859.

Wren, T. R., DeLauder, W. R. & Bitman, J. (1965). Rat mammary gland composition during pregnancy and lactation. *J. Dairy Sci.* **48**: 1517–1521.

Yokoyama, A., Shinde, Y. & Ôta, K. (1969). In *Lactogenesis: the initiation of milk secretion at parturition* : 65–71. Reynolds, M. & Folley, S. J. (eds). Philadelphia: University of Pennsylvania Press.

Zarrow, M. X., Anderson, J. C. & Callantine, M. R. (1963). Failure of progestogens to prolong pregnancy in the guinea pig. *Nature, Lond.* **198**: 690–692.

Zarrow, M. X., Schlein, P. A., Denenberg, V. H. & Cohen, M. A. (1972). Sustained corticosterone release in lactating rats following olfactory stimulation from the pups. *Endocrinology* **91**: 191–196.

Symp. zool. Soc. Lond. (1977) No. 41, 193–210

Comparative Physiology of Milk Removal

B. A. CROSS

*ARC Institute of Animal Physiology,
Babraham, Cambridge, England*

SYNOPSIS

There is great diversity in the gross structure of mammary glands throughout the mammalian class from the small milk patches and club-shaped lobules of monotremes to the massive serried lobules of whales. Suckling habits also vary widely from the brief hourly suckling of the domestic pig to the once a week suckling in the northern fur seal. Nevertheless a common mechanism of milk ejection seems to apply across the whole class, namely the contraction of the myoepithelium investing the secretory alveoli under the humoral influence of oxytocin released from the maternal neurohypophysis during suckling. This milk-ejection reflex is essential for normal removal of milk by the young of species which store most of the secreted milk in the alveoli and smaller lactiferous ducts, but may be dispensed with in ruminants which possess capacious udder cisterns from which the milk can be extracted by the mechanical action of suckling or milking the teats. The species best studied are the rabbit and rat which exemplify contrasting patterns of the milk-ejection reflex. Initial sucking by the litter releases a single pulse of 20–50 mu oxytocin in the rabbit and milk removal is completed in 2–5 min. In the rat multiple pulses of 0·5–1·0 mu oxytocin are released at intervals of 5–15 min throughout suckling periods of 30–60 min. Microelectrode recordings show that a synchronous high frequency discharge of action potentials in half the neurosecretory cells of the paraventricular and supraoptic nuclei precedes each pulse of oxytocin. The rabbit model seems to fit the pig and bottle-nosed whale, while the rat model corresponds more closely with ruminants and man. The central nervous mechanisms that restrain the release of oxytocin between suckling in the rabbit, and determine the cyclicity of oxytocin pulses in the rat, require further study.

INTRODUCTION

Although the ability to lactate is characteristic of all mammals, from the most "primitive" prototherians to the most advanced eutherians, there are wide divergences of suckling pattern. Discharge of milk to the young in aquatic mammals can occur under water (whales, porpoises, sea-cows, sea otters and hippopotamuses) or on land (seals, sea-lions, pigmy hippopotamus) and a variety of erect or recumbent postures are used by different terrestrial mammals. Nursing may be continuous, as in the joey attached to the teat in metatherians, or occur at widely different intervals characteristic of the species. Thus the interval may be a half-hour in the bottle-nosed whale *Berardius bairdii* (Slijper, 1962) or dolphin *Tursiops truncatus* (Harrison, 1969), an hour in the domestic pig *Sus scrofa* (Gill &

Thompson, 1956), a day in the rabbit *Oryctolagus cuniculus* (Zarrow, Denenberg & Anderson, 1965), two days in the tree-shrew *Tupaia belangeri* (see D'Souza & Martin, 1974) and a week in the northern fur seal *Callorhinus ursinus* (Harrison, 1969). Furthermore there are many anatomical distinctions as regards the number, position and gross structure of the mammary glands. In cetaceans the paired teats are recessed to the side of the genital slit while canids, felids and suids, by contrast, have multiple prominent mammae extending from axilla to groin.

It is remarkable that despite this diversity of habit and structure the principal mechanism of milk removal common to all mammals appears to be the contractile response of the mammary myoepithelium under the hormonal influence of oxytocin released from the neurohypophysis. Before examining this mechanism in more detail in the few species best studied, it will be helpful to consider some of the functional variations encountered in the mammary apparatus of the mammalian class.

FUNCTIONAL ANATOMY OF LACTATING MAMMARY GLANDS

The parenchyma of the developed mammary gland in all genera consists of alveolar and duct epithelium comported in lobules, each with a main duct entering with others either into a lactiferous sinus or into a voluminous cistern formed by the fusion of such sinuses as in ruminants (Turner, 1952). The shape and disposition of the lobules, and the amount of stromal tissue, vary widely in different mammals. No eutherian species possess glands with the club-shaped lobules of the platypus and echidna (Griffiths, Elliott, Leckie & Schoefl, 1973) nor the "herring bone" lobular arrangement of the whale mammary gland (Slipjer, 1962). As to the stromal elements in the mouse and rat lactating glands the intralobular connective tissue is extremely sparse while that of the pig is much more prominent (Cross, Goodwin & Silver, 1958).

To help us understand the functional implications of species variations we may suppose that effective control of milk delivery to the young in the right amount and at the appropriate intervals demands a storage system, exit channels, a prehensible appendage, an expulsion mechanism and also a retention mechanism.

Storage System

Milk secreted in the alveoli becomes distributed in varying proportions between the alveolar lumina and small lactiferous ducts (so-called "alveolar milk"), and the collecting ducts, sinuses or cisterns (sinus milk, Cowie, Folley, Cross, Harris, Jacobsohn & Richardson, 1951). Only the sinus milk is accessible to the suckling young without a positive process of alveolar contraction (milk ejection). In most species alveolar milk probably constitutes 70–90% of the stored volume of milk before suckling. In ruminants, however, the presence of capacious udder cisterns alters this ratio so that in the goat, for example, as much as 80% of the milk yield may be stored in the cisterns.

Gross overfilling of the storage sites can give rise to premature involution of the glands as in early weaning, but there appears to be considerable ability to compensate for minor changes in emptying routine such as a missed nursing or milking period. The ducts and sinuses are fairly distensible and accommodate to such temporary overfilling. More drastic distension could impede milk secretion in two ways: by the direct effect of high intra-alveolar pressure upon the lining secretory epithelium and by the reduced capillary blood flow caused secondarily by the alveolar enlargement. Recent evidence in the rat, however, shows that loss of the suckling stimulus is more important that mechanical engorgement in reducing mammary blood flow after 24 hours of non-suckling (Hanwell & Linzell, 1973). At the height of lactation the rat spends about half of each hour with the litter on the teats and the suckling stimulus enhances both mammary blood flow and the rate of milk secretion, probably by neurohormonal mechanisms. It would be interesting to know how storage capacity is adapted to secretory needs in those mammals which exhibit long intervals between suckling, such as tree-shrews (D'Souza & Martin, 1974) and northern fur seals (Harrison, 1969).

Exit Channels

The primary, secondary and tertiary ducts form an uninterrupted channel for the passage of milk from the alveoli to the mammary sinuses or cisterns. Though sphincter mechanisms on ducts have been postulated from time to time they lack convincing anatomical or physiological support (see discussion by Grosvenor & Findlay, 1968) but it is quite possible that when emptied of milk the walls of ducts may collapse together.

The smaller ducts are invested with myoepithelial cells arranged in longitudinal fashion so that upon contraction they cause a shortening and widening of the ducts (Linzell, 1955). Obviously this mechanism, which is activated by oxytocin, in the absence of any direct myoepithelial innervation (Findlay & Grosvenor, 1969), is designed to facilitate outward passage of milk rather than to retard it. The larger ducts possess a small amount of smooth muscle in their walls which may be under sympathetic motor control. Zaks (1962) cites experiments in sheep, goats and guinea-pigs in which stimulation of the peripheral end of the external spermatic nerve supplying sympathetic fibres to the mammary glands interrupted the flow of milk from a cannulated teat. Work in rats on the other hand suggests that resistance to flow in the ducts is very low but some increase in the rate of flow can be detected after denervation (Grosvenor & Findlay, 1968).

Prehensible Appendage

To enable efficient transfer of milk from the mother, nipples or teats that can be grasped by the young are an obvious necessity. Monotremes such as the echidna *Tachyglossus aculeatus* have the least conspicuous prehensible organ but the paired areolar patches in the abdominal pouch possess nipple-like protuberances and the young can be observed vigorously sucking (Griffiths, 1968). By contrast, in some marsupials, for example the red kangaroo *Megaleia rufa*, prehension of the teat is a lengthily protracted affair. The teat enlarges to occupy the oropharyngeal cavity of the joey in the pouch and this intimate contiguity is sustained for weeks (Sharman & Pilton, 1964).

A process of erection of the areolar region, nipple or teat occurs in most mammals, including man, which facilitates prehension by the young during suckling. Little study has been made of this phenomenon though it may involve both a contraction of sympathetically-innervated smooth muscle (as in piloerection) and a vascular engorgement resulting from mechanical stimulation and suction pressure. Prehension by the young would appear to be a special problem in the cetaceans where the teats are normally recessed. According to Slijper (1962), the tongues of young whales are highly muscular and some species have scalloped edges so that the erect teat can be pressed firmly against the palate in such a way as to allow the milk to spout straight into the throat. Erection must be important also in pinnepeds in which the teats are hidden in the pelage.

Expulsion Mechanism

It is now generally agreed that the principal object of the suction produced by the facial musculature of the young is to draw the teat or areolar region into the mouth and retain it there. Positive pressure is used to expel milk from the gland or teat sinuses and this is applied either by the constricting action of tongue against the hard palate (Ardran, Cowie & Kemp, 1957; Ardran, Kemp & Lind, 1958), or by contractile changes in the mammary gland; in many cases of natural suckling both mechanisms occur. The forceful ejection of milk seen in whales (Slijper, 1962) has been ascribed to a voluntary contraction of the cutaneous musculature. Such a mechanism might be of particular value to species suckling under water such as whales, porpoises, manatees and dugongs. However there is no evidence to support such a view, and Enders (1966) testing the same supposition in marsupials by electrical stimulation of the nerve supply to the striated musculature of the pouch in didelphid opossums found that no expulsion of milk occurred.

On the other hand the occurrence of characteristic myoepithelial cells investing the secretory epithelium has been confirmed in all the forms so far studied which include the platypus (Griffiths, Elliott *et al.*, 1973), echidna (Griffiths, McIntosh & Coles, 1969), red kangaroo (Griffiths,

McIntosh & Leckie, 1972), whale (see Slijper, 1962), rat, cat, rabbit, dog (Linzell, 1952), pig (Cross *et al.*, 1958), goat (Richardson, 1949) and man (Richardson, 1951). There seems little reason to suppose that this contractile tissue is unequal to the task of milk ejection. Indeed the observation by Slijper that in the whale milk may continue to flow for some six seconds after the calf lets go of the teat is certainly more compatible with a myoepithelial ejection mechanism than with contraction of voluntary muscle.

Myoepithelium can be induced to contract by rapid mechanical stretching (the "tap response", Cross, 1954; Yokoyama, 1956; Grosvenor, 1965) and in this way vigorous buffeting of the glands by the suckling young may assist expulsion of some of the alveolar milk in the absence of the milk-ejection reflex (see below). This may be especially true for ruminants in which normal milk yields may be obtained under circumstances (udder transplantation or spinally transected animals) that preclude operation of the neuroendocrine milk-ejection mechanism (Denamur & Martinet, 1954; Tverskoi, 1958; Linzell, 1963).

There can be little doubt that in all mammals the mammary myoepithelium is responsive to the arrival of circulating oxytocin whether this be due to release of hormone from the neurohypophysis or to injection of exogenous hormone (see reviews by Cowie & Tindal, 1971 and Bisset, 1974). Nevertheless, considerable variation occurs between the form of the contractile response, i.e. latency, rate of rise to peak pressure and duration of response. In the platypus *Ornithorhynchus anatinus* injection of a large dose of oxytocin elicits ejection of milk after an interval of three to five minutes (Griffiths, Elliott *et al.*, 1973). In the domestic pig the contractile response is very short-iived and would be ended well within three minutes (Braude & Mitchell, 1950; Whittlestone, 1954). Whether such contrasts are the consequence of differing myoepithelial sensitivity or architecture is not clear. Nor can any taxonomic progression be discerned, for the time occupied by withdrawal of milk during suckling appears to be comparably short in the dolphin, rat or pig and comparably longer in the manatee, dog or sheep.

Retention Mechanism

Milk secretion is a continuous process and loss of milk by leakage would be an obvious hazard if the exit channels were not plugged. Gravity, abdominal movements and the rush of water in aquatic forms are all factors tending to oppose retention of milk stored in the lactiferous sinuses or cisterns. The most universal anatomical provision to avoid milk loss is the teat sphincter, composed of smooth muscle, which in cows surrounds the streak canal at the tip of the teat (Turner, 1952). This sphincter is under sympathetic nervous control (Peeters, Coussens & Sierens, 1949). It is not stretching a point unduly to suggest that all teats function as importantly to retain milk between suckling as they do to provide a mechanism whereby milk can be expelled under pressure into the throats of the suckling young. No doubt the relative inaccessibility of the teats in all aquatic mammals assists milk retention under circumstances where diffusional loss would constitute an additional hazard.

We have already noted that the smooth muscle associated with the larger lactiferous ducts has been shown to change its calibre so as to impede the passage of milk. This effect is also mediated by sympathetic nerve fibres (Findlay & Grosvenor, 1969). One further way in which sympathetic nervous activity can operate to oppose milk ejection is by increasing vasoconstrictor tone thus reducing access of circulating oxytocin to the mammary myoepithelium (Cross, 1955a; Hebb & Linzell, 1951). Inappropriate expulsion from milk-filled glands would thus be minimized by sympathetic activity in conditions of apprehension or muscular exertion.

In this connection it is relevant that milk ejection can be blocked by emotional disturbance of reflex excitation of the neurohypophysis (Cross, 1955b; Grosvenor & Mena, 1967). Recent investigations on the central nervous control of milk ejection indeed suggest that several restraining mechanisms exist (Aulsebrook & Holland, 1969; Tindal & Knaggs, 1975) whose function in nature may be to ensure that milk ejection can only occur under circumstances wholly conducive to effective removal of milk by the suckling young.

THE MILK-EJECTION REFLEX

In all species that have been studied a rise of intramammary pressure and/or flow of milk from the teats occurs as a reflex event in suckling or milking (for reviews see Cross, 1961; Denamur, 1965; Benson & Fitzpatrick, 1966; Cowie & Tindal, 1971; Bisset, 1974). The only mechanism that provides an adequate explanation for these observations is an excitation of the neurohypophysis with release of oxytocin which is conveyed in the bloodstream to the mammary capillaries where it evokes contraction of the myoepithelium. In many cases, including the monotremes, some aquatic and many land mammals, the evidence rests solely upon the demonstration of myoepithelium in the mammary glands and the milk-ejecting effect of exogenous oxytocin. Much more convincing is the appearance of endogenous oxytocin in high plasma concentration as shown for a few species by bioassay. The ultimate proof is that of abolishing the reflex by neurohypophysial inactivation, and this still only exists for the rabbit and rat. It is sometimes forgotten that to abolish the reflex by other means, such as anaesthesia, and restore milk-ejection by injecting oxytocin is no proof of neurohypophysial involvement in the natural phenomenon. It simply confirms the occurrence of reflex milk ejection and its simulation by hormone treatment.

By examining what has been discovered about the milk-ejection reflex in goats, pigs, rabbits, rats and man we shall be in a position to make some generalizations on the comparative physiology of milk removal and define some of the remaining problems.

Bovidae—Goat and Cow

The first good scientific account of the milk-ejection reflex was that of Gaines (1915) for the goat in which he compared the natural response with that induced by posterior pituitary extract. The capacity of endogenous neurohypophysial hormone to mimic natural milk ejection was revealed by Andersson (1951) who stimulated the supraoptic nucleus in the hypothalamus by the Hess technique in conscious goats with cannulated udders. Then Tverskoi (1960) reported that surgical division of the pituitary stalk in lactating goats suppressed the milk-ejection reflex for 7–11 days and complete recovery did not occur for several weeks. This was strong support for a physiological role of the neurohypophysis but also indicated that it was not indispensable. More recently Knaggs, McNeilly & Tindal (1972) have delineated an afferent path in the midbrain corresponding to the spinothalamic tract, electrical stimulation of which reproducibly evokes release of oxytocin.

Assays of oxytocin in the jugular blood at milking revealed the presence of raised concentrations of the hormone in only 32% of tests, and normal yields were obtained in the absence of detectable release of oxytocin (Cleverley & Folley, 1970). Coupled with the failure of spinal transection (Denamur & Martinet, 1954), udder transplantation (Linzell, 1963) and pituitary stalk section (Tverskoi, 1960) to eliminate milk ejection in the goat, these data lead to the conclusion that the reflex is a facultative but not an obligatory mechanism for effective discharge of milk in this species. Presumably adequate compensation is secured in its absence by the operation of other processes described on p. 197 above.

Probably the relative contribution of the neurohypophysial mechanism varies between individuals and even in the same goat at different stages of lactation. McNeilly (1972) found that hand-milking more readily evoked release of oxytocin in early than in late lactation. Other factors play a part in the timing and magnitude of the oxytocin pulses. Early release of the hormone occurred more commonly with suckling by the kid than with hand-milking, perhaps because emotional stimuli precede actual contact with the teats. Conditioning of the reflex undoubtedly occurs for Cleverley (1968) detected a release of oxytocin in goats upon the approach and entry of the milker into the pen within 7–10 days of establishing the milking routine.

Similar phenomena to those just mentioned also obtain for cows in which reflex release of oxytocin is a variable event and not always associated with normal milk removal (see Cowie & Tindal, 1971). Evidently the high ratio of cistern milk to alveolar milk minimizes the importance of the neurohypophysical milk-ejection mechanism. Once milk secreted in the alveoli has passed into the cisterns it remains available to the calf, lamb or kid and may be withdrawn by the pumping action of the sucklings or by the milking machine. Late pulses of oxytocin occurring during suckling or milking presumably serve to shunt alveolar milk into the cisterns, replacing that already withdrawn. As we shall see, this state of affairs differs markedly from that observed in non-ruminant species.

Suidae–Domestic Pig

Removal of milk from lactating sows cannot be achieved in the absence of the milk-ejection mechanism (Braude & Mitchell, 1950). A further distinction from ruminants is the abrupt onset and rapid decay of the ejection phase which has an average duration of 14 sec (Gill & Thompson, 1956). Strong circumstantial evidence implicates oxytocin as the mediator of the natural response. Not only is the latter faithfully reproduced by suitable intravenous injections of oxytocin (Braude & Mitchell, 1950; Whittlestone, 1954) but also the hormone can be detected in raised concentration in the circulating blood of suckled sows. Moreover, again in contrast to ruminants, absence of detectable oxytocin in the blood is associated with the inability of the piglets to withdraw milk (Folley & Knaggs, 1966). It would seem that for the sow the rule is for release of the pulse of oxytocin to follow reflexly the initial massaging of the teats by the litter. From the brevity of the withdrawal phase we can assume that only one reflex release of oxytocin occurs. Furthermore, it is unlikely that conditioning plays a significant part in normal lactation in pigs which is characterized by concerted massage of the teats by the whole litter at approximately hourly intervals eliciting a powerful but transient unconditional milk-ejection reflex.

Leporidae—Rabbit

The rabbit, like the pig, has a powerfully developed milk-ejection reflex without which the litter can withdraw only a small fraction of the normal milk yield (<15%, Cross, 1955b). There can be no doubt in this species that neurohypophysial oxytocin mediates the natural milk-ejection response. Destruction of the neurohypophysis abolishes the reflex without directly affecting milk secretion and 20–50 mu oxytocin given intravenously immediately before nursing restores normal milk removal (Cross & Harris, 1952). Vasopressin is not involved since the quantity released by the suckling stimulus is less than the threshold dose for milk ejection (Cross, 1951; Cross & van Dyke, 1953).

Natural suckling in the rabbit occurs only once a day and lasts 2–5 min during which time up to 250 g milk may be transferred to the litter. As in the pig the unconditioned stimulus appears to be the vigorous massaging and sucking activity of the pups which lasts for about 1 min and is followed by rapid gulping for 2 or 3 min (Cross, 1952). Conditioned stimuli appear to play no part for even after establishing a once-daily suckling routine in the laboratory local anaesthesia of the teats completely abolishes the milk-ejection reflex (Findlay, 1968).

Apparently in the rabbit, as in the pig, release of oxytocin occurs as a single spurt. Intramammary pressure recordings made under stressful conditions showed an abbreviated or absent reflex milk-ejection response associated with greatly reduced milk yield, but normal yields were obtained with single injections of oxytocin producing a somewhat larger recorded response (Cross, 1955b). Bioassay of plasma samples withdrawn from suckled does also suggested a single release of hormone with the concentration achieved correlated with milk yield (Bissett, Clark & Haldar, 1970). Other studies indicate that the size of the oxytocin pulse may be determined by the number of pups suckling (Fuchs & Wagner, 1963).

In this species the highly efficient milk-ejection mechanism seems well adapted to sustain a rapid litter growth on a single brief nursing period each day, thus enabling the doe to spend long periods away from the nest and reducing the risk from predators (Zarrow, Denenberg & Anderson, 1965). To this end the retention mechanisms discussed on p. 198 above seems to be highly developed. Experiments involving electrical stimulation in the brain show clearly that in addition to excitatory pathways for release of oxytocin (Tindal, Knaggs & Turvey, 1969; Urban, Moss & Cross, 1971) there are also powerful inhibitory systems (Aulsebrook & Holland, 1969; Tindal & Knaggs, 1975). From work with iontophoretic application of drugs to final common path neurones in the hypothalamic magnocellular nuclei it appears that the excitatory synapses are cholinergic and the inhibitory synapses noradrenergic (Moss, Urban & Cross, 1972). We may conclude therefore that in the natural state the milk-ejection reflex in the rabbit is only likely to be triggered in a particular situation with the doe undisturbed and maternally motivated, and the hungry litter providing a strong spatially-summated sensory excitation from the teats.

Muridae—Rat

It has been known for many years that the neurohypophysis is essential for normal milk removal by the litter from the mammary glands of lactating rats (Harris & Jacobsohn, 1952; Benson & Cowie, 1956), but the essential characteristics of the rat milk-ejection reflex have only recently become clear. We now know that although the pups may remain attached to the mothers' teats for long periods removal of milk is limited to brief intermittent milk-ejections occurring at intervals of 5–20 min (Wakerley & Lincoln, 1971; Lincoln, Hill & Wakerley, 1973), during which the pups all display a characteristic stretch reaction (Vorherr, Kleeman & Lehman, 1967) as they actively imbibe the milk. Unlike other species which have been studied, this normal sequence of regular milk ejections survives anaesthesia provided the glands are well filled and a full litter is attached to the teats (Lincoln et al., 1973). This favourable circumstance enables a much more detailed analysis of the underlying mechanism.

The suckling-induced milk-ejection responses, recorded as intramammary pressure changes from cannulated teats, generally correspond in magnitude and time course to those evoked by intravenous injection of 0·5–1·0 mu oxytocin. The natural response is selectively abolished in glands given a prior intra-arterial injection of the anti-oxytocin agent carbamyl methyloxytocin and less than 0·01 mu vasopressin accompanies the release of 1·0 mu of oxytocin. The natural responses can be mimicked readily by brief stimulus pulses to the infundibular stem. Moreover these electrically induced milk-ejections as well as the suckling-induced responses can be reversibly blocked by interrupting nervous conduction in the neurosecretory axons of the infundibular stem with radio-frequency currents (Cross & Wakerley, 1975—see Fig. 1).

From the foregoing it will be seen that incontestable evidence exists for the essential involvement of neurohypophysial secretion of pulses of oxytocin in the normal transfer of milk from the lactating rat to her litter. Microelectrode studies in the paraventricular and supraoptic nuclei of the hypothalamus in lactating suckled rats have added much deeper insights on the mechanism of the pulsatile release of oxytocin. Approximately half the neurones in each nucleus exhibit a synchronous 20- to 40-fold acceleration of firing rate lasting 2–4 sec, and these explosive episodes precede each pulse of oxytocin (Wakerley & Lincoln, 1973). It can be calculated that about half-a-million action potentials in 2–4 sec are transmitted to the neural lobe to release 1 mu of oxytocin (Lincoln, 1974a). There is now persuasive evidence that the other half of the population of neurones, which do not take part in milk ejection, are vasopressin neurones which can be activated by haemorrhage and osmotic stimuli (Wakerley, Poulain, Dyball & Cross, 1975; Poulain, Wakerley & Dyball, 1977). It is possible to modify the intensity of discharge of the oxytocin

---

FIG. 1. Milk-ejection reflex in the rat.
(a) Left, siting of stimulating (Stim) and radiofrequency (RF) electrodes on neurosecretory fibres of hypothalamo-neurohypophysial(HN) tract. Right, intramammary pressure records showing, similarity of the milk-ejection response to intravenous injection of 1 mu oxytocin (Oxy) and to stimulation of the HN tract, and the blocking effect of a radiofrequency current. In the lower records the stimulating and RF electrodes were transposed and excitation of the peripheral end of the HN tract after proximal blockade of impulse transmission did not prevent milk ejection.
(b) A sequence of suckling-induced milk ejections occurring at 5–8 min intervals (arrows). Note that application of RF current to the hypothalamo-neurohypophysial tract after the 17th successive milk-ejection temporarily blocked the occurrence of ejections.
(c) Left, microelectrode in the magnocellular nuclei for recording impulse activity in single units and the antidromic stimulating electrode for identifying neurosecretory action potentials. Right, polygraph records of action potentials from three supraoptic neurones which display the characteristic burst of activity about 20 sec before suckling-induced milk-ejection response (seen in top trace).

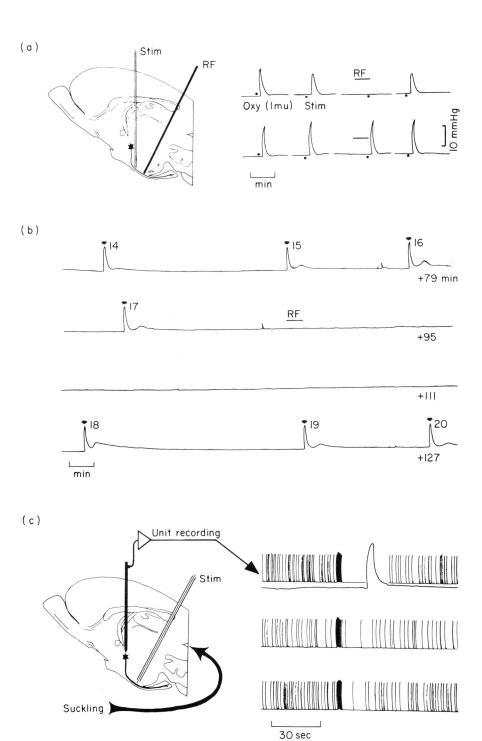

(a)

Stim

RF

Oxy (1mu)   Stim   RF

10 mmHg

min

(b)

14   15   16

+79 min

17   RF

+95

+111

18   19   20

+127

min

(c)

Unit recording

Stim

Suckling

30 sec

neurones by changing the number of pups but this does not seem to affect the number of neurones discharging or the size of the ejection responses (Lincoln & Wakerley, 1975). If the number of pups is reduced below six the reflex milk ejections cease altogether and there are no further synchronous bursts from the oxytocin neurones.

The mechanism that controls the periodicity of the oxytocin pulses remains elusive. The frequency of the pulses can be affected by the degree of mammary distension and the depth of anaesthesia. Milk ejections do not occur when the electroencephalogram (EEG) exhibits an arousal pattern. However, interpolated milk ejections induced either by stimulus pulses to the infundibular stem or by intravenous injections of oxytocin do not disturb the endogenous sequence of milk ejections (Lincoln, 1974b; Wakerley & Deverson, 1975) and this implies that the modulating influence is on the afferent side of the reflex mechanism.

In marked contrast to the pig and the rabbit the initial sucking activity of the litter is not an adequate stimulus for milk ejection in the rat because the first ejection response characteristically occurs some 10–20 min after attachment of the litter to the teats. Furthermore no change in the suckling activity of the pups can be discerned to account for the occurrence of the successive pulses of oxytocin (Wakerley & Drewett, 1975). In this species therefore it is evident that while afferent input from the suckled glands is essential, it is not the only factor in programming the endogenous sequence of recurrent milk ejections.

Hominidae—Man

The occurrence of a process of milk ejection from suckled lactating human breasts has been appreciated for many centuries (see Amoroso & Jewell, 1963); however, there is still remarkably little information on the nature of the reflex, the first modern description of which was that of Newton & Newton (1948). The necessity of the neurohypophysis remains unproven though circumstantial evidence leads to a strong presumption that release of neurohypophysial oxytocin is a vital component of the reflex. Thus the rhythmic intramammary pressure changes occurring in a cannulated breast in a lactating woman when the other breast was suckled could be roughly duplicated in the absence of suckling by infusion of oxytocin intravenously at the rate of 4 mu/min (Sica-Blanco, Mendez-Bauer, Sala, Cabot & Caldeyro-Barcia, 1959). However, later work showed convincingly that the successive ejection pressure-peaks could be duplicated much more accurately by a series of separate oxytocin injections than by the same total dose as a single injection or by a continuous infusion of the hormone (Cobo, Bernal, Gaitán & Quintero, 1967). This clearly implies that oxytocin is released from the neurohypophysis in spurts. Assays of the oxytocin activity of internal jugular blood of suckled women (Coch, Fielitz, Brovetto, Cabot, Coda & Fraga, 1968) gave values of 12–25 μu/ml plasma but the technique used did not reveal whether release occurred in a pulsatile fashion, though Fox & Knaggs (1969) obtained suggestive evidence of this in peripheral venous blood of a suckled woman. If we may assume pulsatile release of oxytocin in man from the work of Cobo *et al.* (1967) this, coupled with evidence that suckling in women does not release detectable amounts of vasopressin (Gaitán, Cobo & Mizrachi, 1964), suggests a parallel with the neuroendocrine system of milk ejection just described for the rat.

CONCLUDING DISCUSSION

Our first generalization shall be that in all mammals the secretion of oxytocin by the neurohypophysis has a role to play in assisting evacuation of milk from the mammary glands during suckling, by its action upon the contractile myoepithelium. The importance of this role can vary widely and probably depends most of all upon the ratio of the alveolar/sinus milk stored in the glands before suckling. Hence we could have the series pig > rabbit > man > cow > goat. Any evolutionary significance of this series is not immediately apparent. However one might conjecture that the grazing habit of the bovids reduces the necessity for a rapid and efficient milk-ejection mechanism, since grazing and nursing are not incompatible activities.

Extensive study of the reflex milk-ejection mechanisms in the rabbit and rat respectively affords us two distinctive models against which to view the other mammals. Disregarding for the moment the difference of nursing interval it is immediately clear from the rat model that a single nursing period does not necessarily mean a single milk-ejection response or a single release of oxytocin from the neurohypophysis. With important qualifications, the ruminants and man can be aligned with the rat model of multiple ejections, whereas the pig and the bottle-nosed whale compare with the rabbit model of a single brief ejection phase during suckling. Probably species that can withdraw milk during suckling periods of more than ten minutes, for example echidna, manatee and bitch, belong to the rat category. The young echidna in the pouch presents an interesting case, for though it may be attached to the teat it apparently receives only one feed of milk a day (Griffiths, 1968). This suggests that central nervous mechanisms in the mother may be crucial in timing the oxytocin release which as we have seen is also true for the rat.

Certainly it should not be supposed that the milk-ejection reflex resembles a classical spinal reflex such as limb flexion or extension. Not only are the sensory components more complex as regards their modalities and the need for temporal and spatial summation, but the response is much more dependent upon fore-brain influences. A closer analogy would be with the micturition reflex which also depends upon a complex afferent input and moreover is subject in different mammals to various ritualizations. These can, as in canids, be sexually dimorphic. Our brief discussion of suckling patterns also suggests a high degree of ritualization which in turn implies a close neural connection between cognitive (behavioural) and vegetative (hormonal) responses. The nature of these linkages has scarcely begun to be studied.

Alone among the species investigated the rat appears to display a remarkably normal milk ejection even under anaesthesia. This does not mean that fore-brain influences are unimportant for the rat. Indeed the absence of reflex milk ejection during EEG arousal proves the contrary. Furthermore, functional decortication by inducing spreading cortical depression with potassium chloride obliterated the inhibition of milk ejection caused by pain in suckled rats (Taleisnik & Deis, 1964). It remains to be discovered whether such corticohypothalamic influences are also implicated in the emotional blockade of milk ejection in the rabbit and how they interact with other excitatory and inhibitory systems.

Another peculiarity of the rat milk-ejection system is that both the size of the oxytocin pulses and the intervals between them are markedly constant during a given period of suckling, corresponding to an unvarying discharge of the oxytocin neurones in the magnocellular nuclei of the hypothalamus. There is no reason to believe that the pulsatile release of oxytocin that has been reported in ruminants and man occurs in such a predictable fashion. It is possible that quite different neurophysiological mechanisms are involved. An intriguing question that needs an answer now is whether inhibitory interneurones are concerned in the patterning of the periodic bursts of firing in the oxytocin neurones and if so whether they are also involved in the tonic restraint exercised between suckling periods and in the inhibition of milk ejection by stress.

From an evolutionary viewpoint it is of great interest that oxytocin and vasopressin neurones, despite their common developmental origin and their intimate juxtaposition in the magnocellular nuclei, perform such different roles. Recent electrophysiological studies (Wakerley et al., 1975; Poulain et al., 1977) show that the two neurosecretory cell types operate in distinctive ways according to physiological state (for example, suckling or dehydration). The vasopressin neurones seldom, if ever, display the synchronous high-frequency bursts of firing that characterize oxytocin neurones in suckled rats. One may conjecture that the gradual perfecting of this synchronous firing may have evolved alongside the development of myoepithelium in the mammary gland as a mammalian specialization for increased lactational efficiency. Relative to the kidney, or even the uterine myometrium, the sensitivity of the myoepithelium to neurohypophysial octapeptide hormone is low. Thus a large increase in plasma oxytocin concentration is needed to evoke a physiologically effective milk-ejection response and this necessitates special provision for the simultaneous activation of thousands of oxytocin neurones. At all events their lactational role is the only one not played in lower vertebrates and it is a cause of wonder that upon these few thousand neurosecretory cells depends the survival of the mammalian class.

ACKNOWLEDGEMENT
I wish to thank my colleague Dr J. B. Wakerley for critical reading of the manuscript and for helpful discussion.

REFERENCES
Amoroso, E. C. & Jewell, P. A. (1963). The exploitation of the milk-ejection reflex by primitive peoples. *Occ. Pap. R. Anthrop. Inst.* **18**: 126–137.

Andersson, B. (1951). The effect and localisation of electrical stimulation of certain parts of the brain stem in sheep and goats. *Acta physiol., Scand.* **23**: 8–23.

Ardran, G. M., Cowie, A. T. & Kemp, F. H. (1957). A cineradiographic study of the teat sinus during suckling in the goat. *Vet. Rec.* **69**: 1100–1101.

Ardran, G. M., Kemp, F. H. & Lind, J. (1958). A cineradiographic study of breast feeding. *Br. J. Radiol.* **31**: 156–162.

Aulsebrook, L. H. & Holland, R. C. (1969). Central inhibition of oxytocin release. *Am. J. Physiol.* **216**: 830–842.

Benson, G. K. & Cowie, A. T. (1956). Lactation in the rat after hypophysial posterior lobectomy. *J. Endocr.* **14**: 54–65.

Benson, G. K. & Fitzpatrick, R. J. (1966). The neurohypophysis and the mammary gland. In *The pituitary gland*: 414–452. Harris, G. W. & Donovan, B. T. (eds). London: Butterworths.

Bisset, G. W. (1974). Milk ejection. In *Handbook of physiology, endocrinology* 4(1): 493–520. Knobil, E. & Sawyer, W. H. (eds). Washington, D.C: American Physiological Society.

Bisset, G. W., Clark, B. J. & Haldar, J. (1970). Blood levels of oxytocin and vasopressin during suckling in the rabbit and the problem of their independent release. *J. Physiol., Lond.* **206**: 711–722.

Braude, R. & Mitchell, K. G. (1950). "Let-down" of milk in the sow. *Nature, Lond.* **165**: 937.

Cleverley, J. D. (1968). *Blood levels of oxytocin, with special reference to lactation.* PhD thesis: University of Reading.

Cleverley, J. & Folley, S. J. (1970). The blood levels of oxytocin during machine milking in cows with some observations on its half life in the circulation. *J. Endocr.* **46**: 347–361.

Cobo, E., Bernal, M. M. de, Gaitán, E. & Quintero, C. A. (1967). Neurohypophyseal hormone release in the human—II. Experimental study during lactation. *Am. J. Obstet. Gynec.* **97**: 519–529.

Coch, J. A., Fielitz, C., Brovetto, J., Cabot, H. M., Coda, H. & Fraga, A. (1968). Estimation of an oxytocin-like substance in highly purified extracts from blood of puerperal women during suckling. *J. Endocr.* **40**: 137–144.

Cowie, A. T., Folley, S. J., Cross, B. A., Harris, G. W., Jacobsohn, D. & Richardson, K. C. (1951). Terminology for use in lactational physiology. *Nature, Lond.* **168**: 421.

Cowie, A. T. & Tindal, J. S. (1971). *The physiology of lactation.* London: Edward Arnold.

Cross, B. A. (1951). Suckling antidiuresis in rabbits. *J. Physiol., Lond.* **114**: 447–453.

Cross, B. A. (1952). Nursing behaviour and the milk-ejection reflex in rabbits. *J. Endocr.* **8**: xiii.

Cross, B. A. (1954). Milk ejection resulting from mechanical stimulation of mammary myoepithelium in the rabbit. *Nature, Lond.* **173**: 450.

Cross, B. A. (1955a). The hypothalamus and the mechanism of sympathetico-adrenal inhibition of milk ejection. *J. Endocr.* **12**: 15–28.

Cross, B. A. (1955b). Neurohormonal mechanisms in emotional inhibition of milk ejection. *J. Endocr.* **12**: 29–37.

Cross, B. A. (1961). Neural control of lactation. In *Milk: the mammary gland and its secretion* **1**: 229–277. Cowie, A. T. & Kon, S. K. (eds). New York and London: Academic Press.

Cross, B. A. & van Dyke, H. B. (1953). The effects of highly purified posterior pituitary principles on the lactating mammary gland of the rabbit. *J. Endocr.* **9**: 232–235.

Cross, B. A., Goodwin, R. F. W. & Silver, I. A. (1958). A histological and functional study of the mammary gland in normal and agalactic sows. *J. Endocr.* **17**: 63–74.

Cross, B. A. & Harris, G. W. (1952). The role of the neurohypophysis in the milk-ejection reflex. *J. Endocr.* **8**: 148–161.

Cross, B. A. & Wakerley, J. B. (1975). Reversible blockade of neurosecretory axons in the hypothalamo–neurohypophysial tract with radiofrequency current. *J. Physiol., Lond.* **245**: 117–118P.

Denamur, R. (1965). The hypothalamo-neurohypophysial system and the milk-ejection reflex. *Dairy Sci. Abstr.* **27**: 193–224, 263–280.

Denamur, R. & Martinet, J. (1954). Enervation de la mamelle et lactation chez brebis et la chèvre. *C.r. Séanc. Soc. Biol.* **148**: 833–836.

D'Souza, F. & Martin, R. D. (1974). Maternal behaviour and the effects of stress in tree shrews. *Nature, Lond.* **251**: 309–311.

Enders, R. K. (1966). Attachment, nursing and survival of young in some didelphids. *Symp. zool. Soc. Lond.* No. 15: 195–203.

Findlay, A. L. R. (1968). The effect of teat anaesthesia on the milk-ejection reflex in the rabbit. *J. Endocr.* **40**: 127–128.

Findlay, A. L. R. & Grosvenor, C. E. (1969). The role of mammary gland innervation in the control of the motor apparatus of the mammary gland: a review. *Dairy Sci. Abstr.* **31**: 109–116.

Folley, S. J. & Knaggs, G. S. (1966). Milk-ejection activity (oxytocin) in the external jugular vein of the cow, goat, and sow, in relation to the stimulus of milking or suckling. *J. Endocr.* **34**: 197–214.

Fox, C. A. & Knaggs, G. S. (1969). Milk-ejection activity (oxytocin) in peripheral venous blood in man during lactation and in association with coitus. *J. Endocr.* **45**: 145–146.

Fuchs, A.-R. & Wagner, G. (1963). Quantitative aspects of release of oxytocin by suckling in unanaesthetised rabbits. *Acta endocr., Copenh.* **44**: 581–592.

Gaines, W. L. (1915). A contribution to the physiology of lactation. *Am. J. Physiol.* **38**: 285–312.

Gaitán, E. E., Cobo, E. & Mizrachi, M. (1964). Evidence for the differential secretion of oxytocin and vasopressin in man. *J. clin. Invest.* **43**: 2310–2322.

Gill, J. C. & Thompson, W. (1956). Observations on the behaviour of suckling pigs. *Br. J. Anim. Behav.* **4**: 46–51.

Griffiths, M. (1968). *Echidnas.* Oxford: Pergamon Press.

Griffiths, M., Elliott, M. A., Leckie, R. M. C. & Schoefl, G. I. (1973). Observations of the comparative anatomy and ultrastructure of mammary glands and on the fatty acids of the triglycerides in platypus and echidna milk fats. *J. Zool., Lond.* **169**: 255–279.

Griffiths, M., McIntosh, D. L. & Coles, R. E. A. (1969). The mammary gland of the echidna, *Tachyglossus aculeatus*, with observations on the incubation of the egg and on the newly hatched young. *J. Zool., Lond.* **158**: 371–386.

Griffiths, M., McIntosh, D. L. & Leckie, R. M. C. (1972). The mammary glands of the Red kangaroo with observations on the fatty acid components of the milk triglycerides. *J. Zool., Lond.* **166**: 265–275.

Grosvenor, C. E. (1965). Contraction of lactating rat mammary gland in response to direct mechanical stimulation. *Am. J. Physiol.* **208**: 214–218.

Grosvenor, C. E. & Findlay, A. L. R. (1968). Effect of denervation on the fluid flow into rat mammary gland. *Am. J. Physiol.* **214**: 820–824.

Grosvenor, C. E. & Mena, F. (1967). Effect of auditory, olfactory and optic stimuli on milk ejection and suckling-induced release of prolactin in lactating rats. *Endocrinology* **80**: 840–846.

Hanwell, A. & Linzell, J. L. (1973). The effects of engorgement with milk and of suckling on mammary blood flow in the rat. *J. Physiol., Lond.* **233**: 111–125.

Harris, G. W. & Jacobsohn, D. (1952). Functional grafts of the anterior pituitary gland. *Proc. R. Soc.* B**139**: 263–276.

Harrison, R. J. (1969). Reproduction and reproductive organs. In *The biology of marine mammals*: 253–348. Anderson, H. T. (ed.). New York and London: Academic Press.

Hebb, C. O. & Linzell, J. L. (1951). Some conditions affecting blood flow through the perfused mammary gland, with special reference to the action of adrenaline. *Q. Jl exp. Physiol.* **36**: 159–175.

Knaggs, G. S., McNeilly, A. S. & Tindal, J. S. (1972). The afferent pathway of the milk-ejection reflex in the mid-brain of the goat. *J. Endocr.* **52**: 333–341.

Lincoln, D. W. (1974a). Dynamics of oxytocin secretion. In *Neurosecretion: the final neuroendocrine pathway*: 129–133. Knowles, F. & Vollrath, L. (eds). Berlin: Springer.

Lincoln, D. W. (1974b). Does a mechanism of negative feedback determine the intermittent release of oxytocin during suckling. *J. Endocr.* **60**: 143.

Lincoln, D. W., Hill, A. & Wakerley, J. B. (1973). The milk-ejection reflex of the rat: an intermittent function not abolished by surgical levels of anaesthesia. *J. Endocr.* **57**: 459–476.

Lincoln, D. W. & Wakerley, J. B. (1975). Factors governing the periodic activation of supraoptic and paraventricular neurosecretory cells during suckling in the rat. *J. Physiol., Lond.* **250**: 443–461.

Linzell, J. L. (1952). The silver staining of myoepithelial cells, particularly in the mammary gland, and their relation to the ejection of milk. *J. Anat.* **86**: 49–57.

Linzell, J. L. (1955). Some observations on the contractile tissue of the mammary glands. *J. Physiol., Lond.* **130**: 257–267.

Linzell, J. L. (1963). Some effects of denervating and transplanting mammary glands. *Q. Jl exp. Physiol.* **48**: 34–60.

McNeilly, A. S. (1972). The blood levels of oxytocin during suckling and hand milking in the goat with some observations on the pattern of hormone release. *J. Endocr.* **52**: 177–188.

Moss, R. L., Urban, I. & Cross, B. A. (1972). Microelectrophoresis of cholinergic and aminergic drugs on paraventricular neurons. *Am. J. Physiol.* **223**: 310–318.

Newton, M. & Newton, N. R. (1948). The let-down reflex in human lactation. *J. Pediat.* **33**: 698–704.

Peeters, G., Coussens, R. & Sierens, G. (1949). Physiology of the nerves in the bovine mammary gland. *Arch. int. Pharmacodyn.* **79**: 75–82.

Poulain, D. A., Wakerley, J. B. & Dyball, R. E. J. (1977). Electrophysiological differentiation of oxytocin- and vasopressin-secreting neurones. *Proc. R. Soc.* (B) **196**: 367–384.

Richardson, K. C. (1949). Contractile tissues in the mammary gland, with special reference to myoepithelium in the goat. *Proc. R. Soc.* (B) **136**: 30–45.

Richardson, K. C. (1951). Structural investigation of the contractile tissues in the mammary gland. *Colloq. Int. Cent. natn. Rech. Scient.* **32**: 167–169.

Sharman, G. B. & Pilton, P. E. (1964). The life history and reproduction of the Red kangaroo (*Megaleia rufa*). *Proc. zool. Soc. Lond.* **142**: 20–48.

Sica-Blanco, Y., Mendez-Bauer, C., Sala, N., Cabot, H. & Caldeyro-Barcia, R. (1959). Nuevo metado para el estudio de la funcionalidad mamaria en la mujer. *Archos urug. Ginec. Obstet.* **17**: 63–72.

Slijper, E. J. (1962). *Whales*. London: Hutchinson.

Taleisnik, S. & Deis, R. P. (1964). Influence of cerebral cortex in inhibition of oxytocin release induced by stressful stimuli. *Am. J. Physiol.* **207**: 1394–1398.

Tindal, J. S. & Knaggs, G. S. (1975). Further studies on the afferent path of the milk-ejection reflex in the brain stem of the rabbit. *J. Endocr.* **66**: 107–113.

Tindal, J. S., Knaggs, G. S. & Turvey, A. (1969). The afferent path of the milk-ejection reflex in the brain of the rabbit. *J. Endocr.* **43**: 603–671.

Turner, C. W. (1952). *The mammary gland.* Columbia, Missouri: Lucas Brothers.

Tverskoi, G. B. (1958). Sekretsiya moloka u koz posle polnoi pererezki spinnogo mozga. *Dokl. Akad. Nauk SSSR* **123**: 1137–1139.

Tverskoi, G. B. (1960). Influence of cervical sympathectomy and pituitary stalk section upon milk secretion in goats. *Nature, Lond.* **186**: 782.

Urban, I., Moss, R. L. & Cross, B. A. (1971). Problems in electrical stimulation of afferent pathways for oxytocin release. *J. Endocr.* **51**: 347–358.

Vorherr, H., Kleeman, C. R. & Lehman, E. (1967). Oxytocin induced stretch reaction in suckling mice and rats in a semi-quantitative bioassay for oxytocin. *Endocrinology* **81**: 711–715.

Wakerley, J. B. & Deverson, B. M. (1975). Stimulation of the supraoptico-hypophysial tract in the rat during suckling: failure to alter the inherent periodicity of reflex oxytocin release. *J. Endocr.* **66**: 439–440.

Wakerley, J. B. & Drewett, R. F. (1975). The pattern of sucking in the infant rat during sequences of spontaneous milk ejection. *Physiol. Behav.* **15**: 277–281.

Wakerley, J. B. & Lincoln, D. W. (1971). Intermittent release of oxytocin during suckling in the rat. *Nature, Lond.*, **233**: 180–181.

Wakerley, J. B. & Lincoln, D. W. (1973). The milk-ejection reflex of the rat in a 20- to 40-fold acceleration in the firing of paraventricular neurones during oxytocin release. *J. Endocr.* **57**: 477–493.

Wakerley, J. B., Poulain, D. A., Dyball, R. E. J. & Cross, B. A. (1975). Activity of phasic neurosecretory cells during haemorrhage. *Nature, Lond.* **258**: 82–84.

Whittlestone, W. G. (1954). Intramammary pressure changes in the lactating sow. 1. The effect of oxytocin. *J. Dairy Res.* **21**: 19–30.

Yokoyama, A. (1956). Milk-ejection responses following administration of "tap" stimuli and posterior pituitary extracts. *Endocr. jap.* **3**: 32–38.

Zaks, M. G. (1962). *The motor apparatus of the mammary gland.* (1st English Edn.) Cowie, A. T. (ed.). Edinburgh: Oliver & Boyd.

Zarrow, M. X., Denenberg, V. H. & Anderson, C. O. (1965). Rabbit: frequency of suckling in the pup. *Science, N.Y.* **150**: 1835–1836.

Symp. zool. Soc. Lond. (1977) No. 41, 211–230

The Aetiology of Mammary Cancer in Man and Animals

R. V. SHORT and J. O. DRIFE[a]

MRC Unit of Reproductive Biology
2 Forrest Road, Edinburgh, Scotland

SYNOPSIS

In order to understand the aetiology of mammary cancer, it is first necessary to have a clear understanding of the physiology of normal mammary development and lactation. The species with the highest spontaneous incidence of mammary cancer are man, the laboratory rat and mouse, and the domestic dog and cat. Mammary cancer is relatively uncommon in primates, even when kept in captivity, and virtually unknown in the cow, which has nevertheless been selected for large udder size and high milk yield for many generations. Why should this be so?

Part of the answer may lie in the fact that mammary cancer is most common in those species that have been ordained by man to spend the greater part of their reproductive lives in the non-pregnant state, undergoing a succession of regular oestrous or menstrual cycles. Another part of the answer may lie in species differences in the hormonal sensitivity of the mammary gland itself. For example, the human breast is unquestionably the most oestrogen-sensitive of all the primates, since it becomes fully developed anatomically at the time of puberty, well in advance of the first ovulation or the first pregnancy. This is not the case in any other primate. The aetiological factors known to predispose to breast cancer in women include early age at menarche, late age at first pregnancy, and late age at menopause. This suggests that it may be harmful to the breasts to be exposed to a succession of menstrual cycles, with high oestrogen levels in the follicular phase unopposed by the protective effect of progesterone.

Absence of pregnancy is also a major aetiological factor in the rat and cat, but is of less importance in the dog. This could be explained by species differences in the endocrinology of the oestrous cycle.

INTRODUCTION

All the epidemiological evidence points to a major role of steroid hormones in the development of mammary cancer; witness for example the low incidence in males, and the protective effect of ovariectomy in females. Although genetical predisposition, infectious agents and chemical carcinogens are also important, especially for the experimental induction of mammary tumours in laboratory animals, the object of this paper

---

[a] Present address: *Eastern General Hospital, Edinburgh, Scotland.*

is to review the hormonal factors that might be involved in the aetiology of the condition.

Before we can begin to appreciate the abnormal, it is first necessary to have a clear understanding of the physiology of normal mammary development and lactation. This may help us to understand why it is that spontaneously occurring mammary tumours are very common in women (cumulative risk, one woman in 20 by the age of 70 in Britain; Doll, 1975), but rare in all other primates (Seibold & Wolf, 1973; Appleby, Keymer & Hime, 1974), very common in dogs (Schneider, 1970), fairly common in cats (Dorn, Taylor, Schneider, Hibbard & Klauber, 1968), and yet virtually unknown in horses and cattle (Priester & Mantel, 1971) (see Figs 1 and 2). At first sight there is no obvious explanation to account for such marked differences in incidence rates between the various species.

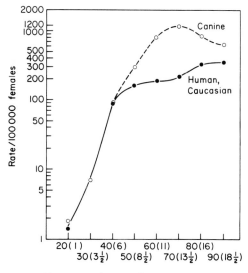

Human age (dog age) at diagnosis in years

FIG. 1. Annual age-specific incidence rates for human and canine mammary cancer. From Schneider (1970).

Although trite, it is nevertheless necessary to remind ourselves that the mammary gland has been specifically designed by evolution to succour the newborn young. Evolution has also always operated to maximize fertility; pregnancy therefore followed hard on the heels of puberty, and the mammary gland became a functional organ soon after the commencement of an animal's reproductive life. It is interesting that man has chosen to suppress this reproductive potential in only a few selected species: his wife, his domestic pets, the cat and dog, and his laboratory animals. In his farm animals on the other hand, fertility has always been at a premium in

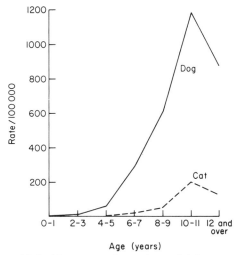

FIG. 2. Annual age-specific incidence rates for canine and feline mammary cancer. From Dorn *et al.* (1968).

order to achieve maximal production of milk or offspring. Could the high incidence of mammary cancer in some species be related to man's self-imposed infertility, resulting in the exposure of the mammary glands to an increased number of sterile oestrous or menstrual cycles before they achieve their ultimate developmental goal of lactation?

Cancer of the mammary gland during the reproductive years would place an individual at a severe disadvantage in evolutionary terms, whereas if the same fate were to befall the individual after reproduction had ceased, it would be beyond the reach of natural selection, and of no significance for the survival of the species. It is therefore interesting to remember that we are the only species in which the female undergoes a true menopause, and the peak incidence of breast cancer is in the post-menopausal years (see Fig. 1).

If we are to attempt to analyse possible endocrine factors that could account for the marked species differences in mammary cancer incidence rates, it is essential to take into account the major differences that are known to occur between species in the hormonal control of mammary development and lactation. These have received scant attention in the past from cancer epidemiologists; for example, Schneider (1970) states that "one event common to man and dog is the same reproductive cycle in the female". Yet an appreciation of these differences must form the basis for any comparative epidemiological study, or any attempt to develop an animal model for human breast cancer. It therefore seems desirable to divide the species up into a number of different lactational categories before attempting to analyse the role of endocrine factors in the genesis of mammary cancer.

THE SIMPLEST SITUATION: MAMMARY DEVELOPMENT AND LACTATION OCCUR AT PUBERTY, AND ARE INDEPENDENT OF PREGNANCY

The most striking examples of this situation are seen amongst the marsupials, of which the red kangaroo (*Megaleia rufa*) will serve as an appropriate case in point.

Male red kangaroos have no pouch and no teats. The four teats first become apparent in the pouch of the female at the time of puberty (Sharman & Calaby, 1964). The length of the oestrous cycle (35 days) is slightly longer than the length of pregnancy (33 days); all the evidence suggests that the hormonal changes during the cycle are identical to those of pregnancy itself, so there has been no need to develop any endocrine mechanism for the maternal recognition of pregnancy (Sharman, 1970). Therefore it is hardly surprising to find that it is unnecessary for female red kangaroos to mate and become pregnant in order to lactate. Sharman & Calaby (1964) successfully removed a newborn joey less than one day old from the teat in its mother's pouch and transferred it to the teat of a virgin, unmated foster mother three days before she herself was due to return to oestrus. The foster mother apparently lactated normally, since the joey had a normal growth rate throughout the duration of its pouch life.

These observations on marsupials may seem a far cry from the aetiology of mammary cancer, especially since there are no reports of such tumours in marsupials. However, the point is made that in some species lactation is normally initiated by the hormonal changes of the oestrous cycle, and no additional endocrine stimulus of a pregnancy is required.

Amongst the eutherian mammals, the domestic dog (*Canis familiaris*) and the ferret (*Mustela furo*) appear to have a pattern of mammary development that is very similar to that of the marsupials. The bitch is a spontaneous ovulator and has a gestation length of about 60 days; she normally comes into oestrus twice a year. The lifespan and secretory activity of the corpora lutea of pseudopregnancy are very similar to those of pregnancy (Jones, Boyns, Cameron, Bell, Christie & Parkes, 1973), suggesting that once again there has been no need for the maternal recognition of pregnancy, and hence no need for the placenta to develop an endocrine function. Thus in any unmated bitch, oestrus is invariably followed by ovulation and the formation of corpora lutea of pseudopregnancy, which persist for about 60 days. During the course of the first pseudopregnancy the mammary glands of the virgin bitch begin to develop and at the end of the pseudopregnancy she may lactate copiously (Asdell, 1964).

Much confusion has arisen in the literature over the misuse of the term "pseudopregnancy" in the bitch. It has tended to lose its scientific connotation (an extended luteal phase in the absence of pregnancy), and has become a subjective description of the animal's behaviour following oestrus. Thus pseudopregnancy may be described as "mild", "moderate" or "severe", depending on the degree of mammary enlargement noticed by the owner. In "severe" cases, the virgin bitch may even go so far as to make a nest and retrieve and "mother" some inanimate object, whilst lactating profusely.

If the endocrine changes of pregnancy in the bitch are so similar to those of pseudopregnancy, we might expect to find little difference in the mammary cancer incidence rates between mated and unmated animals. This still seems to be a point of some dispute. On the one hand, Uberreiter (1966) stated that pregnancy inhibited the growth of mammary tumours while pesudopregnancy stimulated it, and Andersen (1965 and pers. comm.), in a lifespan study of 354 female beagles, reported that about 50–60% of the unmated animals had developed clinical signs of mammary tumours by the age of ten, whereas only 10% of the parous animals had developed tumours. However, in a carefully executed retrospective case-control study, Schneider, Dorn & Taylor (1969) failed to find any association between pseudopregnancy and mammary cancer risk. Parous females were neither over- nor under-represented in the 93 animals with malignant mammary tumours, nor did there appear to be any effect of age at birth of the first litter, litter number, litter size, or total number of offspring per female. In another careful case-control study, Fidler, Abt & Brodey (1967) even reported that tumour-bearing bitches had significantly fewer episodes of pseudopregnancy than controls.

Further research will be required to resolve this point. In view of the human evidence (see below), it may be the time elapsed to first pregnancy, rather than the total number of pseudopregnancies or pregnancies, that is the critical determinant of the mammary cancer incidence rate in the bitch. From the evidence to date, all that can be said is that there is some suggestion that nulliparous bitches are more likely to develop mammary cancer than their parous counterparts.

In contrast to this somewhat equivocal situation there is excellent evidence to show that early ovariectomy of the bitch has a most pronounced sparing effect on the subsequent likelihood of developing mammary cancer. Schneider et al. (1969) showed that bitches that were ovariectomized before their first oestrus had only 0·5% of the mammary cancer risk of intact bitches. If ovariectomy was delayed until after the first oestrus the risk increased to 8%, whereas animals that had two or more oestrous cycles before ovariectomy had 26% of the risk of intact controls. Within this latter group, animals ovariectomized before the age of 2·5 years had a much lower risk than animals ovariectomized after 2·5 years. This evidence would all suggest that mammary cancer is predetermined by the time a bitch is approximately 2·5 years old, and ovarian hormones are crucially important; there is clearly a time-lag of several years between sensitisation and tumour development. Once the tumour is established, however, ovariectomy appears to have no effect on the subsequent course of the disease. As to which ovarian hormones are the culprits, and how they are implicated, this must remain a matter of speculation until some definitive experimental evidence is forthcoming. Although it is fashionable to implicate oestrogens, Jabara (1962) failed to induce mammary tumours in bitches given large doses of diethylstilboestrol, although this treatment did give a high incidence of ovarian tumours; perhaps the animals were not kept for long enough for mammary tumours to become apparent. The beagle bitch has also become notorious with the drug regulatory agencies for producing benign mammary tumours when treated chronically with synthetic progestagens (Nelson, Weikel & Reno, 1973).

Since the majority of mammary cancers in dogs, as in women, are adenocarcinomas (Schneider, 1970), and since the two species show very similar age-specific incidence rates when their ages are adjusted for differences in absolute lifespan (see Fig. 1), it has often been suggested that the dog is the most appropriate animal model for the study of human breast cancer. The great differences in the hormonal control of mammary development between the two species, to say nothing of the profound endocrine differences during the cycle and pregnancy, should make one cautious about making such a naïve assumption. Nothing seems to be known about the spontaneous incidence of mammary cancer in the ferret. Since its reproductive cycle is so similar to that of the dog, it would be worth exploring as a possible experimental animal for mammary cancer studies.

THE INTERMEDIATE SITUATION: MAMMARY DEVELOPMENT AND LACTATION DEPENDENT ON COPULATION

The domestic rabbit (*Oryctolagus cuniculus*) is perhaps the best example of a species in which mammary development is dependent on copulation. The unmated female rabbit is almost constantly in oestrus or sub-oestrus, but there is no mammary development or lactation. Following a sterile mating, ovulation is induced, and this is followed by a pseudopregnancy of about 16 days (the gestation period is 29 days) and mammary development; at the end of pseudopregnancy, when the corpora lutea are beginning to regress, the doe plucks her hair to make a nest, and starts to lactate (Asdell, 1974). The fact that pseudopregnancy is of shorter duration than pregnancy suggests that the placenta does exert some endocrine influence on the mother, although this is clearly not essential for mammary development. Mammary cancer appears to be a rare event in the rabbit, so it is necessary to look at other animals in this group in order to obtain epidemiological evidence about mammary cancer incidence rates.

The domestic cat (*Felis catus*) is a good case in point. The female is an induced ovulator, and in the absence of copulation, the female will undergo a succession of 10-day periods of oestrus every two to three weeks from the spring to the autumn. Ovulation does not occur, so no corpora lutea are formed, and mammary development does not take place. However, a sterile mating is followed by ovulation and a pseudo-pregnancy which does not last as long as pregnancy itself (30 v. 63 days; Asdell, 1964). This, together with the fact that ovariectomy late in pregnancy does not result in abortion (Asdell, 1964), suggests that the placenta has begun to develop an endocrine function towards the end of gestation in the cat. The mammary glands are also said to develop during pseudopregnancy, and lactation may occur when the corpora lutea regress (E. C. Amoroso, pers. comm.).

There is good epidemiological evidence from a retrospective case-control study to show that ovariectomy reduces the risk of mammary cancer in cats about sevenfold (Dorn et al., 1968). In a most extensive uncontrolled investigation, Weijer, Head, Misdorp & Hampe (1972) examined mammary tumours, mostly adenocarcinomas, from 156 cats, of which 114 were intact females, 40 ovariectomized females and two castrated males; the average age of spaying was 5·8 years. One most interesting point was that only 35 of the 114 intact females had ever had kittens.

It is tempting to try and relate these findings to the different endocrine status of the animals. As in the dog, the protective effect of ovariectomy suggests that ovarian steroids are implicated in the aetiology of feline mammary cancer, and it would be interesting to know whether the protective effect is greater the earlier the operation is performed. Normally there are very few opportunities for a cat to become pseudopregnant, since most matings are likely to be fertile. However, there will be a small group of cats that are so jealously guarded by their owners, that although they come into heat regularly, they are never given the opportunity to mate, and so they seldom if ever ovulate. These would appear to be the animals that are most likely to develop mammary cancer. It would require a careful case-control study to establish the point, but it seems possible that in the cat, it is repeated cycles of oestrogenic stimulation of the nulliparous mammary gland that is one of the main endocrine factors predisposing to mammary cancer. It would be particularly interesting to be able to compare the incidence of spontaneous mammary tumours, and the structure of the mammary gland, in three groups of cats, namely those which were never mated, those that had repeated pseudopregnancies following sterile matings, and those that had repeated pregnancies.

In contrast to the rabbit and the cat, the laboratory rat (Rattus norvegicus) is a spontaneous ovulator. However, the stimulus of copulation is necessary to activate the corpus lutem and produce a pseudopregnancy. In the absence of mating, the rat will undergo four-day oestrous cycles, and the corpora lutea formed after ovulation secrete very little progesterone for only two days. Following a sterile mating, the corpora lutea are much more active and last for 12 days, whilst following a fertile mating the corpora lutea are fully active and persist for the duration of pregnancy:

22 days (Hashimoto, Henricks, Anderson & Melampy, 1968). At the time of puberty, ovarian oestrogen secretion results in rapid growth of the duct system within the mammary gland. However, alveolar development requires the additional stimulus of pregnancy, and the rat's placenta is known to produce both luteotrophic and mammotrophic hormones (Cowie & Tindal, 1971). There seems to be little information about the degree of mammary development produced by pseudopregnancy.

The reproductive cycle of the mouse (*Mus musculus*) is very similar to that of the rat. In addition to the oestrogen-induced ductular proliferation of the mammary gland at puberty, Faulkin & DeOme (1960) made the interesting observation that the stroma of the mammary gland had an important regulatory action on ductular development.

There is an enormous literature on the genetic, viral and carcinogenic factors that influence the incidence of mammary tumours in rats and mice, and Welsch & Meites (1974) have recently reviewed the literature on the hormonal control of mammary tumours in these two species. There is abundant evidence to show that chronic oestrogen administration increases the incidence of spontaneous mammary tumours, and ovariectomy diminishes it. Similarly, hypophysectomy decreases tumour incidence, whilst multiple pituitary transplantation increases it. It seems reasonable to conclude that oestrogen exerts its carcinogenic effects both by a direct action on the mammary gland, and indirectly by stimulating the secretion of prolactin from the pituitary. But in spite of all this experimental work, few investigators have addressed themselves to the important question of determining how the animal's own reproductive state (repeated oestrous cycles, repeated pseudopregnancies or repeated pregnancies) influences the incidence of mammary tumours. Howell & Mandl (1961) obtained spectacular results by keeping a group of nulliparous rats until they become senile or died; they found a 100% incidence of mammary tumours. A comparable control group, housed with males and hence allowed to become pregnant on every possible occasion, only had a 4% incidence of mammary tumours. It would have been interesting to know the tumour incidence rate in a group undergoing repeated pseudopregnancies.

Although the evidence is admittedly very incomplete, it does seem that in those species in which mammary development and lactation are dependent on copulation, the risk of developing mammary cancer is highest in animals subjected to repeated oestrous cycles in the absence of an ensuing luteal phase. Whether an initial pregnancy, or even a pseudopregnancy, by temporarily transforming the mammary gland into a secretory structure, can reduce the incidence of mammary cancer even though the animal reverts to a succession of oestrous cycles, remains to be determined. Experiments of this nature should be simple and cheap to perform, particularly in rats and mice, and they could yield much valuable information about the stage of mammary development that is most susceptible to neoplastic change following oestrogen exposure. This might be much more relevant to an understanding of the aetiology of human breast cancer, than all the experiments on "triply operated" rodents (ovariectomized, adrenalectomized and hypophysectomized) given pharmacological doses of exogenous hormones and carcinogenic agents.

THE USUAL SITUATION: MAMMARY DEVELOPMENT AND LACTATION DEPENDENT ON PREGNANCY

Most mammals with which we are familiar fall into this category. They are all, by definition, spontaneous ovulators. The duration of the luteal phase of the oestrous or menstrual cycle is much shorter than the duration of pregnancy, and is seldom referred to as a pseudopregnancy. The extended life of the corpus luteum during pregnancy can be attributed to an endocrine role of the placenta, which may produce a whole variety of hormones. In species like the horse, cow, sheep, goat, pig and guinea-pig, where the life of the corpus luteum is prematurely cut short at the end of the oestrous cycle by the luteolytic action of prostaglandin-$F_{2\alpha}$ secreted from the uterus, the placenta has to exert an initial anti-luteolytic action in order to save the life of the corpus luteum and maintain the pregnancy. In primates, on the other hand, where there is no evidence for a uterine luteolysin, prolongation of luteal life in pregnancy is brought about by the secretion of chorionic gonadotrophin and maybe other luteotrophins from the placenta (Short, 1969). The placenta may also produce a mammotrophic hormone (see the chapter by Forsyth, this volume) and a variety of steroid hormones, including oestrogens and progesterone. These may be manufactured in sufficient quantity to take over the endocrine maintenance of pregnancy, so that ovariectomy of the mother no longer results in abortion.

Thus it can be seen that pregnancy involves major endocrine changes, and it is hardly surprising that nature seems to have used these, rather than the more modest hormonal changes of the shortened oestrous cycle, to initiate lactation. The increased ovarian oestrogen secretion at puberty initiates ductular growth within the mammary gland, but the two- to three-week luteal phase of the oestrous cycle common to all members of this group is barely sufficient to initiate alveolar development; this is well illustrated by the guinea-pig (Turner & Gomez, 1933). If the life of the guinea-pig's corpora lutea is prolonged by hysterectomy, then mammary development occurs and the animal may lactate (Loeb, 1927). If pregnancy is terminated prior to mid-gestation, this is followed by mammary involution, but termination after mid-gestation will initiate lactation (Loeb & Hesselberg, 1917).

The rhesus monkey, *Macaca mulatta*, another member of this group, shows more pronounced mammary changes during the menstrual cycle. Speert (1941, 1948) carried out a most thorough histological investigation of the mammary gland, and noted that variations *between* individuals at a given stage of the menstrual cycle were greater than those *within* an individual at different stages of the cycle; thus it was essential to study serial biopsies from the same animal. When this was done, there was clear evidence of histological changes during ovular cycles. These changes took the form of lobular enlargement, hyperaemia and alveolar dilatation during the luteal phase, with regression following menstruation. During pregnancy, there was little histological change in the mammary gland from the luteal state for the first two months, but by the third month the lobules had become markedly hypertrophied and the alveoli were enlarged and full of secretion; these changes became even more pronounced with advancing gestation. When abortion was induced on the 36th day of pregnancy, it was followed by mammary regression, but abortion on day 60 resulted in immediate although transitory lactation.

Speert (1948) also obtained some most interesting results on mammary involution following ovariectomy; this produced generalized atrophy of the mammary gland and the appearance of discrete hyperplastic nodules which disappeared following oestrogen or progesterone therapy. He also carried out one of the few detailed investigations of mammary involution following parturition, and showed that this was arrested or even reversed in those animals that started to ovulate again soon after delivery. Finally, Speert studied the effects of chronic administration of large doses of oestrogen or progesterone to intact and castrated monkeys. Depending on the dosage, oestrogen took about a month to produce duct, lobule and alveolar development, although it did not initiate secretory activity. This hypertrophied state could be maintained for the duration of treatment (up to 30 months) with no evidence of spontaneous involution. No mammary tumours were produced in any of these animals, perhaps because they were not kept for long enough, although some showed benign metaplastic changes of the alveolar epithelium. Progesterone was also capable of stimulating alveolar growth and development in castrated animals; it is of interest that a ductal carcinoma has recently been described in one out of six rhesus monkeys given physiological doses of an oral contraceptive (Kirschstein, Rabson & Rusten, 1972).

It would be most exciting if one could begin to relate all this histological information to the aetiology of spontaneously occurring mammary cancer in animals within this group. However, such tumours are exceedingly rare in herbivores (Priester & Mantel, 1971) and sub-human primates (Seibold & Wolf, 1973; Appleby *et al.*, 1974), and perhaps we should be seeking an explanation for this fact. It could be argued that in members of this group the oestrogenic phase of the oestrous or menstrual cycle is followed by an abbreviated although fully-functional luteal phase, which might in some way counteract the effects of repeated oestrogenic stimulation of the mammary gland without producing full alveolar development.

THE HUMAN SITUATION: MAMMARY DEVELOPMENT OCCURS AT PUBERTY, BUT LACTATION IS STILL DEPENDENT ON PREGNANCY

We are the only primate in which the breasts undergo a major anatomical development at the time of puberty. Breast enlargement is in fact the first external sign of impending puberty in girls (Marshall & Tanner, 1974), and the breast can achieve its adult size before the first ovulation, which is usually a year or two after menarche (Short, 1976). The reason is not far to seek; breasts are regarded as erotic in most human societies (Ford & Beach, 1952), and it makes obvious sense to develop the organs of sexual attraction before conception occurs. In contrast, the breasts appear to be devoid of erotic significance in all other primates. Much of this pubertal breast growth is due to stromal development, and the fact that the human breast can respond to the minute amounts of oestrogen secreted by the ovaries at the commencement of puberty suggests that this stromal tissue has become extremely sensitive to oestrogen. Since the stromal tissue of the mouse mammary gland has an important regulatory action on ductal growth (Faulkin & DeOme, 1960), it would be interesting to know if the same was true of human breast stroma.

The development of mammary ducts in women at the time of puberty has been described by Dawson (1934) and Haagensen (1971). Coincident with the stromal development of puberty there is a lengthening and branching of the ductular tree, which is rudimentary in the pre-pubertal girl, and the formation of lobules. Following these initial pubertal changes, the amount of glandular tissue shows wide variations between individuals, and indeed between different areas of the same breast (Foote & Stewart, 1945), and this heterogeneity has bedevilled all attempts to analyse the histological changes in the normal human breast.

Rosenburg (1922) claimed that there was budding of the ductal epithelium in the pre-menstrual phase, with the subsequent disappearance of the acini formed from this proliferation after menstruation had occurred. However, Dieckmann (1925) reanalysed Rosenburg's material, and concluded that the differences he had found were due to the patient's age, rather than to the stage of the menstrual cycle. Foote & Stewart (1945) also concluded that the number of lobules increased with age, but in addition they claimed that the number and size of the alveoli increased prior to menstruation.

Curiously, none of these early workers took into account the parity of the subjects they studied, so there exists no information in the literature about possible changes in breast histology as a result of pregnancy. We have attempted to remedy this deficiency.

Several surgical units in Edinburgh have co-operated to provide us with tissue from 170 women of reproductive age undergoing reduction mammoplasty, or breast biopsy for benign disease. Any subjects who were shown to have histological evidence of malignant breast disease were excluded from the study. A full reproductive and menstrual history was taken from each subject, and blood was collected for plasma progesterone assay. Each woman was asked to notify us of the date of her next menstrual period.

In the case of mammoplasty, a piece of tissue was taken at random from the specimen removed by the surgeon; in the case of biopsy, a piece of apparently normal tissue was removed through the same incision as the biopsy, but as far away as possible from the abnormal area. The tissue was immediately fixed in 4% neutral buffered formaldehyde, and 5μm-thickness paraffin-embedded sections were stained and examined.

As in previous studies, we found wide variations within a single breast in the number of lobules per unit area, the number of acini per lobule, and the area of the specimen occupied by lobules; however, the variations between individuals were much greater than those within an individual.

The proportion of glandular tissue varied at random from 0·3% to 29·0% of the area of the section; epithelial height and epithelial cell size were also determined, and an objective assessment was made of the degree of oedema of the intralobular stroma. However, none of these variables showed consistent variations with the stage of the menstrual cycle in either nulliparous or parous women.

The only histological structures that did appear to show a difference between nulliparous and parous women were the vacuolated cells in the basal layer of the alveolar epithelium, as described by Dieckmann (1925) and Bassler (1970). There was a tendency for these cells to be more numerous in nulliparous women, suggesting that the epithelium of the breast might undergo a structural change after the first pregnancy.

Even though histological changes may not be readily apparent in the human breast during the course of the menstrual cycle, structural and functional changes undoubtedly do occur. For example, there are pronounced changes in the total breast volume; nulliparous girls had maximal breast volumes at about the time of menstruation, decreasing by as much as 20% during the early follicular phase of the ensuing cycle; there is even a suggestion that women on the combined oestrogen + progestagen contraceptive pill may show a different pattern of change to women having normal menstrual cycles (Milligan, Drife & Short, 1975). At a functional level, we have found increased IgA synthesis in tissue cultures of human breast during the luteal phase of the cycle in parous women, but not in nullipara (Drife, McClelland, Pryde, Roberts & Smith, 1976), and there are similar changes in DNA synthesis in organ cultures of breast tissue as assessed by the incorporation of tritiated thymidine (Masters, Drife & Scarisbrick, 1977). It seems safe to conclude, from several different lines of evidence, that structural and functional differences do exist between the breasts of nulliparous and parous women.

If the initial growth of the human breast at the time of puberty is for purposes of sexual advertisement only, one would not necessarily expect the volume of the resting breast to give any indication of its potential functional capacity during lactation. Hytten (1954) in fact showed that there was no correlation between initial breast size at the beginning of pregnancy and subsequent lactational performance, whereas there was a good correlation between the degree of secondary breast enlargement during pregnancy and subsequent milk yield.

How can we begin to relate all this fragmentary evidence about the physiology of the human breast to the epidemiology of human breast cancer? We have many tantalizing clues, but little hard proof of cause-and-effect relationships. Thus it has been known for over two centuries that breast cancer is particularly common in nuns (Doll, 1975) and it is now an established fact that nulliparous women, such as nuns, have a much higher incidence of the disease than parous women (Taylor, Carroll & Lloyd, 1959). The evidence initially suggested that the more pregnancies a women had, the less likely she was to develop breast cancer. However, MacMahon and his colleagues (MacMahon, Cole, Lin, Lowe, Mirra, Ravnihar, Salber, Valaoras & Yuasa, 1970) were able to show in five different areas of the world that it was the age of the mother at the birth of her first full-term infant that determined the risk throughout the rest of her life (Fig. 3). Women who gave birth to their first child before the age of 18 had about one-third the breast cancer risk of those whose first birth was delayed until 35 or later. Indeed, women whose first birth was after the age of 35 had a higher breast cancer risk than nulliparous women. The

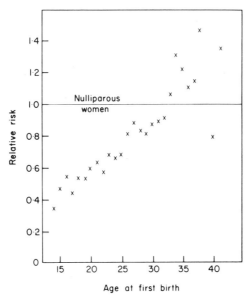

FIG. 3. The relative risk of developing breast cancer in relation to the mother's age at the birth of her first child. The risk in nulliparous women is taken as one. Pregnancy before the age of 35 reduces the risk, and after 35 increases it. From MacMahon, Cole & Brown (1973).

actual number of pregnancies was irrelevant, nor did it matter whether the children were breast or bottle fed. The previous correlation between low cancer incidence and high parity could be explained simply by the fact that the mothers of large families inevitably started reproducing early in life. MacMahon's observations have been confirmed in a large-scale prospective study carried out in New York city (Shapiro, Goldberg, Venet & Strax, 1973), and in a retrospective case-control study in New York State (Lilienfield, Coombs, Bross & Chamberlain, 1975).

The message from MacMahon's work is clear and simple; early first pregnancy protects, whereas late first pregnancy exacerbates. It emphasizes the point that the predisposing factors act early in a woman's reproductive life, but that a pregnancy can protect the breast against these influences. Late first pregnancy may be harmful because it activates a pre-cancerous state that already exists.

Another important factor that has been shown to influence the breast cancer incidence rate is the age of menarche; women with an early menarche have an increased risk (MacMahon, Cole & Brown, 1973; Shapiro et al., 1973). It is well known that the age of menarche varies markedly in different areas of the world, being most advanced in the most developed countries, and most retarded in the poorest (Marshall & Tanner, 1974), whereas the reverse is true for the age at first pregnancy, with the affluent nations having the motivation and the means to defer the birth of the first child, whilst the deprived nations have neither (Short, 1976). It therefore seems significant that the breast cancer incidence rates are highest in the developed countries (MacMahon, Cole & Brown, 1973), which have the longest period of time between menarche and first pregnancy, whereas they are lowest in the developing countries, where the period between these events is the shortest (Fig. 4). Although Mac-Mahon himself rejects it as an explanation, it still seems plausible to imagine that the nulliparous breast may be stressed by a repeated succession of menstrual cycles, whereas following the first pregnancy there is a structural and functional transformation of the breast which protects it from further precancerous change.

It would be fascinating to know the breast cancer incidence rates of women ovariectomized at various ages before and after the onset of puberty, but this information is never likely to become available. However, there is a mass of evidence to show that ovarian hormones do have a major part to play in the genesis of human breast cancer. The classical work of Beatson (1898) showed that ovariectomy could occasionally cause remission of breast cancer in pre-menopausal women. He was led to perform the operation because, he believed, "lactation is at one point perilously near becoming a cancerous process if it is at all arrested", and he asked himself the simple question, "Is cancer of the mamma due to some ovarian initiation, as from some defective steps in the cycle of ovarian changes; and if so, would the cell proliferation be brought to a stand-still . . . were the ovaries to be removed?" There is now abundant

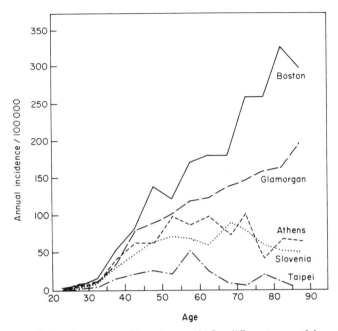

FIG. 4. Age-specific incidence rates of breast cancer in five different areas of the world. Care must be taken in interpreting these data, as there could be "cohort effects" with progressive declines in the age at menarche in the last century. From MacMahon, Cole & Brown (1973).

clinical evidence to show that ovariectomy prior to the menopause reduces the breast cancer risk for the rest of life, and the earlier the operation is performed, the greater the reduction (Trichopoulos, Mac-Mahon & Cole, 1972; MacMahon, Cole & Brown, 1973); ovariectomy before the age of 35 reduces the risk to about a third of that in women having a normal menopause at 45–54, whereas women with a late natural menopause, after 55, have the risk increased by half.

Other aetiological clues are provided by racial differences in breast cancer incidence rates; the Japanese have one of the lowest rates in the world, and yet if Japanese migrate to the United States, their children have incidence rates very similar to those of native Americans (Buell, 1973). This suggests a major influence of environmental factors. There is evidence to suggest that in the early part of this century, Japanese women were probably experiencing a very late menarche and an early first pregnancy, which could account for the low incidence of the disease in Japan in recent decades, whereas American-born Japanese would be expected to conform more to the reproductive norms for the United States, with early menarche and late first birth. Other evidence that also points to a major influence of environmental factors comes from a review of published breast cancer incidence rates in monozygotic and dizygotic

TABLE I

The incidence of malignant breast tumours in female monozygotic and dizygotic twins

	Monozygotic twins		Dizygotic twins	
	One twin affected	Both twins affected	One twin affected	Both twins affected
No. of cases reported in literature	4	4	3	3

Data from Macklin (1940).

twins (Macklin, 1940). These results are summarized in Table I. In spite of the small numbers, and the biases inherent in any survey of the published literature, the 50% non-concordance in monozygotic and also in dizygotic twins would suggest that the environment is much more important than the genotype. For the future, it would be fascinating to see whether differences in reproductive life history could account for these differences between monozygotic twins. However, it is only fair to point out that the female relatives of women with breast cancer are said to have a two- to three-fold increased risk (MacMahon, Cole & Brown, 1973), indicating that genetic predisposition may be quite important. Once again, it would be necessary to establish that this apparent familial trend was not due to a familial similarity in reproductive life histories.

Although all the epidemiological evidence about human breast cancer points to an involvement of the ovaries, it does not tell us which ovarian hormone is likely to be the culprit. Now that many millions of women have been exposed to a combined oestrogen + progestagen oral contraceptive for up to a decade, and many thousands of women have been taking oestrogens in the form of post-menopausal hormone replacement therapy for an even longer period of time, we are beginning to get some answers.

There is general agreement that the combined oral contraceptive has not led to an increased incidence of breast cancer (Arthes, Sartwell & Lewison, 1971; Vessey, Doll & Sutton, 1971, 1972; Boston Collaborative Drug Surveillance Programme, 1973; Royal College of General Practitioners, 1974), although Fasal & Paffenbarger (1975) have suggested that oral contraceptives may hasten the development of pre-existing cancer. However, all the above studies have shown that oral contraceptives do reduce the incidence of benign lesions of the breast. Since women with benign breast lesions have an increased likelihood of subsequently developing a malignant tumour (Davis, Simons & Davis, 1964; Black, Barclay, Cutler, Hankey & Asire, 1972; MacMahon, Cole & Brown, 1973), it is reasonable to hope that the oral contraceptive may ultimately be shown to reduce the incidence of malignant breast disease. We shall have to wait for several more years before women who took the oral contraceptive early in their reproductive years enter the high breast cancer incidence age group. When they do, it will be particularly interesting to know whether taking "the pill" for an extended period of time in the sensitizing years prior to first pregnancy has had a greater sparing effect than when taken after the first pregnancy. Since the pill contains both an oestrogen and a progestagen, it prevents the breast tissue from being exposed to oestrogen alone, whereas during the first half of the normal menstrual cycle, the ovaries are secreting oestrogen and little or no progesterone.

It is only very recently that reliable evidence has become available about breast cancer incidence rates in post-menopausal women on oestrogen replacement therapy (Hoover, Gray, Cole & MacMahon, 1976). The risk of developing breast cancer is increased by this therapy, particularly 10 years or more after first exposure; 15 years after the start of treatment, the risk was doubled. Furthermore, this oestrogen replacement therapy appears to cancel out the normal protective effects of multiparity and oophorectomy, and enhance the chance of malignancy in those with benign breast disease.

All this evidence leaves one with the strong impression that oestrogens may be involved in the genesis of human mammary cancer, and that the breast is in a particularly vulnerable state in the period between menarche and first pregnancy. If we could discover more about the functional changes that the breast undergoes during this time, we might come much closer to an understanding of how breast cancer is caused. Since it seems likely that we shall be using steroid hormones on an ever-increasing scale to regulate human reproduction, it would be comforting to think that we knew enough to use them judiciously. They might enable us to prevent not only pregnancy, but breast cancer as well.

REFERENCES

Andersen, A. C. (1965). Parameters of mammary gland tumours in ageing beagles. *J. Am. vet. med. Ass.* **147**: 1653–1654.

Appleby, E. C., Keymer, I. F. & Hime, J. M. (1974). Three cases of suspected mammary neoplasia in non-human primates. *J. comp. Path. Ther.* **84**: 351–364.

Arthes, F. G., Sartwell, P. E. & Lewison, E. F. (1971). The pill, estrogens and the breast. Epidemiologic aspects. *Cancer* **28**: 1391–1394.

Asdell, S. A. (1964). *Patterns of mammalian reproduction.* 2nd edn. Ithaca, N.Y: Cornell University Press.

Bassler, R. (1970). The morphology of hormone-induced structural changes in the female breast. *Curr. Top. Path.* **53**: 1–89.

Beatson, G. T. (1898). On the treatment of inoperable cases of carcinoma of the mamma; suggestions for a new method of treatment, with illustrative cases. *Lancet* **1898 (ii)**: 104–107, 162–165.

Black, M. M., Barclay, T. H. C., Cutler, S. J., Hankey, B. F. & Asire, A. J. (1972). Association of atypical characteristics of benign breast lesions with subsequent risk of breast cancer. *Cancer* **29**: 338–343.

Boston Collaborative Drug Surveillance Programme (1973). Oral contraceptives and venous thromboembolic disease, surgically confirmed gall-bladder disease, and breast tumours. *Lancet* **1973** (i): 1399–1404.

Buell, P. (1973). Changing incidence of breast cancer in Japanese-American women. *J. natn. Cancer Inst.* **51**: 1479–1483.

Cowie, A. T. & Tindal, J. S. (1971). *The physiology of lactation.* London: Edward Arnold.

Davis, H. H., Simons, M. & Davis, J. B. (1964). Cystic disease of the breast; relationship to carcinoma. *Cancer* **17**: 957–978.

Dawson, E. K. (1934). A histological study of the normal mamma in relation to tumour growth. *Edinb. med. J.* **41**: 653–682.

Dieckmann, H. (1925). Uber die Histologie der Brustdruse bei gestortem und ungestortem Menstruationsablauf. *Virchows Arch. path. Anat. Physiol.* **256**: 321–356.

Doll, R. (1975). The epidemiology of cancers of the breast and reproductive system. *Scott. med. J.* **20**: 305–315.

Dorn, C. R., Taylor, D. O. N., Schneider, R., Hibbard, H. H. & Klauber, M. R. (1968). Survey of animal neoplasms in Alameda and Contra Costa Counties, California. II. Cancer morbidity in dogs and cats from Alameda County. *J. natn. Cancer Inst.* **40**: 307–318.

Drife, J. O., McClelland, D. B. L., Pryde, A., Roberts, M. M. & Smith, I. I. (1976). Immunoglobulin synthesis in the "resting" breast. *Br. med. J.* **1976** (ii): 503–506.

Fasal, E. & Paffenbarger, R. S. (1975). Oral contraceptives as related to cancer and benign lesions of the breast. *J. natn. Cancer Inst.* **55**: 767–773.

Faulkin, L. J. & DeOme, K. B. (1960). Regulation of growth and spacing of gland elements in the mammary fat pad of the C3H mouse. *J. natn. Cancer Inst.* **24**: 953–969.

Fidler, I. J., Abt, D. A. & Brodey, R. S. (1967). The biological behaviour of canine mammary neoplasms. *J. Am. vet. med. Ass.* **151**: 1311–1318.

Foote, F. W. & Stewart, F. W. (1945). Comparative studies of cancerous versus non-cancerous breasts. *Ann. Surg.* **121**: 6–53.

Ford, C. S. & Beach, F. A. (1952). *Patterns of sexual behaviour.* London: Eyre & Spottiswoode.

Haagensen, C. D. (1971). *Diseases of the breast.* 2nd edn. Philadelphia: W. B. Saunders.

Hashimoto, I., Henricks, D. M., Anderson, L. L. & Melampy, R. M. (1968). Progesterone and Pregn-4-en-20α-ol-3-one in ovarian venous blood during various reproductive states in the rat. *Endocrinology* **82**: 333–341.

Hoover, R., Gray, L. A., Cole, P. & MacMahon, B. (1976). Menopausal estrogens and breast cancer. *New Engl. J. Med.* **295**: 401–405.

Howell, J. S. & Mandl, A. M. (1961). The mammary glands of senile nulliparous and multiparous rats. *J. Endocr.* **22**: 241–255.

Hytten, F. E. (1954). Clinical and chemical studies in human lactation. VI. The functional capacity of the breast. *Br. med. J.* **1954** (i): 912–915.

Jabara, A. G. (1962). Induction of canine ovarian tumours by diethylstilboestrol and progesterone. *Aust. J. exp. Biol. med. Sci.* **40**: 139–152.

Jones, G. E., Boyns, A. R., Cameron, E. H. D., Bell, E. T., Christie, D. W. & Parkes, M. F. (1973). Plasma oestradiol, luteinising hormone and progesterone during pregnancy in the beagle bitch. *J. Reprod. Fertil.* **35**: 187–189.

Kirschstein, R. L., Rabson, A. S. & Rusten, G. W. (1972). Infiltrating duct carcinoma of the mammary gland of a rhesus monkey after administration of an oral contraceptive: A preliminary report. *J. natn. Cancer Inst.* **48**: 551–556.

Lilienfeld, A. M., Coombs, J., Bross, I. D. J. & Chamberlain, A. (1975). Marital and reproductive experience in a community-wide epidemiological study of breast cancer. *Johns Hopkins Med. J.* **136**: 157–162.

Loeb, L. (1927). The effects of hysterectomy on the system of sex organs and on the periodicity of the sexual cycle in the guinea pig. *Am. J. Physiol.* **83**: 202–224.

Loeb, L. & Hesselberg, C. (1917). The cyclic changes in the mammary gland under normal and pathological conditions. II. The changes in the pregnant guinea pig, the effect of lutein injections, and the correlation between the cycle of the uterus and ovaries and the cycle of the mammary gland. *J. exp. Med.* **25**: 305–321.

Macklin, M. T. (1940). An analysis of tumours in monozygous and dizygous twins. *J. Hered.* **31**: 277–290.

MacMahon, B., Cole, P. & Brown, J. (1973). Etiology of human breast cancer: A review. *J. natn. Cancer Inst.* **50**: 21–42.

MacMahon, B., Cole, P., Lin, T. M., Lowe, C. R., Mirra, A. P., Ravnihar, B., Salber, E. J., Valaoras, V. G. & Yuasa, S. (1970). Age at first birth and breast cancer risk. *Bull. Wld Hlth Org.* **43**: 209–221.

Marshall, W. A. & Tanner, J. M. (1974). Puberty. In *Scientific foundations of paediatrics*: 124–151. Davis, J. A. & Dobbing, J. (eds). London: Heinemann.

Masters, J. R. W., Drife, J. O. & Scarisbrick, J. J. (1977). Cyclical variations of DNA synthesis in human breast epithelium. *J. natn. Cancer Inst.* **58**: 1263–1265.

Milligan, D., Drife, J. O. & Short, R. V. (1975). Changes in breast volume during normal menstrual cycle and after oral contraceptives. *Br. med. J.* **1975 (iv)**: 494–496.

Nelson, L. W., Weikel, J. H. & Reno, F. E. (1973). Mammary nodules in dogs during four years' treatment with megestrol acetate or chlormadinone acetate. *J. natn. Cancer Inst.* **51**: 1303–1311.

Priester, W. A. & Mantel, N. (1971). Occurrence of tumours in domestic animals. Data from 12 United States and Canadian Colleges of Veterinary Medicine. *J. natn. Cancer Inst.* **47**: 1333–1344.

Rosenburg, A. (1922). Uber Menstruelle, durch das corpus luteum bedingte Mammaveranderungen. *Z. Path.* **27**: 466–506.

Royal College of General Practitioners (1974). *Oral contraceptives and health.* London: Pitman.

Schneider, R. (1970). Comparison of age, sex, and incidence rates in human and canine breast cancer. *Cancer* **26** 419–426.

Schneider, R., Dorn, C. R. & Taylor, D. O. N. (1969). Factors influencing canine mammary cancer development and postsurgical survival. *J. natn. Cancer Inst.* **43**: 1249–1261.

Seibold, H. R. & Wolf, R. H. (1973). Neoplasms and proliferative lesions in 1065 nonhuman primate necropsies. *Lab. anim. Sci.* **23**: 533–539.

Shapiro, S., Goldberg, J., Venet, L. & Strax, P. (1973). Risk factors in breast cancer—a prospective study. In *Host environment interactions in the etiology of cancer in man*: 169–182. Doll, R. & Vodopija, I. (eds). Lyon: International Agency for Research on Cancer.

Sharman, G. B. (1970). Reproductive physiology of marsupials. *Science, N.Y.* **167**: 1221–1228.

Sharman, G. B. & Calaby, J. H. (1964). Reproductive behaviour in the Red kangaroo (*Megaleia rufa*) in captivity. *C.S.I.R.O. Wildl. Res.* **9**: 58–85.

Short, R. V. (1969). The maternal recognition of pregnancy. In *Foetal autonomy*: 2–26. (Ciba Foundation Symposium): Wolstenholme, G. E. W. & O'Connor, M. (eds). London: Churchill.

Short, R. V. (1976). The evolution of human reproduction. *Proc. R. Soc.* (B) **195**: 3–24.

Speert, H. (1941). Cyclic changes in the mammary gland of the Rhesus monkey. *Surgery, Gynec. Obstet.* **73**: 388–390.

Speert, H. (1948). The normal and experimental development of the mammary gland of the Rhesus monkey, with some pathological correlations. *Contr. Embryol.* **32**: 11–65.

Taylor, R. S., Carroll, B. E. & Lloyd, J. W. (1959). Mortality among women in three Catholic religious orders with special reference to cancer. *Cancer, Philadelphia* **12**: 1207–1225.

Trichopoulos, D., MacMahon, B. & Cole, P. (1972). Menopause and breast cancer risk. *J. natn. Cancer Inst.* **48**: 605–613.

Turner, C. W. & Gomez, E. T. (1933). The normal development of the mammary gland of the male and female guinea pig. *Res. Bull. Mo. Agric. exp. Sta.* No. 194.

Uberreiter, O. (1966). Effect of pregnancy and false pregnancy on the occurrence of mammary tumours in the bitch. *Berl. Münch. tierarztl. Wschr.* **79**: 451–456.

Vessey, M. P., Doll, R. & Sutton, P. M. (1971). Investigation of the possible relationship between oral contraceptives and benign and malignant breast disease. *Cancer* **28**: 1395–1399.

Vessey, M. P., Doll, R. & Sutton, P. M. (1972). Oral contraceptives and breast neoplasia: a retrospective study. *Br. med. J.* **1972 (iii)**: 719–724.

Weijer, K., Head, K. W., Misdorp, W. & Hampe, J. F. (1972). Feline malignant mammary tumours. I. Morphology and biology: some comparisons with human and canine mammary carcinomas. *J. natn. Cancer Inst.* **49**: 1697–1704.

Welsch, C. & Meites, J. (1974). Neuroendocrine control of murine mammary carcinoma. In *Mammary cancer and neuroendocrine therapy*: 25–26. Stoll, B. A. (ed.). London: Butterworths.

NOTE ADDED IN PROOF

Recent evidence suggests that lactation itself may have a cancer-sparing effect on the human breast. Ing, Ho & Petrakis (1977) studied a group of Chinese women in Hong Kong who traditionally only fed their babies from one breast. They found that cancer developed only in the unsuckled breast in 79% of such women when postmenopausal, a significantly different distribution from that seen in the normal Chinese population in Hong Kong.

Ing, R., Ho, J. H. C. & Petrakis, N. L. (1977). Unilateral breast feeding and breast cancer. *Lancet* **1977 (ii)**: 124–127.

Symp. zool. Soc. Lond. (1977) No. 41, 231–240

Comparative Studies on Milk Lipids and Neonatal Brain Development

BARBARA M. HALL[a] and JANET M. OXBERRY

Nuffield Institute of Comparative Medicine,
The Zoological Society of London, Regent's Park, London, England

SYNOPSIS

Mammals produce milk of diverse composition and the mammary secretion supports the young during its post-natal phase of most active growth. It is therefore of interest to study composition of milk in relation to the post-natal events in a given species. There is a relationship between the rate of growth and the protein intake in milk and there appears to be a correlation between the maturity of the central nervous system at birth and the mammary supply of the long-chain polyunsaturated fatty acids. These fatty acids, as a part of larger lipid molecules, play an important role in the formation of biological membranes. It has been shown that the chain length and degree of unsaturation of fatty acids affect the activity of membrane-associated enzymes such as Na^+K^+-ATPase and acetylcholinesterase as well as the performance of other metabolic functions dependent on these membranes (for example, cellular recognition). The long-chain polyunsaturated fatty acids are derived from two essential fatty acids, linoleic (18:2w6) and linolenic (18:3w3) in a series of reactions which, at best, are slow. Dietary manipulations taking place during the period of intensive post-natal brain growth have been shown to change fatty acid composition of brain lipids. Human milk supplies some four times the amount of the long-chain polyunsaturated fatty acids found in artificial formulae not supplemented with linoleic acid, and bottle-fed babies show biochemical signs of essential fatty acid deficiency.

MORPHOLOGICAL DEVELOPMENT

A balance is normally maintained between the processes involved in growth and development of a tissue. The formation of cells, their migration, differentiation and death are sequential events and an alteration in any one has an effect on development. In the brain, the adult complement of neurones is almost achieved at an early stage. This corresponds to the first stage of development (Table I) and is completed by approximately 45 days of gestation in the guinea-pig, 170 days gestation in man and by the third day after birth in the rat. The rapid increase in the size of the brain which has been termed the brain "growth spurt" corresponds to stage 2. The morphological events that occur at this time are the outgrowth of axons and dendrites with the establishment of interneuronal connections at the synapses, the extremely rapid multiplication of the oligodendroglial

[a] Present address: *Institute of Child Health, 30 Guildford Street, London WC1, England.*

cells and the formation of the myelin sheaths by these cells. The rate of growth is gradually reduced as the mature adult stage is reached. The sequence of events in the developing brain always appears to follow the same pattern. However, their timing shows regional differences and a significant overlap of the developmental stages is found within the whole organ.

TABLE I

Process in growth and development of brain

---

1. Organogenesis and neuronal multiplication
2. The brain "growth spurt"
   (a) Axonal and dendritic growth glial multiplication and myelination
   (b) Growth in size
3. Mature adult state
4. Senile regression

From Davison & Dobbing, 1968.

The interdependence of the stages of development in the cerebrum and cerebellum is shown by the fact that glial cells provide guidelines for the migrating neurones. In the monkey cerebellar cortex, Rakic (1971) demonstrated that granule cells migrate from the external granular layer to the granular layer and that they utilize the Bergman glial filaments in this migration. In the cerebellar cortex growth occurs later than in other brain regions. In the rat the external granular layer (EGL) increases in size after birth. During the second post-natal week migration of the granule cells reduces the size and by 20 days post-partum there is no detectable DNA synthesis and the EGL disappears (Altman, 1969). Undernutrition has been shown to prolong the time when mitosis occurs in the EGL in the cerebellum of the rat (Gopinath, Bijlani & Deo, 1976; Barnes & Altman, 1973). It has been suggested that this could be due to a prolongation of the cell cycle from day 12 post-partum (Lewis, Balazs, Patel & Johnson, 1975).

RELATIONSHIP BETWEEN MILK COMPOSITION AND MATURITY AT BIRTH

Birth in mammals occurs at different stages of their development, in relation to both their brain and their body growth. This is followed by a characteristic slope of the post-natal growth curve, which again can display an independent pattern for body and brain. The timing of birth and the post-natal growth rate are specific for a given species and both are reflected in the supply of nutrients in the milk.

It was suggested at the end of the last century (Bunge, 1898), and confirmed more recently (Payne & Wheeler, 1968), that the protein intake in the milk of different species is correlated with the rate of growth of their young (Table II). We have focused our attention on lipid because

TABLE II

Growth and protein intake from milk

Species	Protein intake (g/day kg$^{0.73}$)	Growth rate (g/day kg$^{0.73}$)
Man	2·7	8
Rat	7·6	28
Rabbit	12·0	48
Ox (calf)	16·8	61
Pig	17·2	76

From Payne & Wheeler, 1968.

of its abundant presence in the nervous system and its rapid incorporation into the brain during the neonatal period. The lipid content of newborn mouse brain is 18% of dry weight while in the adult this rises to 37% (Folch, 1955).

THE ESSENTIAL FATTY ACIDS

The essential fatty acids (EFA), linoleic (18:2w6) and α-linolenic (18:3w3) cannot be synthesized by animals from carbohydrate or amino acid precursors and have to be obtained in the diet. These two fatty acids subsequently form the long-chain polyunsaturated fatty acids (LCP) of the w6 and w3 family. LCPs are formed in processes involving chain elongation and desaturation. Evidence suggests that the activity of $\Delta6$-desaturase is low in the rat (Hassam, Sinclair & Crawford, 1975). This rate-limiting step may represent a mechanism by which the synthesis of LCP and the formation of prostaglandins are controlled (Hassam & Crawford, 1976). It has been found recently that the cat, along with other carnivores, is unable to desaturate the essential fatty acids and therefore has an absolute requirement for LCP (Rivers, Hassam & Alderson, 1976).

There is substrate competition for the $\Delta6$-desaturase, the affinity for the enzyme increasing with the number of double bonds from oleic to α-linolenic acid (Brenner & Peluffo, 1968). The long-chain polyunsaturated derivatives of the non-essential oleic acid are therefore found only in animals whose diet is deficient in linoleic and α-linolenic acids. Of the derivatives, one in particular, the 20:3w9 fatty acid, accumulates in tissues of essential fatty acid-deficient animals.

LONG-CHAIN POLYUNSATURATED FATTY ACID CONTENT OF MILK

As the supply of protein in milk of different species is different, so the yield of LCP varies (Table III), and it is of interest that the intake of LCP appears to be matched by the maturity of the central nervous system at birth, there being some tenfold difference in the supply of these fatty acids in milk of the rat and the cow.

TABLE III

LCP intake and CNS maturity at birth

Species (arranged in ascending order of magnitude of brain maturity at birth)	LCP intake (g/day kg$^{0.73}$)
Rat	0·44
Marmoset	0·69
Man	0·24
Ox (calf)	0·05
Guinea-pig	0·02

During post-natal development of rat brain, lipids have been shown to be rapidly incorporated into its structure (Sinclair & Crawford, 1972). They are deposited in two phases: arachidonic (20:4w6) and docosahex aenoic (22:6w3) acids during the time when rapid cell division is occurring, and lignocenic (24:0) and nervonic (24:1) acids during the period of myelination.

Changes in the composition of human milk lipid seem to parallel the events taking place in the neonatal brain. The human infant is born with almost its adult complement of neurons. Post-natal mitosis is predominantly due to glial cell multiplication which actively continues for up to three years. Amongst the glial cells are those which form myelin and rapid myelination takes place during this time (Dobbing, 1974). As demonstrated by Crawford, Hall, Laurance & Munhambo (1976) there is a gradual decrease in the amount of LCP in the milk from the beginning of lactation. These fatty acids are in high concentration in brain cells. At the same time there is a slow but constant rise in the content of the long-chain saturated and mono-unsaturated fatty acids characteristic of myelin. However, the rate of fall in the amount of LCP in milk can be increased in the rat by increasing the size of the litter (Hassam & Crawford, 1976). A similar phenomenon has been observed in the milk of undernourished human mothers (Crawford, Stevens, Msuya & Munhambo, 1974). This points to the limitations in the maternal supply of LCP. These limitations

TABLE IV

Blood triglyceride levels of w6 and w3 long-chain polyunsaturated fatty acid in breast- and bottle-fed babies (g/100 g fatty acids)

	Milk	Week 1	Week 6	Week 18	Adult
Arachidonate	Human	4·5 ± 0·31	2·2 ± 0·11	1·7 ± 0·08	0·52
	Cow	4·3 ± 0·24	1·0 ± 0·08	0·07 ± 0·01	± 0·04
Docosahexaenoate	Human	3·8 ± 0·17	2·5 ± 0·16	1·0 ± 0·09	0·44
	Cow	3·5 ± 0·15	0·91 ± 0·11	0·05 ± 0.01	± 0·02

20 : 3w9/20 : 4w6 ratio in blood triglycerides of breast- and bottle-fed babies

Milk	Umbilical cord blood	Week 1	Week 6	Week 18	Adult
Human $n = 10$	0·01	0·008 ± 0·0016	0·015 ± 0·002	0·012 ± 0·001	0·010 ± 0·001
Cow $n = 12$	± 0·001	0·075 ± 0·0063	0·14 ± 0·015	0·28 ± 0·033	

From Crawford, Hassam *et al.*, in press.

are imposed by the metabolism, upon which we have little influence, and by the diet, which can be regulated.

Generally, the composition of human milk lipids is similar throughout the world in the milk of well nourished mothers. However, in extreme dietary conditions such as under- or malnutrition the fatty acid pattern can be altered (B. M. Hall, unpubl. obs.). Analysis of fatty acids in human milk can provide information about the mother's diet since different fatty acids are affected by different conditions. There have been few reports on the LCP content of human milk (Insull & Ahrens, 1959; Insull, Hirsch, James & Ahrens, 1959). However, these fatty acids have been found in milk samples from a socio-economic cross-section of the community (Crawford, Hall et al., 1976)—from Tanzania, Uganda and Copenhagen as well as from London. Much smaller amounts have been found in the artificial formulae used for infant feeding since even in the milks in which the original lipid is modified, the butterfat is replaced by vegetable oil lacking LCP.

The human habit of bottle-feeding provides us therefore with a good model for studying the effect of LCP deprivation in infants. As demonstrated by Crawford, Hassam & Hall (in press), the levels of arachidonic and docosahexaenoic acids in blood triglycerides fall from birth to adulthood, probably reflecting the diminishing dietary supply during this time. However, the fall is much faster in bottle-fed babies (Table IV); after only six weeks it has been found to reach the concentration found in breast-fed babies at 18 weeks. At the same time the blood triglycerides of bottle-fed babies contained the 20 : 3w9 fatty acid found in animal tissues only in the state of the essential fatty acid-deficiency. After one week on artificial milk the triene-tetraene ratio (20 : 3w9/20 : 4w6) in infants' blood triglycerides reached a value seven times higher than that found in the adult and after 18 weeks it was 28 times above that level. This demonstrates that the fatty acid composition of human infants' blood can be altered by nutrition, implying that the post-natal phase of development in humans may be open to dietary manipulation.

FUNCTIONAL DEVELOPMENT IN THE BRAIN

Spontaneous and evoked electrical activity begins to appear early in stage 2 of development (Table II). There is a temporal relationship between the maturation of electrical activity and the optimal synthesis of the enzyme $(Na^+ - K^+)$-ATPase. In rat brain $(Na^+ - K^+)$-ATPase is detected on day 21 of gestation and reaches an adult level by day 12 post-partum; electrical activity is traceable on day 21 of gestation and during the first post-natal week becomes more rhythmic and regular (Côte, 1964; Abdel-Latif, Brody & Ramahi, 1967).

Phospholipids are indicated as having a significant effect on the activity of membrane-associated enzymes such as $(Na^+ - K^+)$-ATPase

(Schwarz, Lindenmayer & Allen, 1975; Wheeler, Walker & Barker, 1975). The activity of this enzyme has also been shown to be altered by variations in chain length and the degree of unsaturation of the fatty acids of the phospholipids (Kimelberg & Popahadajopoulos, 1974).

DIETARY INDUCED CHANGES IN BRAIN LCP

By varying the dietary intake of essential fatty acids, significant changes can be made in the unsaturated fatty acid composition of cell membranes in most tissues (Guanieri & Johnson, 1970). The brain fatty acid pattern is established relatively early in life (Sinclair & Crawford, 1972), and, in the rat, dietary manipulation after weaning must be prolonged in order to effect a change in polyunsaturated fatty acids. By feeding diets containing very low levels of EFA a reduced concentration of 20:4w6, 22:4w6 and 22:6w3 and significantly high levels of 20:3w9 and 22:3w9 are found in the brain. Changes in brain fatty acid pattern are also induced by feeding diets containing extremes in the ratios of w6 to w3 fatty acids, for example sunflower seed oil will cause an increase in the 22:5w6 and decrease in 22:6w3 content of brain (Galli, White & Paoletti, 1970; Galli, Trzeciak & Paoletti, 1971).

EFFECT OF DIET ON MEMBRANE FUNCTION

Mice reared on a diet deficient in EFA for six months of their adult lives were used for the preparation of brain synaptosomal membranes. It was found (Table V) that the activity of the $(Na^+ - K^+)$-ATPase was altered (Sun & Sun, 1974). In another experiment rats were kept on a diet

TABLE V

Comparison of $(Na^+ - K^+)$-ATPase activity between control and essential fatty acid-deficient animals

| Animal | Tissue | $(Na^+ - K^+)$-ATPase activity (μ mol/mg protein/h) | |
		Control	EFA-deficient
Rat	Liver homogenate	5·9*	7·5*
	Liver membranes	57·8*	67·2*
Mouse	Brain homogenate	16·8†	21·2†
	Brain synaptosomal membranes	77·4†	89·3†

* Brivio-Haugland *et al.*, 1976
† Sun & Sun, 1974

deficient in EFA and their isolated liver plasma membranes were found to have an increased V_{max} and K_m for total magnesium $(Na^+ - K^+ - Mg^{2+})$-ATPase (Brivio-Haugland, Louis, Musch, Waldeck & Williams, 1976). Dietary manipulation affects the function of the membrane bound enzyme $(Na^+ - K^+)$-ATPase. It is possible that alterations in membrane structure could have repercussions on the many biological phenomena for which membranes are responsible, namely transport of metabolites and metal ions associated with the processes responsible for energy transduction, information transfer and recognition. Certainly, in the highly specialized membrane of the retina, a change in electrical activity has been found in rats deficient in 22:6w3 fatty acid (Benolken, Anderson & Wheeler, 1973); impairment of learning ability has been reported in rats deficient in 22:6w3 by Lamptey & Walker (1976), and behavioural changes found in EFA-deficient rats by Paoletti & Galli (1972).

NUTRITION OF THE HUMAN INFANT

At present there is little evidence to suggest that the post-natal supply of fatty acids is related to the mental development of the child, but there is a sufficient body of evidence indicating that the milk composition of a particular species represents an optimal adaptation (Widdowson, 1970; Jelliffe & Jelliffe, 1971). Consequently, it appears that in our own species we should aim at supplying this optimum.

REFERENCES

Abdel-Latif, A. A., Brody, J. & Ramahi, H. (1967). Studies on sodium-potassium adenosine triphosphatase of the nerve endings and appearance of electrical activity in developing rat brain. *J. Neurochem.* **14**: 1133–1141.

Altman, J. (1969). Autoradiographic and histological studies of postnatal neurogenesis. III. Dating the time of production and onset of differentiation of cerebellar microneurons in rats. *J. comp. Neurol.* **136**: 269–294.

Barnes, D. & Altman, J. (1973). Effects of two levels of gestational–lactational undernutrition on the postweaning growth of the rat cerebellum. *Expl Neurol.* **38**: 420–428.

Benolken, R. M., Anderson, R. E. & Wheeler, T. G. (1973). Membrane fatty acids associated with the electrical response in visual excitation. *Science, N.Y.* **182**: 1253–1254.

Brenner, R. R. & Peluffo, R. O. (1968). Regulation of unsaturated fatty acids biosynthesis 1. Effect of unsaturated fatty acid of 18 carbons on the microsomal desaturation of linoleic acid into γ-linolenic acid. *Biochim. biophys Acta* **176**: 471–479.

Brivio-Haugland, R. P., Louis, S. L., Musch, K., Waldeck, N. & Williams, M. A. (1976). Liver plasma membranes from essential fatty acid-deficient rats. *Biochim. biophys Acta* **433**: 150–163.

Bunge, T. (1898). *Lehrbuch der physiologischen chemie.* 4th ed. Leipzig.

Côte, L. J. (1964). Mg^{++} dependent Na$^+$K$^+$ stimulated ATPase in the developing rat brain. *Life Sci.* **3**: 899–901.

Crawford, M. A., Hall, B., Laurance, B. M. & Munhambo, A. (1976). Milk lipids and their variability. *Curr. med. Res. Op.* **4**, suppl. 1: 33–43.

Crawford, M. A., Hassam, A. G. & Hall, B. M. (in press). Metabolism of essential fatty acids in the human fetus and neonate. *Nutr. Metab.* **20**, suppl. 2.

Crawford, M. A., Stevens, P., Msuya, P. & Munhambo, A. (1974). Lipid composition of human milk: comparative studies on African and European mothers. *Proc. Nutr. Soc.* **33**: 50–51A.

Davison, A. N. & Dobbing, J. (eds). (1968). *Applied neurochemistry.* Oxford: Blackwell.

Dobbing, J. (1974). The later development of the brain and its vulnerability. In *Scientific foundations of paediatrics*: 565–577. Davis, J. A. & Dobbing, J. (eds). London: Heinemann.

Folch, J. (1955). Composition of the brain in relation to maturation. In *Biochemistry of the developing nervous system*: 121–133. Waelsch, H. (ed.). New York and London: Academic Press.

Galli, C., Trzeciak, H. K. & Paoletti, R. (1971). Effects of dietary fatty acids on the fatty acid composition of brain ethanolamine phosphoglyceride: reciprocal replacement of n-6 and n-3 polyunsaturated acids. *Biochim. biophys. Acta* **248**: 449–454.

Galli, C., White, H. B. & Paoletti, R. (1970). Brain lipid modifications induced by essential fatty acid deficiency in growing male and female rats. *J. Neurochem.* **17**: 347–355.

Gopinath, G., Bijlani, V. & Deo, M. G. (1976). Undernutrition and the developing cerebellar cortex in the rat. *J. Neuropath. exp. Neurol.* **35**: 125–135.

Guanieri, M. & Johnson, R. M. (1970). The essential fatty acids. In *Advances in lipid research* **8**: 115–174. Paoletti, R. & Kitchevsky, D. (eds). London and New York: Academic Press.

Hassam, A. G. & Crawford, M. A. (1976). Influence of maternal dietary γ-linolenic acid on the milk and liver lipids of suckling rats. *Nutr. Metab.* **20**: 112–116.

Hassam, A. G., Sinclair, A. J. & Crawford, M. A. (1975). The incorporation of orally fed radioactive γ-linolenic acid and linoleic acid into the liver and brain lipids of suckling rats. *Lipids* **10**: 417–420.

Insull, W. & Ahrens, E. H. (1959). The fatty acids of human milk from mothers on diets taken *ad libitum*. *Biochem. J.* **72**: 27–33.

Insull, W., Hirsch, J., James, T. & Ahrens, E. M. (1959). The fatty acids of human milk. II. Alterations produced by manipulation of caloric balance and exchange of dietary fats. *J. clin. Invest.* **38**: 443–450.

Jelliffe, D. B. & Jelliffe, E. F. P. (1971). The uniqueness of human milk. *Am. J. clin. Nutr.* **24**: 968–969 & 1013–1024.

Kimelberg, H. K. & Popahadajopoulos, D. (1974). Effects of phospholipid acyl chain fluidity, phase transitions, and cholesterol on (Na$^+$ + K$^+$)-stimulated adenosine triphosphatase. *J. biol. Chem.* **249**: 1071–1080.

Lamptey, M. S. & Walker, B. L. (1976). A possible essential role for dietary linolenic acid in the development of the young rat. *J. Nutr.* **106**: 86–93.

Lewis, P. D., Balazs, R., Patel, A. J. & Johnson, A. L. (1975). The effect of undernutrition in early life on cell generation in the rat brain. *Brain Res.* **83**: 235–247.

Paoletti, R. & Galli, C. (1972). Effects of essential fatty acid deficiency on the central nervous system in the growing rat. In *Lipids, malnutrition and the developing brain*: 121–140 (Ciba Foundation Symposium). Elliott, K. & Knight, J. (eds). Amsterdam: Elsevier.

Payne, P. R. & Wheeler, E. F. (1968). Comparative nutrition in pregnancy and lactation. *Proc. Nutr. Soc.* **27**: 129–138.

Rakic, P. (1971). Neuron-glia relationship during granule cell migration in developing cerebellar cortex. A Golgi and electromicroscopic study in *Macacus rhesus. J. comp. Neurol.* **141**: 283–312.

Rivers, J. P. W., Hassam, A. G. & Alderson, C. (1976). The absence of Δ6-desaturase activity in the cat. *Proc. Nutr. Soc.* **35**: 67–68A.

Schwarz, A., Lindenmayer, G. E. & Allen, J. C. (1975). The sodium-potassium adenosine triphosphatase: pharmacological, physiological and biochemical aspects. *Pharmac. Rev.* **27**: 3–134.

Sinclair, A. J. & Crawford, M. A. (1972). The accumulation of arachidonate and docosahexaenoate in the developing rat brain. *J. Neurochem.* **19**: 1753–1958.

Sun, G. Y. & Sun, A. Y. (1974). Synaptosomal plasma membranes: acyl group composition of phosphoglycerides and $(Na^+ + K^+)$-ATPase activity during fatty acid deficiency. *J. Neurochem.* **22**: 15–18.

Wheeler, K. P., Walker, J. A. & Barker, D. M. (1975). Lipid requirement of the membrane sodium-plus-potassium ion-dependent adenosine triphosphatase system. *Biochem. J.* **146**: 713–722.

Widdowson, E. M. (1970). Harmony of growth. *Lancet* **1970** (i): 901–905.

Symp. zool. Soc. Lond. (1977) No. 41, 241–260

Role of the Mammary Gland and Milk in Immunology

A. K. LASCELLES

CSIRO Division of Animal Health, Melbourne, Australia

SYNOPSIS

In the immunological sense the mammary gland's performance, at least for a limited period of the lactation cycle, is no less striking than its remarkable biosynthetic capacity. In those species which passively transfer immunity to their newborn after birth, massive amounts of the complement-fixing IgG_1 (an immunoglobulin) are transferred into colostrum; in the cow, for example, it has been computed that around 500 g of serum-derived IgG is transferred into secretion during the week prior to parturition. This transport system is also remarkable for its high degree of selectivity which would appear to be associated with the possession of specific receptor sites located on the basal or intercellular membrane of the glandular epithelium. These receptors can distinguish between molecules as similar as IgG_1 and IgG_2.

Mammary glands are also able to make antibody locally. The extent to which this occurs varies among the species, being most highly developed in those species such as the human and rabbit which transfer immunity to their young before birth. Most of the antibody made locally is secretory IgA. Even in ruminants, significant quantities of local antibody are secreted provided that the glands are locally stimulated with antigen prior to parturition when the gland possesses the cellular machinery required to mount an immune response.

While the importance of colostral antibody in the protection of the new-born of animals normally born agammaglobulinaemic is well established, the importance of colostral secretory IgA antibody in species such as man and the rabbit acquiring passive immunity pre-natally is less clear. In these latter animals protein is not absorbed from the gut lumen of the neonate and any beneficial effect of IgA would be confined to the lumen itself. Recent evidence suggests that agglutination of bacteria and the prevention of their adherence to the mucosa, brought about by secretory IgA antibody, plays an important role in the elimination of pathogens by peristaltic action.

There is also evidence that the local immune response plays a role in protection of the mammary glands against bacterial mastitis.

INTRODUCTION

The biosynthetic performance of the mammary gland, both in terms of the repertoire of substances synthesized and the sheer magnitude of the total synthetic activity, continues to intrigue the most experienced biologist. Another impressive activity of the mammary gland which, until recently, has received rather less than its fair share of attention concerns the secretion of immunoglobulins, although the significance of this activity in relation to the protection of the newborn against infectious agents and intoxications in early life has been appreciated in general terms for a long time (Brambell, 1970).

During the course of lactation the concentration of immunoglobulin in the mammary secretion varies enormously; concentrations in colostrum of ruminants may exceed 10 g/100 ml but fall quickly after parturition to values around 50–100 mg/100 ml. The levels remain relatively stable during the period of copious milk production but increase again at the end of lactation, especially so after regular milking ceases. Before considering this subject further, it may be useful to briefly discuss the nomenclature and properties of immunoglobulins in mammals.

The term "immunoglobulin" was first proposed by Heremans (1959) to describe the group of serum proteins closely related chemically, immunologically and functionally which appeared to be carriers of antibody activity. The immunoglobulins are a heterogeneous group of glycoproteins of high molecular weight with γ- or β-electrophoretic mobility, which are synthesized by cells of the lymphocyte plasma series. The basic unit of all immunoglobulins comprises two identical heavy and two identical light polypeptide chains covalently linked by interchain disulphide bridges. Immunoglobulin is a "Y"-shaped molecule (Fig. 1), the two arms of the Y constituting the $F(ab)_2$ fragment while the Fc fragment is contributed by the stem of the Y. On the basis of isotypic variation, immunoglobulins have been classified into classes and subclasses. The different classes of immunoglobulin have distinct antigenic properties which are carried in the Fc portion of the molecule and each class may be further subdivided into subclasses on the basis of additional antigenic variation in the Fc fragment. Biological properties of the molecule, such as complement binding and transport across epithelial membranes, can also be attributed to the Fc portion of the molecule. In man, five major classes of immunoglobulin have been recognized; these are IgG, IgM, IgA, IgD and IgE. Immunoglobulin classes have also been described in many animal species and shown to be analogous to the above human classes. IgG, the most plentiful immunoglobulin in the serum of mammals, exists in monomeric form, being composed of only two heavy and two light chains, whereas IgM is a pentamer of this basic structure, and IgA a dimer. IgA is the so-called secretory immunoglobulin because it is found in highest concentration in mucous secretions. IgA dimers are capable of specific combination with another glycoprotein called "secretory component" (SC) which is synthesized in the glandular epithelium to form a complex called "secretory IgA" (SIgA) (Fig. 2). This is the form in which virtually all the IgA occurs in external secretion whereas the relatively small amounts found in serum of most species lack SC. Details of the structure and various properties of the four major immunoglobulins found in secretion and serum of milk are tabulated in Table I.

Details of immunoglobulins in mammary secretion from a number of species are given in Table II. It can be seen that the major immunoglobulin in mammary secretion of the ewe is IgG_1 which is generally regarded as

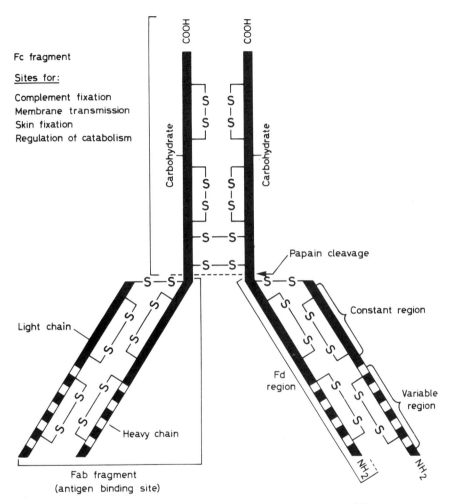

FIG. 1. Model of IgG molecule. This molecule is composed of one basic 7S unit.

the complement fixing subclass of IgG in ruminant animals. A similar situation holds for the cow and goat but in the pig, horse and dog IgA predominates in milk, although IgG is the most abundant immunoglobulin in the colostrum. In marked contrast to ruminants, IgA is the major immunoglobulin in colostrum of man, the rabbit and guinea-pig.

Immunoglobulins in mammary secretion are either derived from the bloodstream or made locally by cells of the lymphocyte plasma cell series situated close to the glandular epithelium. The relative importance of these two sources differs strikingly between species and the generalization can be made that the former is quantitatively much more important in

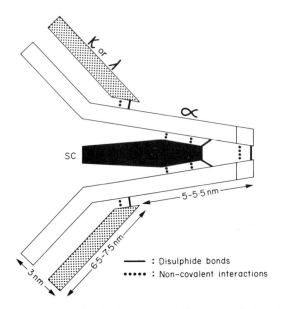

FIG. 2. Model of secretory IgA molecule. This molecule is composed of two basic 7S units and secretory component (SC).

TABLE I

Details of immunoglobulin classes in sheep and cattle

	IgG$_1$	IgG$_2$	IgM	IgA
Molecular weight	150 000	150 000	1 030 000	IgA = 330 000
				SIgA = 420 000
Sedimentation	6·5–7·0	6·5–7·0	19	IgA = 10
coefficient				SIgA = 10·8–11·2
Half-life (days)				
Sheep	9·5	10·6	6	3
Ox	9·6	17·7		
Carbohydrate (%)	2–4	2–4	10–12	SIgA = 8
Reaginic activity	?	–	–	–
Complement fixation	+	–	+	–
Placental transfer	–	–	–	–
Antibody activity	+	+	+	+

TABLE II

*Concentrations of immunoglobulins (mg/100 ml) in mammary secretions of several species**

Species	Secretion	IgG§ IgG$_1$	IgG$_2$	IgA	IgM
Sheep	Colostrum	6000		200	410
	Milk	30		6	3
	Milk whey–primiparous ewe	72	8	8	3
	Milk whey–immunized gland†	103	13	21	7
Cow	Colostrum	7900		440	490
	Colostrum	7500	190	440	490
	Colostral whey	4824	398	469	710
	Milk	40		5	4
	Milk	35	6	5	4
	Milk whey	40	6	11	15
Goat	Colostrum	5800		170	380
	Milk	25		6	3
Pig	Colostrum	5870		1070	320
	Colostrum	5609		883	240
Horse	Colostrum	6000		?	?
	Milk	62		?	?
	Milk—1 day post-partum	310		20	22
	Milk—16 days post-partum	36		80	7
Dog	Colostrum	1453		313	217
Guinea-pig	Milk	14	50	76	11
Rabbit	Colostrum	240		450	10
	Colostrum-immunized gland‡	380		650	20
Man	Fat-free colostrum	17		1788	79

* From Lascelles & McDowell (1974).
† Glands infused 3 weeks before parturition with 10^9 killed *Brucella abortus* in saline.
‡ Glands immunized prior to parturition intradermally and subcutaneously with live *Salmonella typhimurium*.
§ In the ruminant IgG$_1$ fixes homologous and heterologous complement, whereas IgG$_2$ fixes low doses of homologous complement only (Feinstein & Hobart, 1969).

those species which transmit immunity from mother to young after birth by way of the colostrum (Brambell, 1970).

IMMUNOGLOBULINS DERIVED FROM THE BLOODSTREAM

On the basis of similarities in size, physicochemical and precipitation characteristics of the immunoglobulin fractions of serum and colostrum described by earlier workers (for example, Little & Orcutt, 1922; Smith, 1946), it was generally considered that the immunoglobulin in colostrum was derived from blood serum. This view, however, was seriously questioned for a time when Campbell, Porter & Petersen (1950) reported the presence of a large number of plasma cells in mammary tissue of cows during colostrum formation and went on to suggest that the large amounts of antibody in colostrum were made locally. This interpretation has not been substantiated and the following unequivocally points to the humoral origin of most of the antibody in colostrum. Askonas, Campbell, Humphrey & Work (1954) injected antibody to type-3 *Pneumococcus*, biosynthetically-labelled with sulphur-35, into goats shortly before parturition and observed that the specific activities in serum and secretion were similar. Subsequently ^{131}I-labelled homologous γ-globulin was observed to be transferred from serum into colostrum of cows, reaching concentrations in colostrum 13 times higher than in serum (Blakemore & Garner, 1956; Garner & Crawley, 1958). In conformity with these findings, Larson & Kendall (1957) reported a substantial decrease in the blood serum concentration of γ-globulin as the cows approached calving. In response to earlier claims by Campbell *et al.* (1950) regarding the suggested local origin of colostral immunoglobulin, Dixon and his colleagues reported a virtual absence of plasma cells in the mammary tissue of cows immediately before parturition (Dixon, Weigle & Vazquez, 1961). However, they did report a substantial decrease in serum concentration of γ-globulin and presented evidence suggesting that the decrease in serum concentration could be quantitatively accounted for by the corresponding rise in concentration in colostrum. Therefore, the observations of Dixon and his colleagues indicated that plasma cells were not a normal inhabitant of the gland and it would seem therefore that the mammary glands examined by Campbell *et al.* (1950) may have been chronically infected.

Thus by 1960 it was firmly established that most of the immunoglobulin in colostrum was derived from serum and that its transfer was affected by a selective process relative to other serum proteins. The next major step in our understanding of this process can be attributed to Richards & Marrack (1963) who were the first to recognize that it was the electrophoretically fast-migrating immunoglobulin (IgG$_1$) which selectively accumulated in colostrum of sheep.

FACTORS AFFECTING TRANSFER OF IMMUNOGLOBULIN INTO MAMMARY SECRETION

The ability of the mammary gland of ruminants to accumulate IgG$_1$ was considered, until relatively recently, the property of the colostrum-forming gland and possibly also the involuting gland (Carroll, 1961; Murphy, Aalund, Osebold & Carroll, 1964; Carroll, Murphy & Aalund, 1965; Pierce & Feinstein, 1965; Aalund, 1968). However, by measuring the distribution of radio-labelled IgG$_1$ and IgG$_2$ between milk and blood following intravenous injection, Mackenzie & Lascelles (1968) were able to show that selective transport of IgG_1 continued into lactation, although at a much lower magnitude than occurs during colostrum formation. In other words, the ability to selectively transport IgG_1 was shown to be a characteristic of the mammary gland and was not confined to the colostrum-forming period. In an attempt to seek a satisfactory explanation for the substantial elevation in the selective transfer of IgG_1 during colostrum formation, Lascelles (1969, 1971) noted that selective transfer appeared to be greatest when synthetic activity was lowest and resorption of milk was occurring.

Subsequent studies have been concerned with examining the relationship between milk production (taken to be essentially synonymous with *de novo* synthetic activity) and selective transfer into mammary secretion. In these studies quantitative expression of selective transfer was obtained by computing an index of selective transfer of IgG_1. This index was computed by correcting the ratios of $IgG_1 : IgG_2$ in secretion for differences in their respective concentrations in serum as follows:

$$\frac{\text{Conc. of } IgG_1 \text{ in secretion}}{\text{Conc. of } IgG_2 \text{ in secretion}} \times \frac{\text{Conc. of } IgG_2 \text{ in serum}}{\text{Conc. of } IgG_1 \text{ in serum}}.$$

Watson, Brandon & Lascelles (1972) observed in sheep that there was a transitory increase in the selective index for IgG_1 during the early stages of involution when milk production was declining sharply but before substantial degradation of the glandular epithelium had taken place (Lascelles & Lee, in press). In a further attempt to determine whether local influences affected transfer of IgG_1 into colostrum of cows a comparison was made of selective transfer indices between glands milked continuously during the period preceding calving and those allowed to undergo normal involution (Brandon & Lascelles, 1975). In contrast to the striking increases in selective index for IgG_1 in unmilked glands as parturition approached, the increases in milked glands were reduced in magnitude and delayed in time of onset (Fig. 3). It was clear that the maintenance of milk production during the colostrum-forming period was associated with an inhibition of the massive selective transfer of IgG_1 which normally characterizes this period. Finally, attention is drawn to recent experiments on the effects of corticosteroid-induced lactogenesis on the transfer of immunoglobulin into mammary secretion of cows (Brandon, Husband & Lascelles, 1975). Injections of long-acting synthetic glucocorticoids are characteristically followed by the production of copious quantities of milk and, in contrast to the situation in untreated cows, are accompanied by sharp decreases in the concentration of IgG_1 in the secretion and in the selective index of IgG_1.

While the above studies do not actually exclude the possibility of central endocrine influences playing a direct role in the selective transfer of large amounts of IgG_1 into colostrum, they certainly support the notion

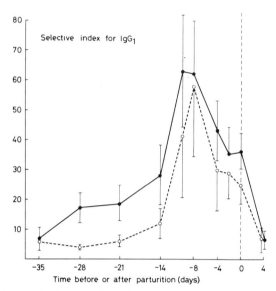

FIG. 3. Changes in the selective index for IgG$_1$ for milked and unmilked glands of cows from 35 days before until four days after parturition. Values presented are means ± standard errors, each plotted point being the mean value of selective indices for individual cows. O---O: Selective index for milked glands. ●——●: Selective index for unmilked glands.

that selective transfer of IgG$_1$ is a physiological characteristic of the mammary gland and that its magnitude is influenced by local factors such as synthetic activity. This view is at variance with that of Smith, Muir, Ferguson & Conrad (1971) who concluded that ovarian steroid hormones exert at least some control over selective transfer of IgG$_1$. These conclusions were based on the observation that increases occurred in selective transfer of IgG$_1$ following injection of ovarian steroids into non-pregnant, non-lactating cows. Lascelles & McDowell (1974) have suggested that the changes observed by Smith and his colleagues are simply a consequence of the known mammogenic effects of these hormones. Under their influence glandular epithelial cells proliferate (see Cowie, 1961) and, after a period, a fully-developed lobulo-alveolar system is elaborated. However, in the absence of a lactogenic stimulus, the developed gland does not produce copious quantities of secretion and consequently closely resembles the colostrum-forming gland of the pre-parturient cow.

It has long been known that inflammatory changes in the mammary gland are associated with marked compositional changes in milk which can be attributed to decreases in synthetic activity and also to increases in the permeability of the glandular epithelium of the mammary gland and its associated capillary endothelium. The effect of inflammation on the transfer of radio-labelled IgG$_1$ and IgG$_2$ from blood into milk was studied by Mackenzie & Lascelles (1968). Their results demonstrated that inflammation increased the permeability of the glandular epithelium and also, presumably, of the capillary endothelium. Since the whey:serum ratios for radio-labelled IgG_2 approached those for IgG_1 as the inflammation developed, it was concluded that the selective mechanism for the transfer of IgG_1 was severely inhibited. Mackenzie & Lascelles (1968) also observed that the selective transport recovered, indeed tended to operate at a higher level of activity than in the uninflamed gland, after the acute inflammation subsided. Essentially similar conclusions were reached by Harmon, Schanbacher, Ferguson & Smith (1976) who observed that levels of IgG and serum albumin increased dramatically during acute inflammation induced experimentally with *Escherichia coli* but that IgG levels remained elevated after serum albumin levels had returned to normal.

LOCALLY-SYNTHESIZED IMMUNOGLOBULIN

Considerable difficulty has been experienced in establishing the local origin of immunoglobulin in external secretions. Satisfaction of the following two criteria represents strong but not unequivocal evidence that most of the immunoglobulin in external secretion is of local origin. First, the secretion:serum concentration ratio for the particular immunoglobulin is higher than that for other immunoglobulins and serum albumin. Second, a substantial number of cells of the lymphocyte plasma cell series are present in the tissue near the glandular epithelium, and a significant proportion can be demonstrated to contain immunoglobulin of the same class as that in secretion. These criteria have been used to demonstrate the local origin of IgA in a number of external secretions; IgA is the major immunoglobulin in mammary secretion in man, the rabbit and guinea-pig. Clearly, however, a selective transfer mechanism similar to that responsible for the accumulation of IgG in the mammary secretion of ruminants will also result in a high secretion:serum ratio and accordingly additional evidence will also be required to determine the relative importance of local synthesis and selective transfer. This may be obtained by comparing specific radioactivities of immunoglobulin in serum and secretion following intravenous injection of radio-labelled immunoglobulin.

The question of whether the mammary gland is capable of synthesizing antibody following local infection or infusion of antigen has been the subject of many investigations (Pierce, 1959; Lascelles, 1963). In retrospect, the experiments of Batty & Warrack (1955), who claimed to demonstrate local antibody production in the mammary gland of the rabbit, were convincing whereas studies by others (for example, Kerr, Pearson & Rankin, 1959), who reported local antibody formation in lactating mammary glands following a series of antigen infusions, were not. Such a procedure may give rise to chronic inflammatory lesions which would be associated with increases in the permeability of the capillary endothelium and of the glandular epithelium to plasma proteins. Moreover, it is likely that such chronic inflammatory foci would become infiltrated with cells of the lymphocyte plasma cell series which could contribute significant quantities of antibody to the secretion.

More recent experiments have demonstrated that infusions of killed bacteria and flagellar antigens into involuting and dry mammary glands give rise to a persisting local production of antibody, whereas there is no suggestion of a local production following infusion into lactating glands (Lascelles, 1963; Outteridge, Rock & Lascelles, 1965; Lascelles, Outteridge & Mackenzie, 1966; Outteridge, Mackenzie & Lascelles, 1968; McDowell & Lascelles, 1969, 1971) (Figs 4 and 5). More recently, workers

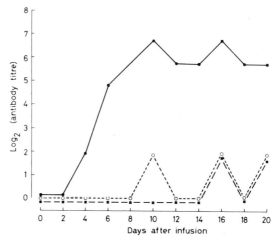

FIG. 4. Brucella agglutination titre in samples of serum and whey following infusion of killed *Brucella abortus* organisms into one side of the udder of a ewe during lactation. ●——●: Blood serum. ○– – –○: Whey–*Brucella*-infused side. ■– – –■: Whey–non-infused side.

in other laboratories have confirmed these findings using a variety of antigens in cows and sheep (Plommet, 1968; Porter, 1968; Wilson, 1972; Wilson, Duncan, Heistand & Brown, 1972). The virtual absence of IgA in the mammary secretion of primiparous ewes, which contrasts with the presence of significant quantities in secretion from glands infused with antigen before parturition, lead to the suggestion (Lascelles & McDowell, 1970) that local antigenic stimulation awakens an almost dormant secretory IgA system in the mammary gland of the sheep. The quantitative importance of the IgA system in the intestine of ruminants, as in other species a region subject to continuous antigenic stimulation, supports this concept. It is of interest to mention that pre-partum local antigenic stimulation of the mammary gland of the guinea-pig, one of those species

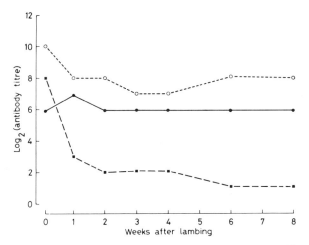

FIG. 5. Brucella agglutination titre in samples of serum and whey collected from both sides of the udder of a ewe during the first eight weeks of lactation. Four weeks before lambing the ewe was infused with killed *Brucella abortus* into one side and killed *Salmonella typhi* "O" into the other side of the udder. ●——●: Blood serum. ○ - - - ○: Whey–*Brucella*-infused side. ■ - - - ■: Whey–*Salmonella*-infused side.

in which IgA predominates in colostrum and milk, gives rise to persisting production of IgA antibody (McDowell, 1973).

CELLULAR BASIS FOR LOCAL IMMUNITY

With a view to establishing the cellular basis for the immune responsiveness of the mammary gland, Lee & Lascelles (1969) followed the histological changes in the glands of ewes undergoing involution. Very few lymphocytes were observed in lactating mammary tissue. However, during the course of involution there was a progressive increase in the concentration of lymphocytes, and in the fully involuted gland lymphocytes appeared to outnumber epithelial cells. The lymphocytes were located next to the epithelium of the alveolar remnants and small ducts (Fig. 6). Numerous large cells with basophilic or foamy cytoplasm, previously identified as macrophages, (Lascelles, Gurner & Coombs, 1969; Lee, McDowell & Lascelles, 1969) were also seen in tissue and secretion. It was evident that the concentration of lymphocytes and macrophages in the tissue at this time was sufficient to account for the immune responsiveness of the involuted gland.

In the rabbit at least, the Peyer's patches are an enriched source of potential IgA-producing cells (Craig & Cebra, 1971) and presumably the IgA-specific lymphoid cells of the mammary gland are derived from the same source. It is apparent from the results of Lascelles & McDowell

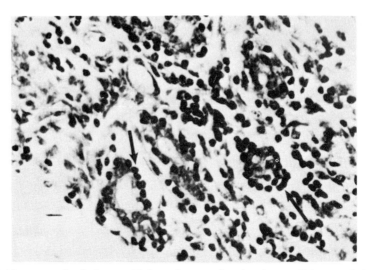

FIG. 6. Mammary gland of a ewe 32 days after weaning showing small ducts and alveolar remnants surrounded by a layer of lymphoid cells (arrows). Alcian blue methyl green-pyronin Y. ×585.

(1970) for the sheep that local instillation of antigen causes proliferation of this population of cells.

The occurrence of large numbers of lymphoid cells in human colostrum capable of producing antibody to the O-antigens of commonly occurring strains of *Escherichia coli* was recently reported by Ahlstedt, Carlsson, Hanson & Goldblum (1975). Up to 8% of colostral lymphoid cells were shown to produce antibody of this specificity. Using a sensitive immunoglobulin class-specific plaque assay, these workers found that most of the reactive cells were IgA-specific. In this context it is of interest to mention that recent observations of Husband & Gowans (pers. comm.) suggest that initial antigenic stimulation of IgA antibody-producing cells occurs in the Peyer's patches from whence cells, after entering the circulation, ultimately seed the intestinal lamina propria, being retained in those areas harbouring specific antigen. The high frequency of colostral lymphoid cells forming antibodies to *Escherichia coli* suggests that these cells also arise in the gastrointestinal tract. In the functional sense, the homing of these cells into the gland ensures that milk is provided with antibody of appropriate specificity for protection of the gastrointestinal tract of the newborn.

MECHANISM OF TRANSFER OF IMMUNOGLOBULINS

Electron microscopic evidence, together with the fact that lactating mammary glands (unless allowed to become grossly distended with milk) are essentially impermeable to water-soluble molecules as small as lactose, suggests that there would be virtually no transfer of serum proteins between epithelial cells. It is evident therefore that transfer of proteins from interstitial fluid into mammary secretion occurs through the cell itself, presumably within the vesicles described by Bargmann & Welsch (1969) as originating from the basal border of the epithelial cell. Some years ago I proposed that selective transfer of IgG_1 into colostrum of ruminants required the presence of receptors on the basal or intercellular membrane of glandular epithelial cells (Lascelles, 1969, 1971). This proposal was different from the one offered by Brambell (1966, 1970) to account for selective transport of immunoglobulin across various membranes in laboratory rodents. Brambell suggested non-selective uptake of protein in vesicles and degradation of those protein molecules not attached to receptors for a given immunoglobulin class (Fig. 7).

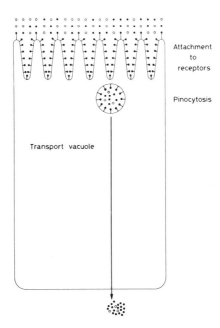

FIG. 7. Theoretical model for the selective transfer of IgG_1 (relative to IgG_2) across the glandular epithelium of the ruminant mammary gland. ●: IgG_1 molecule. ○: IgG_2 molecule.

An attempt was made to test the validity of the above hypothesis by examining alterations in the concentrations of IgG_1 and IgG_2 in blood serum of cows during colostrum formation, when large quantities of IgG_1 are known to be selectively transported from blood into mammary secretion (Brandon, Watson & Lascelles, 1971). It was predicted that if the Brambell degradation scheme is correct, serum IgG_1 and IgG_2 levels

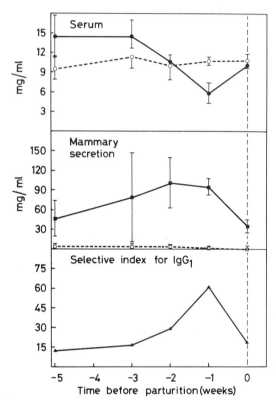

FIG. 8. Changes in concentration of IgG₁ and IgG₂ in blood serum and mammary secretion of cows before and after parturition. Values presented are means ± standard errors. Computation of the selective index for IgG₁ is described in the text, each plotted point being the mean value of selective indices for individual cows. ●———●: Concentration of IgG₁ in blood serum. ○---○: Concentration of IgG₂ in blood serum. ■———■: Concentration of IgG₁ in mammary secretion. □---□: Concentration of IgG₂ in mammary secretion. ▲———▲: Selective index for IgG₁.

would decrease in the same proportion whereas only IgG₁ would decrease if transfer were effected by the alternative mechanism. As a sharp fall (Fig. 8) in the concentration of IgG₁ in serum occurred in the absence of a simultaneous decrease in IgG₂ levels, it was concluded that the degradation hypothesis proposed by Brambell was invalid, at least for the mammary gland of the cow, and the data were fully consistent with the proposal involving the occurrence of surface receptors and an absence of degradation. However, the experiments of Brandon, Watson *et al.* (1971) are open to the criticism that the failure to observe a fall in level of plasma IgG₂ may have been due to an increased rate of synthesis of IgG₂ equal to the rate of degradation of this immunoglobulin. The final resolution of this aspect came from recent unpublished experiments in sheep by

Cripps, Fulkerson, Griffiths, McDowell & Lascelles (1976) using radio-tracer techniques to compare the metabolism of IgG_1 and IgG_2 in late pregnant and non-pregnant ewes. They reported similar rates of synthesis of IgG_1 and IgG_2 in both groups but the irreversible loss of IgG_1 was significantly greater than the loss of IgG_2 in pregnant ewes and IgG_1 in non-pregnant ewes.

An attempt was made to determine whether the glandular epithelium of the mammary gland has the capacity to selectively transfer IgA and IgM into secretion in a way similar to that described for IgG_1 (Watson & Lascelles, 1973). This study was undertaken by comparing the concentrations of the various immunoglobulins in simultaneously-collected samples of secretion and regional lymph. Experiments were carried out in ewes in which one of each pair of mammary glands was infused with antigen three weeks before parturition in order to stimulate the local immune system (IgA and IgM) in the gland. During the subsequent lactation, samples of milk, blood and lymph were collected and concentrations of immunoglobulins and serum albumin were measured (Fig. 9). The results illustrated clearly that the magnitude of preferential transfer was related to

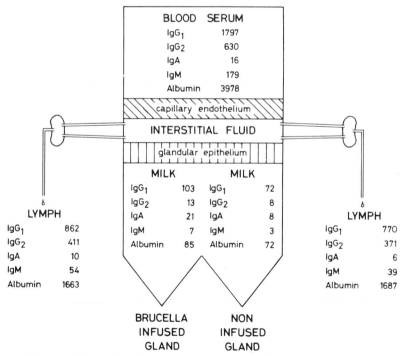

FIG. 9. Diagram of the udder of the ewe. It should be noted that the lymphatic drainage of the two sides of the udder is separate. Concentrations (mg/100 ml) of the various immunoglobulins and albumin in blood serum, lymph (representing interstitial fluid) and milk from *Brucella*-infused and non-infused glands are mean values for five ewes.

TABLE III

Whey : lymph ratios* for IgG₁, IgA and IgM, relative to those for IgG₂, for whey from immunized and non-immunized mammary glands of primiparous ewes

Gland	Whey : lymph ratio for		
	IgG_1	IgA	IgM
Immunized	3·12	81·43	4·86
Non-immunized	2·54	28·11	2·87

$$* \frac{(\text{Conc. of } IgG_1, IgA \text{ or } IgM \text{ in whey})}{(\text{Conc. of } IgG_2 \text{ in whey})} \times \frac{(\text{Conc. of } IgG_2 \text{ in lymph})}{(\text{Conc. of } IgG_1, IgA \text{ or } IgM \text{ in lymph})}.$$

the activity of the local immune system (see Table III). This was strikingly borne out by the results from two of the animals which were lambing for the first time. In these animals the whey : lymph ratios for IgA and IgM for non-immunized glands, in which the results indicated that there was virtually no local production of IgA and IgM, were considerably less than those for IgG_2 and albumin. It was concluded that local production of immunoglobulins close to the glandular epithelium created a concentration gradient with the highest concentration immediately adjacent to the basal membrane of the epithelial cells. In this situation preferential transfer of IgA and IgM across the glandular epithelium could occur without the necessity for specific receptor sites as postulated for the selective transfer of IgG_1.

FUNCTIONAL CONSIDERATIONS

While the importance of colostral antibody in the protection of the new-born of those species normally born without circulating antibody is quite clear, its significance in those species such as the rabbit and man, which acquire passive immunity before birth, is not entirely resolved. In these latter species IgA is predominant and the question arises as to whether secretory IgA ingested in the colostrum has a protective effect. At this point it is important to stress that protein is not absorbed from the gut of new-born human infants and rabbits (Lascelles, 1963) and thus any beneficial effect of the IgA would be manifested in the gut lumen itself. While there is considerable evidence that secretory IgA antibody possesses virus neutralization properties in the absence of complement, the significance of this class of antibody in protection against bacterial diseases is less clear. It was reported by Adinolfi, Glynn, Lindsay & Milne (1966) that human colostral IgA in the presence of lysosome and complement was bactericidal for *Escherichia coli*: these results were largely substantiated in a more recent study by Burdon (1973). It is of interest that similar properties have been attributed to secretory IgA derived from colostrum (Hill & Porter, 1974). On the other hand, Eddie, Schulkind & Robbins (1971) reported slight or no complement-dependent activity for secretory IgA from rabbit colostrum in bactericidal tests with *Salmonella typhimurium* and that lysosome failed to potentiate activity.

In an important series of studies Rowley and his colleagues (Steele, Chaicumpa & Rowley, 1974; Bellamy, Knop, Steele, Chaicumpa & Rowley, 1975; Steele, Chaicumpa & Rowley, 1975) have demonstrated the protective effect of secretory IgA antibody derived from colostrum of immunized rabbits, as well as serum-derived IgG and IgM antibody, against *Vibrio cholerae* in infant mice. The evidence indicated that agglutination and the prevention of adherence of *Vibrio cholerae* to epithelial cells were significantly correlated with protection. It was postulated that the agglutinated, non-adherent organisms were readily eliminated from the small intestine by peristaltic action. These results are generally in accord with earlier results of Freter (1969) and Williams & Gibbons (1972).

Finally, mention is made of the various attempts to immunize mammary glands of domesticated ruminants against bacterial mastitis. Local infusion of staphylococcal vaccines into mammary glands of sheep, goats and cows during the dry period a few weeks before parturition confers some degree of protection against challenge with virulent staphylococci (see Lascelles & McDowell, 1974). Systemic immunization gives a reasonably high level of protection only when live vaccines are administered (Derbyshire, 1962) and in recent studies Watson (1976) has drawn attention to the correlation between abscess formation at the site of vaccination and the level of protection on challenge. This worker has also presented strong evidence that antibody belonging to the IgG_2 sub-class plays an important role in protection probably by enhancing phagocytosis by polymorphs to which it is cytophilic.

REFERENCES

Aalund, O. (1968). *Heterogeneity of ruminant immunoglobulins*. Thesis: Copenhagen.

Adinolfi, M., Glynn, A. A., Lindsay, M. & Milne, C. M. (1966). Serological properties of γA antibodies to *Escherichia coli* present in human colostrum. *Immunology* 10: 517–526.

Ahlstedt, S., Carlsson, B., Hanson, L. A. & Goldblum, R. M. (1975). Antibody production by human colostral cells. 1. Immunoglobulin class specificity and quantity. *Scand. J. Immun.* 4: 535–539.

Askonas, B. A., Campbell, P. N., Humphrey, J. H. & Work, T. S. (1954). The source of antibody globulin in rabbit milk and goat colostrum. *Biochem. J.* 56: 597–601.

Bargmann, W. & Welsch, U. (1969). On the ultrastructure of the mammary gland. In *Lactogenesis: the initiation of milk secretion at parturition*: 43–52. Reynolds, M. & Folley, S. J. (eds). Philadelphia: University of Pennsylvania Press.

Batty, I. & Warrack, G. H. (1955). Local antibody production in the mammary gland, spleen, uterus, vagina and appendix of the rabbit. *J. Path. Bact.* 70: 355–363.

Bellamy, J. E. C., Knop, J., Steele, E. J., Chaicumpa, W. & Rowley, D. (1975). Antibody cross-linking as a factor in immunity to cholera in infant mice. *J. infect. Dis.* **132**: 181–188.

Blakemore, F. & Garner, R. J. (1956). The maternal transferrence of antibodies in the bovine. *J. comp. Path.* **66**: 287–289.

Brambell, F. W. R. (1966). The transmission of immunity from mother to young and the catabolism of immunoglobulins. *Lancet* **1966 (ii)**: 1087–1093.

Brambell, F. W. R. (1970). *The transmission of passive immunity from mother to young.* Amsterdam: North-Holland.

Brandon, M. R., Husband, A. J. & Lascelles, A. K. (1975). The effect of glucocorticoid on immunoglobulin secretion into colostrum in cows. *Aust. J. exp. Biol. med. Sci.* **53**: 43–48.

Brandon, M. R. & Lascelles, A. K. (1975). The effect of pre-partum milking on the transfer of immunoglobulin into mammary secretion of cows. *Aust. J. exp. Biol. med. Sci.* **53**: 197–204.

Brandon, M. R., Watson, D. L. & Lascelles, A. K. (1971). The mechanism of transfer of immunoglobulin into mammary secretion of cows. *Aust. J. exp. Biol. med. Sci.* **49**: 613–623.

Burdon, D. W. (1973). The bactericidal action of immunoglobulin A. *J. med. Microbiol.* **6**: 131–139.

Campbell, B., Porter, R. M. & Petersen, W. E. (1950). Plasmacytosis of the bovine udder during colostrum secretion and experimental cessation of milking. *Nature, Lond.* **166**: 913.

Carroll, E. J. (1961). Whey proteins of drying-off secretions, mastitic milk, and colostrum separated by ion-exchange cellulose. *J. Dairy Sci.* **44**: 2194–2211.

Carroll, E. J., Murphy, F. A. & Aalund, O. (1965). Changes in whey proteins between drying and colostrum formation. *J. Dairy Sci.* **48**: 1246–1249.

Cowie, A. T. (1961). The hormonal control of milk secretion. In *Milk: the mammary gland and its secretion* **1**: 163–203. Kon, S. K. & Cowie, A. T. (eds). New York and London: Academic Press.

Craig, S. W. & Cebra, J. J. (1971). Peyer's patches: An enriched source of precursors for IgA-producing immunocytes in the rabbit. *J. exp. Med.* **134**: 188–200.

Cripps, A. W., Fulkerson, W. J., Griffiths, D. A., McDowell, G. H. & Lascelles, A. K. (1976). The relationship between the transfer of immunoglobulins, sodium and potassium into mammary secretion of the parturient ewe. *Aust. J. exp. Biol. med. Sci.* **54**: 337–348.

Derbyshire, J. B. (1962). Immunity in bovine mastitis. *Vet. Bull.* **32**: 1–10.

Dixon, F. J., Weigle, W. O. & Vazquez, J. J. (1961). Metabolism and mammary secretion of serum proteins in the cow. *Lab. Invest.* **10**: 216-237.

Eddie, D. S., Schulkind, M. L. & Robbins, J. B. (1971). The isolation and biologic activities of purified secretory IgA and IgG anti-*Salmonella typhimurium* 'O' antibodies from rabbit intestinal fluid and colostrum. *J. Immunol.* **106**: 181–190.

Feinstein, A. & Hobart, M. J. (1969). Structural relationship and complement fixing activity of sheep and other ruminant immunoglobulin G subclasses. *Nature, Lond.* **223**: 950–952.

Freter, R. (1969). Studies of the mechanism of action of intestinal antibody in experimental cholera. *Texas Rep. Biol. Med.* **27** (Suppl. 1): 299–316.

Garner, R. J. & Crawley, W. (1958). Further observations on the maternal transference of antibodies in the bovine. *J. comp. Path.* **68**: 112–114.

Harmon, R. J., Schanbacher, F. L., Ferguson, L. C. & Smith, K. L. (1976). Changes in lactoferrin, immunoglobulin G, bovine serum albumin, and α-lactalbumin during acute experimental and natural coliform mastitis in cows. *Infect. Immun.* **13**: 533–542.

Heremans, J. F. (1959). Immunochemical studies on protein pathology: the immunoglobulin concept. *Clin. chim. acta* **4**: 639–646.

Hill, I. R. & Porter, P. (1974). Studies of bactericidal activity to *Escherichia coli* of porcine serum and colostral immunoglobulins and the role of lysozyme with secretory IgA. *Immunology* **26**: 1239–1250.

Kerr, W. R., Pearson, J. K. L. & Rankin, J. E. F. (1959). The bovine udder and its agglutinins. *Br. vet. J.* **115**: 105–119.

Larson, B. L. & Kendall, K. A. (1957). Changes in specific blood serum protein levels associated with parturition in the bovine. *J. Dairy Sci.* **40**: 659–666.

Lascelles, A. K. (1963). A review of the literature on some aspects of immune milk. *Dairy Sci. Abstr.* **25**: 359.

Lascelles, A. K. (1969). Immunoglobulin secretion into ruminant colostrum. In *Lactogenesis: the initiation of milk secretion at parturition*: 131–143. Reynolds, M. & Folley, S. J. (eds). Philadelphia: University of Pennsylvania Press.

Lascelles, A. K. (1971). Mechanisms of milk synthesis and secretion. *Proc. 18th Int. Dairy Congress, Sydney* **2**: 514–524.

Lascelles, A. K., Gurner, B. W. & Coombs, R. R. A. (1969). Some properties of human colostral cells. *Aust. J. exp. Biol. med. Sci.* **47**: 349–360.

Lascelles, A. K. & Lee, C. S. (in press). Involution of the mammary gland. In *Lactation* 4. Larson, B. L. & Smith, V. R. (eds). New York and London: Academic Press.

Lascelles, A. K. & McDowell, G. H. (1970). Secretion of IgA in the sheep following local antigenic stimulation. *Immunology* **19**: 613–620.

Lascelles, A. K. & McDowell, G. H. (1974). Localized humoral immunity with particular reference to ruminants. *Transplant Rev.* **19**: 170–208.

Lascelles, A. K., Outteridge, P. M. & Mackenzie, D. D. S. (1966). Local production of antibody by the lactating mammary gland following antigenic stimulation. *Aust. J. exp. Biol. med. Sci.* **44**: 169–180.

Lee, C. S. & Lascelles, A. K. (1969). The histological changes in involuting mammary glands of ewes in relation to the local allergic response. *Aust. J. exp. Biol. med. Sci.* **47**: 613–623.

Lee, C. S., McDowell, G. H. & Lascelles, A. K. (1969). The importance of macrophages in the removal of fat from the involuting mammary gland. *Res. vet. Sci.* **10**: 34–38.

Little, R. B. & Orcutt, M. L. (1922). The transmission of agglutinins of *Bacillus abortus* from cow to calf in the colostrum. *J. exp. Med.* **35**: 161–171.

Mackenzie, D. D. S. & Lascelles, A. K. (1968). The transfer of [^{131}I]-labelled immunoglobulins and serum albumin from blood into milk of lactating ewes. *Aust. J. exp. Biol. med. Sci.* **46**: 285–294.

McDowell, G. H. (1973). Local antigenic stimulation of guinea-pig mammary gland. *Aust. J. exp. Biol. med. Sci.* **51**: 237–245.

McDowell, G. H. & Lascelles, A. K. (1969). Local production of antibody by ovine mammary glands infused with salmonella flagellar antigens. *Aust. J. exp. Biol. med. Sci.* **47**: 669–678.

McDowell, G. H. & Lascelles, A. K. (1971). Local production of antibody by the lactating mammary gland of the ewe and the effect of systemic immunization. *Res. vet. Sci.* **12**: 113–118.

Murphy, F. A., Aalund, O., Osebold, J. W. & Carroll, E. J. (1964). Gamma globulins of bovine lacteal secretions. *Arch. Biochem. Biophys.* **108**: 230–239.

Outteridge, P. M., Mackenzie, D. D. S. & Lascelles, A. K. (1968). The distribution of specific antibody among the immunoglobulins in whey from the locally immunized gland. *Arch. Biochem. Biophys.* **126**: 105–110.

Outteridge, P. M., Rock, J. D. & Lascelles, A. K. (1965). The immune response of the mammary gland and regional lymph node following antigenic stimulation. *Aust. J. exp. Biol. med. Sci.* **43**: 265–274.

Pierce, A. E. (1959). Specific antibodies at mucous surfaces. *Vet. Rev. Annot.* **5**: 17–36.

Pierce, A. E. & Feinstein, A. (1965). Biophysical and immunological studies on bovine immune globulins with evidence for selective transport within the mammary gland from maternal plasma to colostrum. *Immunology* **8**: 106–123.

Plommet, M. (1968). Origine des anticorps du lait. *Annls Biol. amim. Biochim. Biophys.* **8**: 407–417.

Porter, R. M. (1968). Anti-O chaining titre in colostral serum from cows infused prepartum with *Salmonella pullorum*—H. *J. Dairy Res.* **35**: 7–12.

Richards, C. B. & Marrack, J. R. (1963). Sheep serum γ-globulin. *Protides biol. Fluids* **10**: 154–156.

Smith, E. L. (1946). The immune proteins of bovine colostrum and plasma. *J. biol. Chem.* **164**: 345–358.

Smith, K. L., Muir, L. A., Ferguson, L. C. & Conrad, H. R. (1971). Selective transport of IgG_1 into the mammary gland: role of estrogen and progesterone. *J. Dairy Sci.* **54**: 1886–1894.

Steele, E. J., Chaicumpa, W. & Rowley, D. (1974). Isolation and biological properties of three classes of rabbit antibody to *Vibrio cholerae*. *J. infect. Dis.* **130**: 93–103.

Steele, E. J., Chaicumpa, W. & Rowley, D. (1975). Further evidence for cross-linking as a protective factor in experimental cholera: properties of antibody fragments. *J. infect. Dis.* **132**: 175–180.

Watson, D. L. (1976). The effect of cytophilic IgG_2 on phagocytosis by ovine polymorphonuclear leucocytes. *Immunology* **31**: 159–165.

Watson, D. L., Brandon, M. R. & Lascelles, A. K. (1972). Concentrations of immunoglobulin in mammary secretion of ruminants during involution with particular reference to selective transfer of IgG_1. *Aust. J. exp. Biol med. Sci.* **50**: 535–539.

Watson, D. L. & Lascelles, A. K. (1973). Mechanisms of transfer of immunoglobulins into mammary secretion of ewes. *Aust. J. exp. Biol. med. Sci.* **51**: 247–254.

Williams, R. C. & Gibbons, R. J. (1972). Inhibition of bacterial adherence by secretory immunoglobulin A: a mechanism of antigen disposal. *Science, N.Y.* **177**: 697–699.

Wilson, M. R. (1972). The influence of preparturient intramammary vaccination on bovine mammary secretions. Antibody activity and protective value against *Escherichia coli* enteric infections. *Immunology* **23**: 947–955.

Wilson, M. R., Duncan, J. R., Heistand, F. & Brown, P. (1972). The influence of pre-parturient intramammary vaccination on immunoglobulin levels in bovine mammary secretions. *Immunology* **23**: 313–320.

Symp. zool. Soc. Lond. (1977) No. 41, 261–276

A Comparative Study of Plasma Cells in the Mammary Gland in Pregnancy and Lactation

R. S. H. PUMPHREY

Department of Bacteriology, Royal Infirmary, Glasgow, Scotland

SYNOPSIS

In the past there has been controversy about how much of the immunoglobulin in milk is produced locally in the mammary gland, and how much transported from the serum. It was clear that there must be species differences, but there was no systematic information on this point. There is overwhelming indirect evidence to suggest that all local synthesis of immunoglobulin must occur in plasma cells.

Mammary tissue from about 600 individuals from 60 species has been examined for the number and location of plasma cells during lactation. Certain patterns emerge, and are discussed. It will be suggested that there are at least three different reasons for plasma cells collecting in the mammary gland, which can be distinguished by their location and the other cells accompanying them. These causes are as follows.

(a) In all species, as a natural consequence of involution. The plasma cells are accompanied by lymphocytes and histiocytes, and disappear once the gland reaches its resting state.

(b) In some individuals of all species, following local reaction to local antigen, usually as a consequence of the resolution of micro-abscesses. This leads to local dense lymphoid collections, usually with comparatively few plasma cells, and usually not in the centrilobular position.

(c) In about a third of the species examined, during active lactation (or pregnancy and lactation), there seems to be a special mechanism for trapping cell precursors, often preferentially in the centrilobular zone. There is indirect evidence that these are cells which outside lactation would have homed to the gut, but this has so far proved difficult to demonstrate directly.

INTRODUCTION

Milk is a major route for passive immunization of newborn mammals. The kind of immunity transferred, however, varies greatly from one species to another. It is now clear that there are two categories of passive humoral immunity transferred, the mucosal and the systemic.

During the evolution of the vertebrates, the primitive immunoglobulin molecule has undergone its own form of adaptive radiation, with selection of successful variants for protection of the different compartments of the body. By the time of the basal mammalian stock, there was a variety of heavy chains of different sizes and properties. Besides a modified version of the primitive μ-chain, there were γ- and α-chains.

Thus the role of IgM was being taken over by IgG in the protection of interstitial tissue spaces, and by IgA in the protection of mucosae.

Lactation evolved before placentation in mammals. While it is likely that the yolk sac of the monotremes contains maternal IgG, as is typical of reptiles and birds, it is highly probable the IgA is transferred in the milk. With the adoption of viviparity and the microlecithal egg, the newborn mammal was left in a parlous state, with no maternal IgG. The scanty evidence suggests that little IgG is normally transferred in the milk in marsupials, and that IgA is the major whey protein (Bell, Stephens & Turner, 1974).

During the eutherian radiation, different solutions were found to this problem. Several groups independently devised means of transferring maternal IgG to the fetus *in utero*, either via the haemochorial placenta of higher primates, or via a modification of the yolk sac, as in rabbits (Brambell, 1970). An alternative route came to be used by other groups, which managed to transfer maternal serum IgG to their milk, and again from the milk in the gut of the neonate into the circulation. The principal exponents of this are found in the carnivores, horses and ruminants. While these methods were evolving for the transfer of IgG, in most groups the old secretory immune system persisted for IgA, with local plasma cells in the mammary gland secreting antibodies which protected the gut of the young, but which were not absorbed. The few studies of local synthesis of immunoglobulins by the mammary gland show that IgA is, indeed, the major class produced, though smaller amounts of IgG and IgM can be detected (Hochwald, Jacobson & Thorbecke, 1964; Bourne & Curtis, 1973; Drife, McClelland, Pryde, Roberts & Smith, 1976). Similarly, immunofluorescent and immunoperoxidase staining show that IgA is the predominant immunoglobulin in the plasma cells in the healthy mammary gland.

The source of these plasma cells has not been definitely shown. Circumstantial evidence, however, indicates that they represent a population which normally homes to the gut, but which in lactation is diverted to the mammary glands (McClelland, Samson, Parkin & Shearman, 1972; Kenney, Boesman & Michaels, 1967; Bohl, Gupta, Olquin & Saif, 1972).

This investigation was undertaken to see if there was any general pattern in the mammary plasma cell populations during lactation in different groups of mammals, as a first step towards following the evolutionary history of transfer of passive mucosal immunity from mother to young, and as part of a study to try to understand the mechanism of trapping of plasma cell precursors by the mammary gland.

THE PLASMA CELLS

Plasma cells are the short-lived end product of B-lymphocyte differentiation. Each one secretes antibody of a single class and single antigen binding specificity. A large force of immunological manpower has been directed over the last few years to understanding the stimuli which make B cells mature along the various paths open to them, and a fair amount is known about the ways in which antigen can induce cellular co-operation between phagocytes, T cells and B cells. However, little is known as yet about the triggers which cause the B cells to select which class of immunoglobulin is to be synthesized. We can observe that most of the plasma cells in the lamina propria of the gut, and in the bronchi and lachrymal and salivary glands, are IgA or IgM-secreting, whereas everywhere else they tend to be IgG or IgM. Cell-tracing studies have indicated that a complex sequence of interactions may be necessary to induce the formation of IgA cells, with at least one stage of circulation via the thoracic duct and blood stream between induction by antigen in the gut, and secretion of IgA by the plasma cell in the lamina propria (Gowans & Knight, 1964).

The resting mammary gland has in many species a resident population of plasma cells, which, in man at any rate, are mostly IgA secreting (Drife *et al.*, 1976). Local infection or cancer may change this population towards a majority of IgG cells (McClelland, Roberts & Pryde, pers. comm.) (though there is evidence that in some animals, with some antigens, at some stages of the lactation cycle, the response to local challenge may be predominantly IgA — Bohl *et al.*, 1972; Lee & Lascelles, 1970; Genco & Taubman, 1969; Norcross, 1971).

Plasma cells are highly specialized. They are devoted to rapid synthesis of a single protein, their own particular antibody. As a result of this specialization they have a characteristic morphology, which means that they can easily be recognized by light or electron microscopy. Most methods of fixation lead to peripheral clumps of chromatin in the nucleus, and the uniform intense staining of the cytoplasm of pyronin, or fainter staining by haematoxylin, means that they can readily be identified in the mammary gland of any species. More sophisticated staining techniques, such as immunoperoxidase or immunofluorescence, enable one to see which class of immunoglobulin is being secreted. These techniques, however, need highly specific antisera (which could not be obtained for the majority of species studied here), and may also require special fixation of the tissue (which would have made it much more difficult to obtain the specimens). I have therefore confined my immunoperoxidase studies to human mammary tissue, as the technique is in routine clinical use, and the specificity of the staining well established (Burns, 1975).

THE COMPARATIVE SURVEY

A brief summary of the data collected is presented in Table I.

TABLE I

Summary of comparative data

	Numbers examined				Findings			
	R	P	L	W	R	P	L	W
Monotremata								
Echidna			2				+ +	
Ornithorhynchus anatinus			1				+	
Marsupialia								
Trichosurus vulpecula			5				+	
Perameles nasuta			3				+ +	
Vombatus hirsutus			1				+	
Macropus rufogriseus	1		3				+	
Wallabia bicolor				1				+ + +
Setonix brachyurus			3				+	
Megaleia rufa			1				(+)	
Insectivora								
Sorex araneus		2	1	1		−	−	(+ + +)*
Talpa europaea			1				−	
Chiroptera								
Pipistrellus pipistrellus		2	2			−	−	
Hipposideros diademata			1				−	
Cheiromeles torquatus			1				−	
Macroglossus lagochilus			1				−	
Cynopterus brachyotis	1	2	1		−	−	−	
Tupaiidae								
Tupaia minor			1				−	
Tupaia belangeri	1					−		
Lyonogale tana			1				−	
Primates								
Aotus	1				(+)			
Macaca mulatta		24	8			+	+	
Homo sapiens	12	5	30	3	+	+	+ + +	+ +
Lagomorpha								
Oryctolagus cuniculus	2	6	16	2	+	+	+ + + +	+ +
Rodentia								
Sciurus carolinensis	3	2	3		(−)	(+)	−	
Mus musculus	5	15	45	30	(−)	(−)	(+)	+ +
Rattus norvegicus	4	6	20	11	(−)	(−)	(+)	+ +
Cricetomys auratus	2	2	5		(−)	(+)	(+) + +	
Proechimys guairae	2	1	12	1	−	+	+ + +	+ +
Cavia porcellus		3	5	3		+ +	+ +	+ + +
Galea musteloides	3	2	8	3	(+)	+	+ + + + +	+ + +
Cetacea								
Physeter catodon	1	1	5	2	(+)	+ + +	+ + +	+ + +
Tursiops truncatus				1				+

continued

TABLE I—*continued*

	Numbers examined				Findings			
	R	P	L	W	R	P	L	W
Tursiops aduncus			1				+++	
Balaenoptera musculus	9	9	5	1	(+)	++	+++	
Balaenoptera physalis	20	13	10	2	(+)	+++	+++	+
Balaenoptera borealis	7	1	1		(+)	++	++	
Carnivora								
Felis catus		1	6	3	−		(−)	+
Canis familiaris	2	5	9	8	(+)	−	−.++	++.(+)
Mustela vison		1	3	1		−	(−)	+
Mustela putorius furo	1	3	3		(+)	(+)	(−)	
Halichoerus grypus		3	21			++	++++	
Phoca vitulina			4	1			+++	++
Otaria flavescens			1				−	
Arctocephalus australis			1				(−)	
Arctocephalus gazella			2				−	
Perissodactyla								
Equus caballus	1	2	7		+	(−)	−.+	
Artiodactyla								
Sus scrofa		1	1		+		++++	
Muntiacus reevesi			2				−	
Cervus elaphus			1				−	
Redunca fulvorufula			3	2			(+)	+
Damaliscus albifrons				1				+
Addax nasomaculata			1				(−)	
Gazella subguttorosa			1				−	
Ovis aries	5	11	10	4	++	+.−	−	++
Capra aegragus		1				(+++)*		
Capra hircus	1	7	5	1	++	+.−	−	++
Capra falconeri			1				−	
Syncerus caffer			1				(+)	
Bos taurus			6				(+++)*	

R = resting; P = pregnant; L = lactating; W = weaning.

Plasma cells per 1000 alveolar cells
− ≤10
(−) 11–20
(ǀ) 21–50
+ 51–100
++ 100–200
+++ 200–300
++++ 300–500
+++++ >500
−.+ Change during one stage of lactation.
* Evidence of mastitis.

Numbers

For many of the species only one or two specimens of lactating tissue have been obtained, so the question arises, how representative are these of their species? As far as absolute numbers of plasma cells are concerned, this question is unanswerable without obtaining more specimens, when a statistical analysis would be possible. However, there are two reasons why I think that even single specimens give valuable information.

(a) In species with a moderate average number of plasma cells (i.e. 50–200 per 1000 alveolar cells) there seems to be a very wide range of numbers when 10 or more individuals are examined, but in those species where there are very low numbers (i.e. less than 10 per 1000 alveolar cells) the range is much smaller, and large numbers of plasma cells are only seen where there is frank evidence of mastitis. Many of the species for which only one lactating individual has been examined are of this type, with very few plasma cells.

(b) Even where there is only one specimen per species, there are many lobules in the specimen, and if there is a clear *pattern* of plasma cell distribution in enough of these, it is likely to be representative of the species, even though the actual number may be far from the mean of the species.

Other Interstitial Cells

Table I only shows the numbers of plasma cells. In fact, other cell types were also enumerated, including lymphocytes, histiocytes, mast cells and neutrophil polymorphs. While mast cells and neutrophils were easy to identify and count in all species, histiocytes and lymphocytes were more difficult, particularly when they were squeezed in between distended alveoli. In localized mastitis all the elements of the immune response could be seen, with histiocytes, lymphocytes, plasma cells and polymorphs. If the infection had persisted long enough in one place, the chronic inflammatory elements had organized to form lymphoid nodules, sometimes with germinal centres. In and around these collections plasma cells may be seen but, in man at least, these are mostly IgG-secreting cells.

Plasma cells are not mobile but their immediate precursors, lympho-blasts, are (Wilkinson, Russell, Pumphrey, Sless & Parrott, 1976). These probably have a very short life once they have entered the tissues before they turn into mature plasma cells, but a certain amount of cell division does occur, and occasionally binucleate plasma cells are seen.

During the weaning period, especially if there was a sudden stop to lactation, there was a marked ingress of plasma and other cells. Histio-cytes and mast cells were prominent at this stage, and in some species (dogs and mice, for example), the mast cells were particularly numerous in middle to late involution. Apart from this, there was, in most species, an almost complete absence of polymorphs from the interalveolar spaces during pregnancy and lactation. In one or two species, for example the rhesus monkey, there was a constant low background of widely scattered polymorphs throughout pregnancy and lactation, without any sign of mastitis.

The Plasma Cells of Lactation

It is well known that some species, such as pigs, man and rabbits, have many plasma cells in the mammary gland in lactation, whereas others, such as sheep and goats (Lascelles, 1963) have very few. I have not, however, seen any account of the distribution of these cells, either within the gland as a whole, or within each lobule.

Distribution throughout the gland

In the guinea-pigs I have examined there was a very obvious difference between the centre of the gland, where the alveoli were tightly packed and almost tubular in shape, and the outside, where the alveoli were spherical and less tightly packed. The plasma cells were almost entirely confined to this outer region.

In the reedbucks, where I had complete udders, there were lymph nodes at the base of the gland, and for some way into the substance of the gland there were islands of lymphoid tissue, which looked very much as if the gland had grown into the lymphoid tissue as it developed. It may be that with some of the dense lymphoid collections seen in other species where I only had a small piece of gland, the collections arose in this way, for example in the fur seal, *Arctocephalus australis* (Fig. 1). In other species,

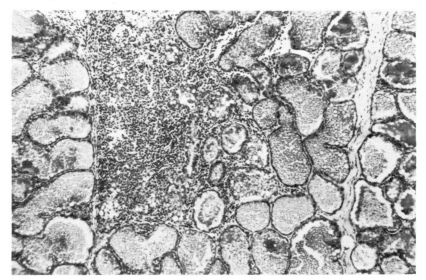

FIG. 1. A dense lymphoid collection within a lobule of the mammary gland of *Arctocephalus australis*.

such as the Nubian goat, *Capra aegragus*, the lymphoid tissue appeared to have replaced some of the gland, suggesting organized chronic inflammation. These lymphoid collections persist even when the gland involutes, so that next lactation, when the gland grows out again, islands of lymphoid tissue are seen in healthy looking lobules. This provides an alternative explanation for the appearance in Fig. 1.

Distribution within the lobules

In the cuis, *Galea musteloides*, the plasma cells are so numerous that they appear to be present in equal numbers throughout the lobule (Fig. 2).

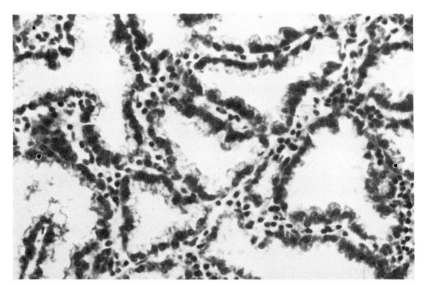

FIG. 2. The uniform high number of plasma cells in the mammary gland of *Galea musteloides*.

This is also true of the rabbit, and most of the whales (Fig. 3). These cells are almost pure plasma cells, with a few lymphoid cells, but no signs of polymorphs or histiocytes, which makes it most unlikely that they are there as the result of inflammation.

In man, on the other hand, the plasma cells are usually accompanied by equal numbers of lymphoid cells (but no polymorphs or histiocytes), and are very clearly localized to the centrilobular region and around the collecting ductule where the blood drains from the lobule. The plasma cells stain specifically for IgA (Fig. 4), with perhaps one-tenth of the number of IgM-staining cells, and very few IgG. This localization may be so complete that lobules that are not favourably sectioned may appear to have no lymphoid or plasma cells. In other lobules, while the lymphoid cells are still largely located round the collecting ductule, the plasma cells are found throughout the lobule. While this centrilobular pattern is most

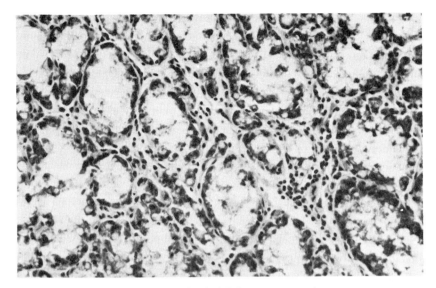

FIG. 3. Plasma cells in the mammary gland of *Balaenoptera musculus.*

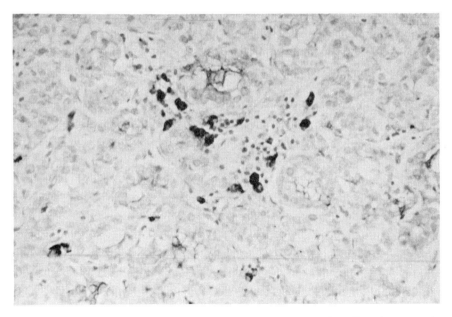

FIG. 4. Human lactating mammary gland, with a centrilobular collection of lymphocytes and plasma cells. The plasma cells are darkly stained by the immunoperoxidase technique, using a specific anti-human IgA. The control, IgM and IgG sections of this same area showed no stained cells.

noticeable in man, it is also seen in many other species, including rhesus monkeys, dogs, seals, pigs, Cape buffalo and whales, where it is clearest in pregnancy, when the numbers of plasma cells are lower.

There are several possible explanations for this centrilobular distribution.

(a) There might be a specialized vascular structure for attracting the plasma cell precursors out of the circulation in this region, analogous to the post-capillary venule which induces T-lymphocytes to emigrate from the blood to the paracortical region of the lymph node (Gowans & Knight, 1964; Farr & de Bruyn, 1975). Certainly it is true that the plasma cells are found by the post-capillary venule in the mammary gland, but the endothelium of these vessels has, by light microscopy at least, a quite normal appearance.

(b) The collecting ductule might secrete a specific chemoattractant. If this were the only region secreting the attractant, it is difficult to understand how the plasma cells are often found out to the periphery of the lobule when so few of the mobile plasma cell precursors are to be seen.

(c) All the alveolar epithelial cells might secrete a specific chemoattractant, which, because of the arrangement of the blood supply, reaches its highest concentration in the centrilobular zone. This would explain how, if the rate of secretion of the attractant is higher, the plasma cells can be found further out towards the periphery. In the cuis and the whales the level might be supposed to be high enough to achieve optimal migration right out to the edge of the lobule, causing the uniform high number throughout the lobule.

Of these three, I favour the third, which gains some additional support from observations in mice.

Mice have rather few plasma cells in the mammary gland during mid-lactation. The few that they have (around 20 per 1000 alveolar cells) are mainly found in little collections round capillaries or venules. The lobules in the mouse mammary gland are small, and the blood supply is not so clearly arranged with a centrilobular drainage as in man, so these collections may not be in the centre of the lobules. If the drainage of milk from one of the glands is obstructed, while the young are left with their mother to maintain the lactation, milk secretion continues in the blocked gland, and it becomes distended. By the end of the first four hours, there is a clear increase in the number of plasma cells, which double their numbers by 12 hours. Beyond 48 hours, when the number of plasma cells is up to 10 times the starting level, involutional changes begin to appear, with frank leakage of milk into the tissues where it is quickly taken up by histiocytes. Polymorphs also start to appear. A reasonable explanation of this might be that the retention of the milk leads to an increased level of the chemoattractant in the venular blood. There is an alternative explanation which, on the face of it, might seem more attractive. Murillo (1971) showed that the cells from human colostrum could incorporate labelled amino acids into IgA. Perhaps when the duct is blocked plasma cells no longer move out into the milk. However, plasma cells are never seen in the milk, or within the alveoli in these mice, so I feel that the first explanation is the more likely.

The phases of lactation

In about a third of the species studied there was enough material from different stages of lactation to be able to understand something of the timing of the appearance of the plasma cells in lactation. There were clear differences; for example, in man the number of plasma cells is maximal around the time of birth, whereas in the series of beagle bitches studied, at this stage there were virtually no plasma cells to be seen. The different patterns seen are shown in Fig. 5. One feature common to all species was

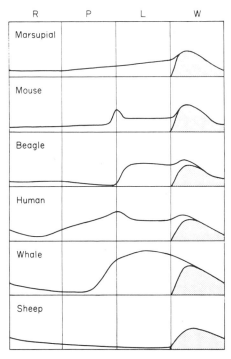

FIG. 5. The abscissa shows the phase of lactation: R = resting, P = pregnant, L = lactating, W = weaning. The ordinate indicates the relative number of plasma cells in the mammary gland of each species.

the presence of plasma cells in the involuting gland. They were always accompanied by histiocytes and polymorphs, and were scattered throughout the lobule, rather than being centrilobular.

In the resting gland there may or may not be any plasma cells, and to a certain extent this may be due to the persistence of the cell population after involution. Unfortunately, I have no way of knowing how many cycles of lactation most of my specimens had been through, but from the few that I do know, I get the impression that the more cycles, the more plasma cells.

In man there is known to be a cyclical variation in the numbers of plasma cells with the menstrual cycle. I have not been able to assess whether a similar change takes place after oestrus in any of these animals, apart from pseudopregnant rabbits (after HCG injection), when the number of plasma cells appears to be the same as in truly pregnant rabbits with the same stage of mammary development.

The phylogenetic sequence

A few plasma cells are normally found around apocrine sweat glands. The monotremes and marsupials examined have low numbers of plasma cells in their mammary glands; some, for example *Perameles nasuta*, have rather more than others, for instance *Megaleia rufa*, but none shows the clear centrilobular pattern that is seen in many of the eutherians. It is not possible to say more than that this *might* represent the primitive condition. It was therefore surprising to find that the insectivores, tree-shrews and bats had glands devoid of plasma cells. Perhaps it will turn out that those species examined are, by chance, atypical and that the other more primitive eutherians have the same number as the marsupials.

Among the rest of the Eutheria there seems little consistency in the numbers of plasma cells, but perhaps slightly more in the pattern of numbers according to the phase of lactation. A striking feature is the absence of plasma cells in most of the ruminants, except the cow (but I have seen no cow tissue that does not show signs of mastitis) (Fig. 6).

A very much larger number of species will have to be examined before any clear evolutionary sequence can be deduced.

Behavioural correlations

From the mouse experiments described on p. 270, it might have been supposed that those species with a long interval between periods of nursing might have more plasma cells. The rabbit, with its daily nursing and plentiful plasma cells might have lent support to this idea. However, when the tree-shrews, with their 48- or 72-hour cycle, and the fur seals with their weekly nursing were examined, the reverse was true. Certainly the fur seals have far fewer plasma cells than the common or grey seals. These species must have a special mechanism to prevent involution, which in most mammals would be rapid if they were suckled so infrequently. Maybe this same mechanism prevents the plasma cells from homing to the mammary gland.

Another possible correlation could be made by comparing the precocious with the altricial mammals. Within the rodents, the precocious seem to have more plasma cells than the altricial, but the ruminants, all precocious, have low numbers. It is very possible that IgA is playing a

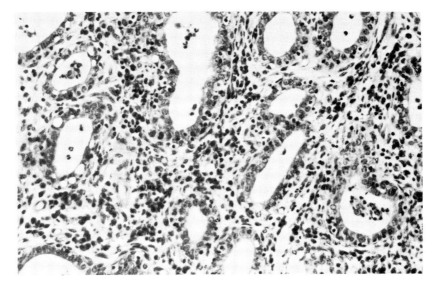

FIG. 6. Mastitis in cow mammary gland. There are numerous polymorphs, lymphocytes and plasma cells between the alveoli, and polymorphs and macrophages in the lumen.

different role in the precocious animals; being directed towards food antigens as well as bacterial, it may play an important part in preventing the establishment of gut allergies, whereas in the altricial the antibacterial role is likely to be the most important. It may be important that the mother commonly eats the faeces of the nestlings in altricial mammals. Her gut immune system is much more mature than theirs, and so in this way she can respond to their gut bacteria, and provide them with passive mucosal immunity.

Variation between individuals

In the series of human and grey seal lactating glands the variation between individuals was very striking. If they really constitute an important physiological mechanism for protecting the young, there must be some explanation of this variability.

(a) There may be regional variations within the mammary gland. The blocks of tissue only constitute a very small part of the gland, and they may not contain representative numbers of plasma cells.

Perhaps relevant to this idea is the observation of variation between lobules. Some have many plasma cells scattered throughout the lobule; others have few, and mostly centrilobular. An attempt was made to find a correlation between the numbers and other quantifiable variables with the lobule, such as the average radius of the alveoli, height and base area of the alveolar cells, number of alveolar cells per alveolus etc. A weak correlation was found with higher numbers in those lobules with more

distended alveoli. To do this study properly serial sections of the gland
will have to be examined to establish the total number of plasma cells in
each lobule, as in single sections it is not possible to see if there is a dense
collection in a different part of the lobule.

(b) There may be variations in the flux of plasma cell precursors to the
mammary gland. These precursors are produced in response to antigenic
stimulus; if the stimulus is temporarily high, owing for example to
inflammation in the gut mucosa allowing a higher antigen load to reach
the immune mechanism, the plasma cell precursor flux will also be high,
and the numbers that the mammary gland can trap will increase.

Other mechanisms of IgA transfer

Species of mammal vary greatly in their plasma level of dimeric IgA. In
those species with high levels, there may be no need for local IgA plasma
cells in the mammary gland, as the dimer can reach the secretions from
the blood. Thus the levels of milk and serum dimeric IgA will have to be
measured in the insectivores, bats and tree-shrews before any conclusions

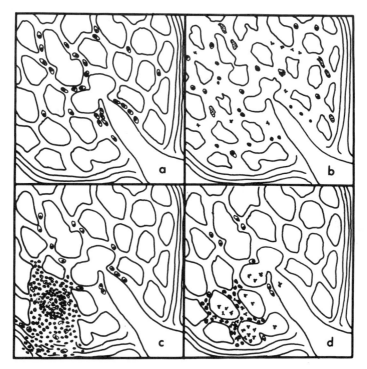

FIG. 7. The distribution of the interstitial cells in the lobule. (a) Centrilobular plasma cells.
(b) The uniform mixed infiltrate of involution. (c) Localized mastitis. (d) A dense lymphoid
collection.

about the passive transfer of mucosal immunity in these groups can be reached.

CONCLUSIONS

Even though 60 species of mammal have been examined, no clear evolutionary sequence emerges. The monotremes and marsupials have low numbers of plasma cells, which may be the same population as those found in the apocrine sweat glands in most mammals. In the Eutheria these numbers have either been suppressed, as in most ruminants, or exaggerated, as in the whales. In those eutherians with moderate numbers of plasma cells in the mammary glands there seems to be a special mechanism to trap them, possibly due to a specific chemotactic agent secreted by the alveolar cells which leads to centrilobular collections of plasma cells (Fig. 7a). These are IgA cells where the immunoglobulin class has been tested for.

Local infection may lead to local plasma cells, which may be of the IgG, IgM or IgA type (Fig. 7c). Local lymphoid collections are found, with few associated plasma cells. These may arise through the resolution of micro-abscesses, or the growth of the mammary tissue through pre-existing lymphoid collections during pregnancy (Fig. 7d). During involution after lactation, plasma cells are seen in all species, together with histiocytes, polymorphs, lymphocytes and commonly mast cells (Fig. 7b).

ACKNOWLEDGEMENTS

I would like to thank all those who gave me the material on which this work is based, including R. G. Bell, K. Benirschke, P. B. Best, British Antarctic Survey, British Museum (Natural History), O. Dansie, E. K. Dawson, R. A. Dieterich, F. D'Souza, Forestry Commission, I. A. Forsyth, R. Gambell, F. Hammond Jr., R. J. Harrison, Huntingdon Research Centre, J. L. Linzell, A. G. Lyne, A. McNeil, G. Pearson, H. Platt, P. A. Racey, G. Schoefl, J. D. Skinner, R. E. Stebbings, C. F. Summers, M. R. M. Warner and I. E. West, I would also like to thank all the departments of histology across the country who cut the thousands of sections involved, and in particular Mr K. Reid for the immunoperoxidase staining.

The author was in receipt of Wellcome grant No. 5798/I.5.

REFERENCES

Bell, R. G., Stephens, C. J. & Turner, K. J. (1974). Marsupial immunoglobulins: an immunoglobulin molecule resembling eutherian IgA in the serum and secretions of *Setonix brachyurus. J. Immunol.* **113**: 371–378.

Bohl, E. H., Gupta, R. K. P., Olquin, M. V. F. & Saif, L. J. (1972). Antibody responses in serum, colostrum, and milk of swine after infection or vaccination with transmissable gastroenteritis virus. *Infection & Immunity* **6**: 289–301.

Bourne, F. J. & Curtis, J. (1973). Transfer of immunoglobulins IgG, IgA, & IgM from serum to colostrum and milk in the sow. *Immunology* **24**: 157–162.

Brambell, F. W. R. (1970). *The transmission of passive immunity from mother to young.* Amsterdam: North-Holland.

Burns, J. (1975). Background staining and sensitivity of the unlabelled antibody-enzyme (PAP) method. Comparison with the peroxidase labelled antibody sandwich method using formalin fixed paraffin embedded material. *Histochem.* **43**: 291–294.

Drife, J. O., McClelland, D. B. L., Pryde, A., Roberts, M. M. & Smith, I. I. (1976). Immunoglobulin synthesis in the resting breast. *Br. Med. J.* **1976** (ii): 503–506.

Farr, A. G. & de Bruyn, P. P. H. (1975). The mode of lymphocyte migration through the post-capillary venule endothelium in the lymph node. *Am. J. Anat.* **143**: 59–92.

Genco, R. J. & Taubman, M. A. (1969). Secretory IgA antibodies induced by local immunisation. *Nature, Lond.* **221**: 679–681.

Gowans, J. L. & Knight, E. J. (1964). The route of recirculation of lymphocytes in the rat. *Proc. R. Soc.* (B) **159**: 257–282.

Hochwald, G. M., Jacobson, E. B. & Thorbecke, G. J. (1964). ^{14}C amino-acid incorporation into transferrin and β_{2A} globulin by ectodermal glands *in vitro. Fedn Proc. Fedn Am. Socs exp. Biol.* **23**: 557.

Kenney, J. F., Boesman, M. I. & Michaels, R. H. (1967). Bacterial and viral coproantibodies in breast-fed infants. *Pediatrics* **39**: 202–213.

Lascelles, A. K. (1963). A review of the literature on some aspects of immune milk. *Dairy Sci. Abstr.* **25**: 359–364.

Lee, C. S. & Lascelles, A. K. (1970). Antibody producing cells in antigenically stimulated mammary glands and in the gastro-intestinal tract of sheep. *Aust. J. exp. Biol. med. Sci.* **48**: 525–535.

McClelland, D. B. L., Samson, R. R., Parkin, D. M. & Shearman, D. J. C. (1972). Bacterial agglutination studies with secretory IgA prepared from human gastrointestinal secretions and colostrum. *Gut* **13**: 450–458.

Murillo, G. J. (1971). Synthesis of secretory IgA by human colostral cells. *Southern Med. J.* **64**: 1333–1337.

Norcross, N. L. (1971). Immune response in the mammary gland. *J. Dairy Sci.* **54**: 1880–1885.

Wilkinson, P. C., Russell, R. J., Pumphrey, R. S. H., Sless, F. & Parrott, D. M. V. (1976). Studies of chemotaxis of lymphocytes. *Agents and actions* **6**: 1–3. Basel: Birkhäuser Verlag.

Symp. zool. Soc. Lond. (1977) No. 41, 277–284

Rearing Human Infants: Breast or Bottle

MAVIS GUNTHER

Esher, Surrey, England

SYNOPSIS

Asepsis and other forms of modern technology have provided mothers in the western world with a relatively safe choice in their method of feeding their babies. The composition of human milk is now recognized to include a diversity of substances regulating bacterial and viral growth and the low concentrations in it of buffering substances are believed to help control the intestinal flora. The anti-infective qualities and its immediate transfer from the gland to the baby's mouth make it life-saving in populations in hot countries without piped water. Even in such circumstances, as well as in temperate countries, there are babies who do not take the breast and mothers who do not make enough milk. For them and for mothers to whom breast feeding is not emotionally acceptable bottle feeding is necessary.

Breast feeding delays ovulation and hence conception but is not so reliable a method of contraception as "the pill". Those who advise mothers should have a proportionate view of the relative advantages and disadvantages of both methods and maintain a sensitivity to the circumstances of each individual mother.

INTRODUCTION

Over the recent decades, beliefs and persuasions about the relative merits of breast and bottle feeding have changed frequently and radically. The "it's natural" argument is too facile and yet it is a hardy growth which has returned many times. Assessment of bottle feeding is difficult because what goes into the bottle can vary from an over-diluted milky-looking liquid to the most recently acclaimed products. Improvements in the past have so many times needed years of use to reveal their drawbacks. Experimental trials of new recipes for artificial milks are up against the difficulties of justification where known recipes work well; mothers cannot be commanded to keep to a trial even if persuaded; captive babies in paediatric units can justly be fed on modifications for their good but investigators are still left with a fear that any change can bring unforeseen disadvantages and the long-term effects remain unknown. Has the recent phase of high sodium content in artificial milks made more hypertensives? Or have cows' milk fats made different brains? I shall confine myself mainly to explaining where I think breast feeding stands.

CHOICE

From a biologist's point of view, lactation is a characteristic part of the mammalian life cycle, so well tried out that further experimentation would have little chance of finding a better way of infant feeding. This assessment is not, to my mind, a justification for ruling on other people's choices between breast and bottle feeding. There is no reason to believe that evolution has reached a stage where "natural" processes ensure the survival of every good member of a new generation. I would maintain too that our human endowments also brought by evolution include rational thought, compassion and choice. We want to aid survival, and some babies cannot live without benefit of substitute milk. It is also a fact that babies can grow, and grow well, on substitute milk, as anyone who saw the great beauty of British babies between 1940 and 1960 will vouch for. So we ask what qualities breast milk brings, how real are the risks to a baby receiving it or deprived of it, and how and in what circumstances will substitute milk be as good, or better, or even essential? If there is choice, whose is it and on what criteria? For each woman with an infant to feed there are many people without. Even if only a small proportion of them are concerned, the woman is likely to be outnumbered by those with their (several) opinions on matters which considerably affect intimate parts of her life and her body. Since breast feeding modifies conception rate there is reason for some public concern to do with its relation to overpopulation. But the decisions made must be those of the individuals concerned and should be based on fair representation.

QUALITIES OF MILK

In the past half-century the analysts, relying on the biologist's invention of asepsis, have offered their services confident of being able to make a matching milk. But they, and for the moment I am classing immunologists as analysts, have recently built much of the knowledge which makes us understand the difficulties involved in providing the equivalent of one species' milk from that of another species. The baby probably has some latitude in dealing with the differences in chemical composition of substitute milks; the need for defence against infection leaves less room for manoeuvre. A baby, from birth, must be able to live symbiotically with bacteria and viruses and yet avoid massive multiplication of unchosen bacteria within the alimentary canal and penetration of his body proper by infection.

Breast milk, having passed over the complicated surfaces of the nipple, is not sterile. This observation has been used in argument by paediatricians who preferred a baby to be bottle fed. But the milk brings with it a formidable package of diverse protective components. Considering the immeasurable range of potential infecting organisms and their mutability one can appreciate the diversity.

There are macrophages and other white cells in colostrum and milk, clinging, probably phagocytic and possibly bringing interferon. There are specific antibodies recapitulating the mother's experience against infection. One pictures these specific components working where infection is most likely to get in, in the nose and mouth and all the way down the gut, controlling multiplication of viruses and bacteria on a day-to-day local basis. Antibodies in the forms in sera are not absorbed intact if given by mouth to neonates (Boorman, Dodd & Gunther, 1958), but the concentration of IgA in the circulation is increased after the ingestion of colostrum (Iyenagar & Selvaraj, 1972), and it is probable that some antibodies are also transferred from milk to the baby's passive store. The local action plays the important part; oral immunization against poliomyelitis does not result in live virus appearing in the baby's stool (the evidence of a good "take") if the baby is breast fed within six hours before or after the dose (John & Devarajan, 1973). Breast feeding also reduces the risk of infection with respiratory syncytial virus (Downham, Scott, Sims, Webb & Gardner, 1976).

The antibodies to bacteria also act locally and do so at least partly by interaction with other substances in milk; *in vitro* experiment has shown that anti-*Escherichia coli* antibodies together with lysozyme, also present in milk, and complement can lyse *E. coli* (Glynn, 1969).

Then, in our list of protective substances, there are the carrier proteins taking iron, vitamin B_{12} and folate almost quantitatively from the milk, by gut absorption, to the circulation so that the baby benefits, the gut bacteria are denied and the mother's resources are used with great economy (Bullen, Rogers & Leigh, 1972; Ford, 1974). The iron-binding protein, lactoferrin, inhibits bacterial growth by its own action but is made much more potent by the specific antibodies in the milk. Although these proteins serve as nourishment to the baby a portion survive the passage through the alimentary tract undigested, limiting bacterial growth as they go.

The bacterial flora of the gut may also be selected by the effects of their own metabolic products. Human milk is not so highly buffered against acid–base changes as is cows' milk and it is thought that the higher phosphate and protein content of cows' milk may cause the difference in faecal flora of bottle- and breast-fed babies; milks for infant feeding are now being modified accordingly. Bullen and her colleagues (Bullen, Tearle & Willis, 1976) have shown by *in vitro* experiments that bifido bacilli and *E. coli* produce much acetic acid, acetate buffer and lactic acid by metabolism of sugar in breast milk; this results in a pH of 5 or less. They have found too that many usual gut bacteria would not survive in such acid media. In the modern jargon breast milk provides for elitism among gut bacteria.

This catalogue of the special qualities of human milk as a food for human babies gives us some understanding of protective mechanisms. The list is far from complete. The addition to bottle milk of corresponding substances derived from animal sources is not a simple matter for, being proteins, they are likely to be antigenic as other known whey proteins are (Gunther, Cheek, Matthew & Coombs, 1962).

The details of the actual nutrient provision of breast milk are also now better understood. For instance, nutritionists used to shake their heads over the low iron content of it and of folate too. In fact the breast-fed full-term baby is less likely than the bottle-fed one to become anaemic and this finding has become understandable with the recognition of the carrier proteins. At present a proportionate view seems to be that if the baby is normal and the mother well fed, the milk is unlikely to be bettered. Indeed the urge to improve can betray us into mistakes. Over the years when mothers got "cereals" into their babies in the first weeks and months some babies grew too heavy for their mothers to carry them; some weighed 12·5 kg at seven or eight months.

RISKS TO THE BABY

However, the practical outcome, the probability of healthy growth and survival, is the real measure of what the choice between breast and bottle implies. The risks to the baby depend first and foremost on the likelihood of infection and this mostly on where he lives. If his mother has the benefit of a cool climate, western hygiene, basic health education, sufficient food and vitamin intake, and access to antibiotics, the difference made by breast or bottle feeding is small. In the Swedish survey (Mellander, Vahlquist & Mellbin, 1959) of 402 babies there were fewer respiratory infections amongst the breast-fed but slightly more gastroenteritis among those who were breast-fed the longest (perhaps because the babies in that group had twice as many siblings of school age as the others had). The differences were not sufficient to make one or other method compellingly advantageous.

In places where the hot climate makes a bacterial culture of bottle milk, where clean water is a scarce commodity, antibiotics hard to obtain, nutritional deficiencies widespread and the mothers have little traditional learning about asepsis, the risks of bottle feeding are very great; one is tempted to make the rule of, "no piped water, no bottle feeding". Nevertheless, in all climates there can still be circumstances where breast feeding is impossible or where the baby's nutrition would be better with additions. The baby born prematurely will not get from breast milk alone as much of some substances, notably copper and zinc, as he would had he continued *in utero* until term (Dauncey, Shaw & Urman, in prep.). Modification of breast milk is beneficial and specialized bottle milks have their place.

DIFFICULTIES IN BREAST FEEDING

Even with a full-term baby, breast feeding can be inadequate, difficult or impossible for some women and totally repugnant to others. The breasts do not necessarily make enough milk for a baby and lactation may come to an end too early; one of the disadvantages of breast feeding can be that the mother or her advisers do not recognize this fact.

Even where the mother almost surely knows how unlikely her baby is to survive without breast feeding, difficulties can make it impossible. Gordon and his co-workers (Scrimshaw, Taylor & Gordon, 1968) in remote villages in the Punjab, observed the feeding practices and their outcome among 775 mothers and their babies. Of these, 16 babies died owing to obstetric difficulties, congenital deformities or prematurity. In addition there were 20 babies who had to be artificially fed from the start owing to maternal deaths (two), the mother having no milk (two), the child being ill (one), the child being too weak (10) and five who would not take the breast. Only one of the 20 survived for a year. This death rate gives us one of the clearest measures of the hopelessness of the lot of the bottle-fed baby in such circumstances, but I want to emphasize that the choice of bottle feeding is not necessarily the result of the mother wishing not to breast feed, nor of seductive advertising. Almost all normal full-term babies who will not take the breast will feed from a bottle. I have seen too often the look of disbelief on the face of doctors and the defensive inventions of mothers where breast feeding has failed but its success had been assumed to need only the will.

As this is a zoological symposium might I describe again one of the common ways in which breast feeding fails? I knew for some years that many mothers and many babies too did not know how to achieve suckling. Meryl Middlemore's book, *The nursing couple*, drew my attention to the difficulties experienced by chimpanzees in captivity. I had seen and had become convinced that a human baby usually needed only one satisfying sucking of the breast to know how for the future. This was instant learning and could only be seen by me as the evoking of an instinctive response with a lasting change in behaviour from one experience—in the baby. The mother's part is different. Through the kindness of Dr G. M. Vevers and Mr A. Budd at the London Zoo I learnt that among chimpanzees only those coming late to captivity had known how to feed their young (Wyatt & Vevers, 1935; Budd, Smith & Shelley, 1943). From this I concluded that instinct could indeed fail and that the mother's part should be maintained from generation to generation by mimicry. Many human mothers have to be taught. Incidentally I have seen no film so far which would help a woman, or those who seek to help her, by providing a basis for mimicry. The accidental finding of good positioning is helped by nakedness or only very thin clothes between mother and child; again perhaps we should learn from animals. The work of the Zoological

Society of London has made a practical contribution enabling some mothers to achieve their choice (Gunther, 1955).

Clinical experience has taught me that in the western world a mother's delight in her baby is worth more to him than her milk—and these can be alternatives. Even if her repugnance to breast feeding is not intense (and it can be) her complicated, often ill-expressed attitude to the modern dual functions of the breast has to be taken into account (Gunther, 1976). In addition, the days and weeks after birth can be bewilderingly difficult for the mother. Many previous ways and habits are disrupted; at a rough estimate a mother with a new baby has to spend five hours a day feeding and tending him. She is likely to lose sleep, to be very tired and to have too much to do. She may have to give up something and there is more likely to be a person about (perhaps the husband) who is delighted to give a bottle than someone willing and skilled to take over the washing and shopping.

The relation of lactation to conception is complicated and very important. Suckling reduces the likelihood of ovulation (and hence conception) and the resultant family spacing affects the health of the individual family and world population. But inhibition of ovulation during lactation may cease and lactation may fail before the mother recognizes that it is failing; suckling is not such a reliable means of contraception as the oestrogen-containing "pill". The progestogen-only pill appears to be intermediate in its effectiveness and a good compromise. Conceiving while the baby is small can be more than a matter for the mother's regret; it is widely believed to risk the health of the mother and both babies. The oestrogen doses in oral contraceptives are not far above those which are known to inhibit lactation and they almost certainly curb breast feeding. This cycle of effects, where the mother can afford the pill, provides an uncertain choice. Many prefer the greater certainty, feeling that bottle feeding is always available. The situation is, however, totally different in developing countries.

FEEDING AND COT DEATH

While the foresight born of fear seems rational, it is not a good basis for argument in discussing the choice of method of feeding with an expectant mother. This is seen when mention of cot death comes up. All mothers, whether they have heard of sudden infant death or not, find themselves, often to their surprise, feeling their babies from time to time to see if they are still alive; the fear of life disappearing seems to be part of the state of mind of a new mother. To them thought of cot death does not come proportionately and fear does not make mothering easier or breast feeding more likely to be successful; it can be inhibiting. Nevertheless it is a fact that unexpected and unexplained sudden infant death is more common among the bottle-fed. The causes are not known; whether one favours attributing the death to infection or respiratory reflexes or both, or to stages of brain maturation, conduction failure in the heart, mineral insufficiency, an anaphylactic type of reaction, hypoglycaemia, botulism or whatever, one can usually see that, in theory, the differences between breast and bottle milks could be involved. The ingenious statistical methods used by Carpenter, Emery and their co-workers (Carpenter, Gardner, McWeeny & Emery, in prep.) leave no doubt that while breast feeding is not a total protection against cot death it is a considerable one, regardless of social class.

CONCLUDING COMMENTS

The news value and interest coming from the growing science of galactology tends to make one speak of the advantages of breast milk. It must not be forgotten, as I have said earlier, that babies can grow up well on modified cows' milk. Group behaviour, fashion, social climate—whatever you call it—can swing, putting persuasion and pressure on a mother, maintaining that breast feeding is, at one time, indecent or, at another, obligatory. In truth, breast feeding is a psychosomatic activity and the method of feeding needs to correspond to the unpersuadable part of the mother's mind. For this reason advisers should consider themselves to be working on a one-to-one basis, sensitive to the needs of the individual mother, not spokesmen for their group, and they should have the humility to go on learning how to help and how to accept in each separate instance the advantages of either method.

One may start with a biologist's view and there is good reason to accept the safety of breast feeding including the safety from society's and the mother's notions. But the stage we have reached provides us with choice. To return to our present stage of evolution: humans now have the mental provision of concepts and fears and some understanding to see when substitute methods of feeding will bring little risk. Within the next decade the available milks should be even better. The basis of choice may change but I hope that choice, part of the proper endowment of present-day humans, will remain and be respected.

REFERENCES

Boorman, K. E., Dodd, B. E. & Gunther, M. (1958). A consideration of colostrum and milk as sources of antibodies which may be transferred to the newborn baby. *Arch. Dis. Childh.* **33**: 24–29.

Budd, A., Smith, L. G. & Shelley, F. W. (1943). On the birth and upbringing of the female chimpanzee "Jacqueline" with an appendix showing some developmental comparisons of four chimpanzees born in the Zoological Gardens from 1935 to 1937. *Proc. zool. Soc. Lond.* **113** (A): 1–20.

Bullen, C. L., Tearle, P. V. & Willis, A. T. (1976). Bifidobacteria in the intestinal tract of infants: an *in-vivo* study. *J. med. Microbiol.* **9**, 325–333.

Bullen, J. J., Rogers, H. J. & Leigh, L. (1972). Iron binding protein in milk and resistance to *Escherichia coli* infection in infants. *Br. med. J.* **1972** (i): 69–75.

Carpenter, R. G., Gardner, A., McWeeny, P. M. & Emery, J. L. (in prep.). A *multistage scoring system of identifying infants at risk of unexpected death*.

Dauncey, N. S., Shaw, J. C. L. & Urman, J. (in prep.). *The absorption and retention of magnesium, zinc and copper by low birth weight preterm infants and light for dates full term infants fed pasteurised human breast milk*.

Downham, M. A. P. S., Scott, R., Sims, D. G., Webb, J. K. G. & Gardner, P. S. (1976). Breast feeding protects against respiratory syncytial virus infections. *Br. med. J.* **1976** (ii): 274–276.

Ford, J. E. (1974). Observations on the possible nutritional significance of vitamin-binding proteins in milk. *Proc. Nutr. Soc.* 33: 15A.

Glynn, A. A. (1969). The complement lysozyme sequence in immune bacteriolysis. *Immunology* 16: 463–471.

Gunther, M. (1955). Instinct and the nursing couple. *Lancet* **1955** (i): 575.

Gunther, M. (1976). The new mother's view of herself. In *Breast feeding and the mother. Ciba Foundation Symposium* 46: 145–152.

Gunther, M., Cheek, E., Matthews, R. H. & Coombs, R. R. A. (1962). Immune responses in infants to cow's milk proteins taken by mouth. *Int. Archs Allergy appl. Immun.* 21: 257–278.

Iyenagar, L. & Selvaraj, R. J. (1972). Intestinal absorption of immunoglobulins by newborn infants. *Arch. Dis. Childh.* 47: 411–414.

John, T. L. & Devarajan, L. V. (1973). Poliovirus antibody in milk and sera of lactating women. *Indian J. med. Res.* 61: 1009–1012.

Mellander, O., Vahlquist, V. & Mellbin, T. (1959). Breast feeding and artificial feeding. The Norbotten study. *Acta paediatr. Stockh.* 48: Suppl. 116.

Middlemore, M. P. (1941). *The nursing couple*. London: Hamish Hamilton.

Scrimshaw, N. S., Taylor, C. E. & Gordon, J. E. (1968). Interaction of nutrition and infection. *WHO Monogr.* 57: 230.

Wyatt, J. M. & Vevers, G. M. (1935). Remarks on the recent birth of a female chimpanzee in the Society's gardens. *Proc. zool. Soc. Lond.* **1935**: 195–197.

Symp. zool. Soc. Lond. (1977) No. 41, 285–296

Dairying:
Past, Present and Future

C. C. BALCH and J. W. G. PORTER

National Institute for Research in Dairying, Shinfield,
Reading, England

SYNOPSIS

From a simple farmhouse occupation the production and sale of milk in the United Kingdom has grown over the past 125 years to a mighty industry, returning annually almost £1000 million to home milk producers who provide us with all our needs of liquid milk and cream, about two-thirds of our cheese and one-tenth of our butter, and, as an important by-product, much of our beef and veal.

The initial development of milk production was stimulated by the building of railways which enabled milk to be brought to the major cities from those areas where grass grows best. Collection and transport is now by road tanker, though distribution is still completed by door-to-door deliveries, which remain the key to maintained levels of consumption of liquid milk.

The present efficient industry is due to radical improvements in cow nutrition, health and management, in the milk-producing potential of cows through breeding, and in equipment, technology and hygiene in the handling of milk at all stages from the milking machine to its final packaging in liquid or manufactured form.

Recognition of the particular value of milk as a food in the 1920s was followed by Government-sponsored schemes to encourage increased consumption of liquid milk. This was accompanied by statutory control of the compositional and hygienic quality of milk and milk products and by the establishment of Milk Marketing Boards.

The present Government has indicated clearly its intention of encouraging milk production, the major enterprise of British agriculture. On grounds of human nutrition, agricultural profitability and national economy it is desirable to continue to increase efficiency in the dairy industry. Such developments must be based on continued provision of funds for research in dairying in its widest sense, but certainly including the physiology and biochemistry of lactation.

INTRODUCTION

The papers presented at this Symposium demonstrate the vast amount of research, fundamental and applied, that has been carried out, and the minute detail in which lactation and its product have been investigated. This interest in turn reflects the great importance that has been attached in many countries to milk as a food. The work has been concerned with both human milk and with the milk of the cows and other animals that provide, in most human lifetimes, thousands of times more milk than is obtained from the human mother. It is therefore appropriate to spend a little time considering the mighty industry that has become necessary to provide this milk for the town-dwelling population. How did it happen, what does it now encompass, what is its future?

Dairying is the process of handling milk, of transferring it from the producing animal to the human consumer, fresh or as a product that can be stored for a shorter or longer time. Dairying in this sense is concerned worldwide with the milk of cows, sheep, goats and buffaloes, in that order of numerical importance. Although space will only allow us briefly to survey the cow-based dairy industry of the UK with a glance at our European future within the European Economic Community (EEC), it must be remembered that there are large regions of the world where milk production and dairying are not concerned solely with the cow.

HISTORY

For centuries, dairying in the UK was a farmhouse occupation. The farmer's wife and her helpers milked out cows that were suckling calves. Some of the milk was consumed fresh, the remainder was made into butter and cheese; amounts surplus to the requirement of the farm were sold at the door or at market. In many economies, pig husbandry was dependent on the by-products—skim milk, buttermilk and whey—of the dairy.

In this simple form, milk production was essentially seasonal, the cows calving in the spring, with the flush of grass in April, May and June sustaining lactation. This system continued to be a major source of butter and cheese for towns until well into the twentieth century. The needs of the growing urban population for liquid milk were, however, steadily growing. These needs have dictated the pattern of development of the dairy industry.

The needs of the cities led first to the increase of dairy farming in the areas immediately around, in the Home Counties, Essex, Cheshire and Lancashire. It also led to the now-forgotten phenomenon of the town dairy; the cow was taken to the customer, sometimes quite literally. The demand for liquid milk throughout the year imposed great agronomic problems.

The growth of the dairy industry can be counted from the building of the railways. This made possible vast increases in milk production in the great grass-growing areas of the West, South and Midlands. Milk was transferred rapidly to London by passenger-train and eventually by special milk trains. Though road transport has now replaced rail, the influence of the railway affected development until the 1930s.

Grass and transport do not provide winter milk. The need to feed lactating cows in the winter months led to improved methods of growing and conserving grassland herbage, including silage-making. It was soon realized that residues from the milling and oil-seed industries were useful animal feeds and this led to the development of the animal feedingstuffs industry, which now also serves pigs and poultry.

Major bacteriological problems had to be overcome as the industry grew. Transport of milk demanded improvements in all aspects of dairy hygiene and in the methods used in handling milk. The nineteenth century had seen the development of small items of apparatus for the dairy industry, and especially for butter- and cheese-making. The early twentieth century saw the gradual overthrow of resistance to approved equipment and practices for operations such as udder washing, hand-milking, milk straining and cooling, milk transporting and bottling. We now accept that our bottle, carton or plastic bag of milk is safe to drink and will keep for two or three days without a refrigerator. It was not always so. This position has been achieved only by continual legislation and the devoted care of a noble band of dairy advisors and instructors, braving the natural resistance of farmers and herdsmen to alterations in long-established routines. It says much also for people working at all levels in the dairy industry that these changes have been accepted, steadily and successfully. The latest milestone in this story of inexorable change is the announcement that collection of milk in cans, or churns, will be abandoned in 1978; from then on milk will be kept in refrigerated bulk tanks on farms and collected by bulk tankers.

Another great problem was the hazard of bovine tuberculosis. The long saga of the contest between the protagonists of "tuberculin testing" and pasteurization ended in victory for both. The country has been clear of bovine tuberculosis for about 10 years, but milk has been made doubly safe by pasteurization for a much longer period. Brucellosis is now being eradicated area by area.

The more intractable problem of mastitis remains a source of serious loss to the farmer. Hygiene routines and antibiotic therapy are reducing losses, but this disease, in its many forms, is unlikely to be eliminated; the price of a low incidence is eternal vigilance.

Throughout the last 60 years the development of the dairy industry has been supported by research. The National Institute for Research in Dairying, founded in 1917, the Hannah Research Institute, founded in 1928, other government-funded institutes, in addition to universities, colleges, farm institutes and commercial establishments have contributed to the development of the dairy industry of today.

TRENDS IN MILK PRODUCTION

The extent to which the various developments have affected milk producers and milk production in the UK are summarized in Table I. By the end of the war the number of milk producers had risen by some 45 000 compared with 1939, with a peak of 196 000 in 1950; since then they have fallen to 77 000.

This greatly reduced number of herds has contained a steady 3·2 million cows over the last 15 years; herd size has trebled and milk yields, as a result of improved feeding and management, have increased by 60%. In consequence our farms now sell two and a quarter times as much milk as in 1939. In the current year it will be worth in the region of £1000 million.

One of the most impressive examples of the adaptability of the dairy industry is provided by the changes in breed that have occurred in recent years. In 1939 the most numerous breed was the Shorthorn or, inaccurately, the Dairy Shorthorn. Figure 1 shows how rapidly and completely the Friesian breed has come to predominate in the nation's dairy herd during the last 20 years.

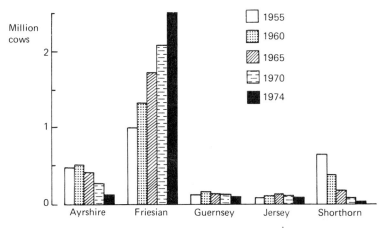

FIG. 1. Changes in numbers of cows of the five main dairy breeds during the last 20 years.

Further dramatic change has been in milking methods. By 1939 some progressive farmers were beginning to use milking machines. Man-power shortages in wartime hastened the change until by about 1950 hand-milking was confined to the smallest holdings; it must now have virtually disappeared. On many farms one worker is expected to milk 100 cows twice daily.

Government policy has, until recently, favoured cheap food and prices have done nothing to encourage the production of milk for manufacturing. The final column of Table I shows that home production has averaged about 50% of the milk that would be required to make the UK self-supporting in dairy products as it already is in liquid milk; in 1973 we produced 59% of the required milk. About three-quarters of our production for beef is met by animals bred, as crosses or pure breeds, in the dairy herds.

The modern dairy cow is not only a high producer of desirable human food; she is also very efficient. The values given in Table II show that

TABLE I

Milk producers and milk production in the UK 1939–1975

Year	Total milk producers (thousands)	Total dairy cows (millions)	Mean herd size	Mean milk yield (gallons)	Milk sales (thousand million gallons)	Producers' receipts (£ million)	Production‡ (% of requirements)
1939	156	(2·5)*	15†	560	1·3	—	—
1945	191	(2·6)*		518	1·4	—	—
1950	196	(2·9)*	15	623	1·8	—	51
1955	175	(3·0)*	17†	675	2·0	306	48
1960	152	3·2	21	765	2·2	315	47
1965	125	3·2	26	775	2·4	369	52
1970	101	3·2	33	825	2·6	429	53§
1975	77	3·2	42	890	2·9§	837	

* Calculated from total of dairy and beef. † England and Wales. ‡ Milk equivalent of UK production as % of equivalent for total requirements. § 1974 (59% in 1973).

TABLE II

Efficiency of production of food protein and energy by domesticated animals

Food product	Production of protein (g/Mcal of DE)	energy (as % of DE)
Pork	6·1	18
Eggs	10·1	12
Broiler	11·9	12
Milk (3600 kg/year)	10·5	22
Milk (5400 kg/year)	12·8	27
Beef	2·9	6

DE = Digestible energy.
From Reid, J. T. (1970). The future role of ruminants in animal production. In *Physiology of digestion and metabolism in the ruminant*: 1–22. Phillipson, A. T. (ed.). Newcastle-upon-Tyne: Oriel Press. (With permission.)

when considering complete livestock production systems the efficiency with which cows convert energy into the energy of product compares very favourably with other farm livestock, especially in high-yielding cows. The energy cost of producing protein also compares very favourably in cows. It must be emphasized that these calculations by Reid (1970) apply to complete production systems.

Inflation has made the comparison of costs difficult. It is interesting to note, however, that in 1974 the public was sold a litre of milk for the cost of 4·4 min of work by an average male industrial worker; in 1938 20 min of work were required.

The animal feedingstuffs industry now handles some 11 million tons of feed, of which over 4 million tons are consumed by cattle; in 1974 feedingstuffs imports totalled £336 million, about as much as was spent on the import of dairy products and eggs.

Milk production in the UK is therefore making a substantial contribution to reducing imports. It could do much more if political pressures which encourage us to purchase dairy products from our traditional suppliers and new EEC partners were relaxed. Farmers need only a fair return and an assured future to increase milk production.

MILK MARKETING

All dairy farmers must sell their milk through one of the Milk Marketing Boards. These Boards are milk producers' organizations which were set up in the 1930s to stabilize and rationalize milk sales. They have been extremely successful not only in their trading, by ensuring that milk is effectively utilized, but also on the farm, in promoting and implementing milk recording schemes and artificial insemination services which have done much to increase production and to ensure that the product is of good compositional quality.

Milk from the farms is transported either to processing dairies or to depots where it is tested for hygienic quality before being accepted. Payment to the farmer is based on compositional quality; the basic price is paid for milk containing 8·40% or more solids-not-fat (SNF) and 12·4–12·5% total solids; additions are made to the price for milk containing more total solids and deductions for milk containing less. It is of interest that there is no legally-defined lower limit to the solids content of milk, though under the Food and Drugs Act of 1955 milk containing less than 3% fat or 8·5% SNF is presumed to have been adulterated unless it can be proved that it is genuine and that nothing has been added or removed since it left the cow.

UTILIZATION OF MILK PRODUCED IN THE UK

Representative figures for the amounts of milk sold as liquid and used for the manufacture of the more important products before and at intervals since the last war are shown in Table III.

As milk output has increased, substantial capital investments have had to be made by both the distributive and the manufacturing sides of the industry to handle the vast amounts of a highly perishable commodity that are produced each day.

TABLE III

Utilization of milk (in million gallons) produced in the UK, 1939–1975

	1939	1945	1970	1975
Liquid sales	830	1260	1600	1700
(pints/head/week	<3	4	4·8	4·9)
Manufactured:	420	170	1000	1200
Butter	130	50	320	260
Cheese	100	45	310	500
Condensed	90	45	140	120
Dried whole	20	35	40	50
Cream	65	—	160	210
Other	16	—	50	50
By-products				
Dried skim	180			445
Whey	80			400

Liquid Milk

The first call on milk supplies in the UK has always been to meet the demand for liquid milk. This demand built up steadily over the years but has now probably reached a peak at just under 5 pints/head/week. Recognition of the particular value of milk in the diet of young children in the 1920s was followed by the introduction of the first school milk scheme in 1927. Subsequent school milk and welfare milk schemes during and after the war helped to emphasize the importance of milk as a food but some sales were lost when the school milk scheme was modified in 1971. Further impetus to sales has been, and continues to be, provided by skilful advertising from the Milk Publicity Councils on behalf of the Milk Marketing Boards and the dairy trade.

Milk sold by retail in the UK must be specially designated according to the heat-treatment used. Over 90% of the milk destined for the liquid market is pasteurized, usually by the high-temperature short-time (HTST) process ($71-73°C$ for 15 sec); the remainder is sterilized either in-bottle at $110-115°C$ for 15 min or by one of the ultra-high temperature (UHT) processes at $140°$ for 1–2 sec and aseptically dispensed into suitable containers.

It is generally agreed within the industry that maintenance of sales of liquid milk depends to a large extent on the regular doorstep delivery service that has come to be accepted as a part of national life in England and Wales and in Ireland but that is virtually unknown in other EEC countries, except to a limited extent in the Netherlands. It is impressive to note that this service involves 40 000 roundsmen visiting $17·5$ million homes daily to deposit 32 million bottles of milk (and to collect the empties). The burden of distributive costs has been eased by rationalizing rounds, by the introduction of lighter bottles and by the development of deliveries of other products (some of them dairy, but often including poultry, bread and potatoes).

Milk for Manufacturing Purposes

During the last few years steadily increasing amounts of milk have become available for manufacturing purposes; this trend seems likely to continue though it was arrested temporarily by the dry weather in 1976. It is evident from Table III that particular emphasis is being given to increasing the outputs of cheese and cream.

An important secondary aspect to the production of cheese, butter and cream is to ensure effective utilization of the by-products: whey (from cheese) and skim milk (from butter and cream). Both are valuable foods; skim milk contains about two-thirds and whey nearly a half of the nutrients of whole milk. Most of the skim milk is dried and much of the product is used in the food industry, but the use or disposal of whey presents greater problems. At present much of the whey produced is used for animal feeding and either sold at a very low price to pig farmers, who use it as a liquid feed, or spray- or roller-dried and used as an ingredient of diets for young calves. Perhaps not surprisingly liquid whey is also a good feed for dairy cows and recent work has indicated that it may have a particular value in stimulating milk production in the early stages of lactation. Whey is also used as a source of lactose and dried whey has uses in the food industry. Nevertheless the scale of current production is such that each year up to 100 million gallons of whey are not used at all and have to be discarded. One of the pressing problems of the industry is to find efficient and economic means of utilizing what, in the future, may be even larger volumes of whey.

IMPORTS OF MILK PRODUCTS IN THE UK

For at least the last 20 years the milk equivalent of UK imports of milk products has been roughly equal to the total production of milk in the UK. Butter and cheese comprise practically the whole of these imports (Table IV).

TABLE IV

Imports (milk equivalents) and home production, 1974

	Imports	Home production (million gallons)	Home as % of total	Import cost (£m)
Total	2538	2882	53	333
Butter	2260	270	11	235
Cheese	250	480	66	76

ROLE OF MILK IN DIETS IN THE UK

At present milk and milk products contribute about 20% of the total dietary intake in the UK, but they are of paramount importance as sources of high quality protein, calcium and certain vitamins. The principal contributions are summarized in Table V, from which it is evident that besides energy they supply about one-quarter of the protein, one-third of the riboflavin and vitamin A and nearly two-thirds of the calcium consumed.

Until comparatively recently, the results of most experimental investigations have emphasized the nutritional virtues of dairy products. However, during the last decade several major components of milk have

Table V

Percentage contribution of dairy products to nutrient intakes in UK

	Energy	Protein	Calcium	Riboflavin	Vitamin A
Liquid milk	11·3	18·1	46·8	38	13
Cheese	2·3	5·3	11·3	–	5
Butter	7·3	0·1	0·3	–	18·6
Other	1·2	1·6	3·7	8	1·8
Total	22·1	25·1	62·1	46	38·4

been cited as potential hazards to health. In the UK particular prominence has been given to the possible implication of milk fat in the aetiology of heart disease; in other countries attention has also been given to the genetic or familial incidence of lactose intolerance and to the apparently increasing occurrence, particularly in infants, of allergic reactions to milk proteins.

THE FUTURE OF THE DAIRY INDUSTRY IN THE UK

The Government White Paper *Food from our own resources* (1975) looks for continued expansion of milk production with an increase of up to 20% in output by 1980. Even if such an increase is achieved there will still be a substantial gap between consumption and home production, provided that sales of milk and cheese can be maintained at present levels. Prices will no doubt be a determining factor, but with the complicated interactions of inflation, subsidies and the value of the EEC's "green pound" it would be imprudent to make predictions about future trends.

Most of our needs for butter will continue to be met by imports and it seems likely that the industry would not be seriously embarrassed by a fall in demand due either to high prices or to increasing public awareness of the supposed virtues of polyunsaturated fatty acids. Recent work has shown that polyunsaturated fatty acids can be introduced into butter fat either by feeding protected fats to cows, or during processing.

On the manufacturing side of the industry particular attention may be given in the future to the methods of concentrating milk by processes such as reverse osmosis and ultrafiltration. The use of concentrated milks in cheesemaking improves yields and reduces both the volume of whey to be handled and the loss of nutrients in it. Retail sale of concentrated milks could reduce very appreciably the quantities of water that have to be carried by roundsmen, and may be an intermediate stage towards the ultimate product for home use—an instantized whole milk powder that reconstitutes to a liquid indistinguishable from fresh milk. Attention will also be given to the preparation and utilization of the individual constituents of milk as raw materials for the manufacture of other foods.

DAIRYING WITHIN THE EEC

Estimates of production and consumption of milk, butter and cheese in the EEC countries in 1973 are shown in Table VI. Taken as a whole, these countries are at present producing about 10% more milk than they need, and production is still rising. In 1974 the EEC countries were almost self-sufficient in butter, had a small surplus of cheese and substantial surpluses of condensed and dried milks. When account is taken of the continuing importation of New Zealand butter there was also a surplus of this product.

TABLE VI

Production, utilization and consumption of milk and milk products in EEC, 1973

	Production million gallons)	Utilization liquid sales	butter	cheese	Consumption liquid milk (pints/ head/wk)	cheese (lb/yr)	butter (lb/yr)
		(million gallons)					
B/L*	820	160	390	35	2·5	19	20
Denmark	1000	140	620	160	4·0	22	18
France	6300	800	2500	1300	2·3	33	19
Germany	4500	910	2400	400	2·3	13	16
Eire	880	130	430	90	7·1	6	29
Italy	2100	870	390	820	2·3	22	4
Netherlands	2000	420	770	510	4·1	21	4·5
UK	3100	1600	490	370	4·7	13	16

* Belgium and Luxembourg.

The long-term prosperity of the dairy industry within the Common Agricultural Policy of the EEC countries depends on maintaining a more satisfactory balance between production and consumption than has been achieved in recent years. Efforts to restrain production are to continue, but have so far proved of little avail. At the same time a more vigorous drive will be made to increase sales, particularly of liquid milk, and proposals are being made for the establishment of school milk schemes.

CONCLUSIONS

Whether viewed from the standpoints of human nutrition, agricultural profitability or national economy there are strong grounds for continuing the development of efficiency in the British dairy industry. To bring about that development it will be necessary to continue research on many facets of the dairy industry; these must certainly include the physiology and biochemistry of the lactation process.

ACKNOWLEDGEMENT

In preparing this paper information provided in the annual publication *Dairy Facts & Figures* (Federation of UK Milk Marketing Boards) has been of the greatest value.

Symp. zool. Soc. Lond. (1977) No. 41, 297–312

Physiological Effects of Lactation on the Mother

ANN HANWELL and M. PEAKER

ARC, Institute of Animal Physiology,
Babraham, Cambridge, England

SYNOPSIS

The metabolic demands of lactation in terms of the requirements of nutrients for milk secretion and mammary energy metabolism are described. Dimensional analysis using allometric equations, or "scaling", has proved valuable in assessing the metabolic effects of lactation on the mother in different species. In general, milk yield, energy output in milk and mammary weight are scaled in a similar manner to metabolic rate. Therefore small animals face a relatively short period of intense metabolic demand during lactation while large animals face a longer period of lesser demand. Some implications of these relationships and the effects of artificial selection for high milk yields in dairy animals are discussed.

The cardiovascular system is also affected by lactation. Blood flow through the mammary glands and alimentary canal increases, and cardiac output is also raised. Possible factors inducing these changes are considered.

INTRODUCTION

Lactation is a process which affects the whole maternal organism and in this contribution we shall discuss two aspects, firstly the metabolic demands of lactation on the mother in different species, and secondly, some of the cardiovascular changes that occur in the mother.

The demands of lactation are well recognized, particularly by human and agricultural nutritionists, and in many respects the functioning mammary gland can be regarded as being a parasite on the mother. For example, direct measurement has shown that for some metabolic substrates the proportion of the amount available to the whole body which is removed by the lactating mammary glands can be very great. In studies involving the simultaneous measurement of uptake by the mammary glands (arterio-venous concentration difference multiplied by the rate of mammary blood flow) and the rate of entry of a substrate into the circulation (determined by isotope-dilution techniques) it has been found in cows, goats, horses and pigs that the mammary glands take 50–80% of the available glucose, 15–40% of the acetate and most of the amino acids (see Linzell, 1974).

The demand for particular substrates may affect some animals more than others. For example, the necessity of glucose for lactose synthesis means that in herbivorous animals which rely on the synthesis of glucose from volatile fatty acids (formed by fermentation processes in parts of the alimentary canal), glucose is at a premium and hepatic gluconeogenesis is a key feature of lactation. The alimentary tract and liver are clearly important in supplying the additional nutrients for milk formation and mammary energy metabolism, and indeed these organs grow during lactation. These and other effects of lactation, for example, increase in food consumption, breakdown of body stores, changes in response to stress and on the reproductive system, are considered in the recent Ciba Foundation Symposium, *Breast-feeding and the mother* (Ciba, 1976).

METABOLIC DEMANDS OF LACTATION IN RELATION TO BODY WEIGHT: A QUESTION OF SCALING

Basic Relationships

In view of the well known relationship between metabolic rate and body size, it is hardly surprising that dimensional analysis using allometric equations, or "scaling", has proved to be of value in considering the metabolic effects of lactation. Long before scaling became a common tool of the comparative physiologist, Brody & Nisbet (1938) and Brody (1945) compared the milk yield of cows, goats and rats, and concluded that energy output in milk and milk yield are, like metabolic rate, related to body weight to approximately the three-quarter power; when data from other species were considered later, similar relationships emerged (Payne & Wheeler, 1968; see Blaxter, 1971).

The most complete study of the relations between milk yield, energy loss in milk, mammary gland weight and body weight is that of Linzell (1972). We have extended this treatment to 19 species of various breeds (Table I) and have slightly modified the methods of analysis. Linzell realized that the inclusion of data from animals which have been subject to artificial selection by man for high milk yields could give a misleading impression and he therefore excluded certain breeds from his mathematical analysis. We have followed a similar course except that our list of animals not artificially selected is slightly different from his (Table I). In considering the data and the conclusions drawn, which incidentally are for eutherian mammals only, there are important caveats to bear in mind.

(i) While species or breeds that have been overtly selected for high milk yields (dairy breeds of cows, goats and sheep) or for their ability to rear large litters in a short period (domestic pigs) have been excluded from the mathematical analysis, a number of domesticated forms are included in which some selection for lactational performance may have occurred indirectly.

(ii) Compared with the number of extant mammalian species, the number included in the analyses is small and, as Blaxter (1971) has remarked, differences between groups of animals will probably emerge as more data accrue. Thus we might expect to see some deviation from the typical pattern in marine mammals, which have extremely high concentrations of fat in their milk, when measurements of milk yield are made. Although the collection of data for such comparative studies can be criticized by quoting such notable aphorisms as "We need no longer record the fall of every apple" (Medawar, 1967), scaling is but one field in which the accumulation of information leads to a better understanding of the similarities and differences between animals.

(iii) The possibility should be considered that some measurements, particularly of milk yield, are wrong. This is not the place to consider the methodology but while Linzell (1972) excluded some data, there is the chance that the calculations are skewed by errors, particularly in small animals (some estimates of milk yield in the literature would defy the law of conservation of energy). Furthermore, data on milk yield, milk composition (for calculation of energy output) and body weight were often not obtained in the same study.

(iv) The figures represent typical milk yields (the maximum recorded values are also shown for some domestic breeds) in established lactation; they do not take into account variations in the rate of milk secretion at different stages of lactation.

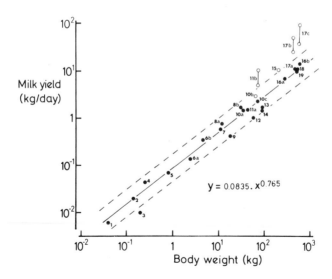

FIG. 1. Relation between milk yield and body weight. The regression line and 95% confidence limits are for those animals not artificially selected for high milk yields (●); for the breeds that have been selected (○) some maximum values are also shown. See Table I for details of the species and breeds.

TABLE I

List of eutherian mammals used in dimensional analysis of milk yield, energy output in milk, mammary weight and body weight

Number on graphs in Figs 1, 2, 3, 5*	Species and breed	Forms selected for milk yield marked "S"	Reference†
1	House mouse (laboratory strain) *Mus musculus*		
2	Golden hamster (laboratory strain) *Mesocricetus auratus*		See also Martin (1968)
3	Tree-shrew *Tupaia* sp.		
4	Brown rat (laboratory strain) *Rattus norvegicus*		
5	Guinea-pig (laboratory strain) *Cavia porcellus*		
6	Rabbit *Oryctolagus cuniculus*		
a	(Dutch breed)		
b	(New Zealand white breed)		
7	Arctic fox *Alopex lagopus*		
8	Domestic dog *Canis familiaris*		
a	(beagle)		
b	(alsatian)		
9	Yellow baboon *Papio cynocephalus*		

No.	Name	Species	S	Reference
10	Sheep	*Ovis aries*	S	Unpubl. obs.
a	(Merino breed)			
b	(Friesland breed)			
c	(Hampshire breed)			
11	Goat	*Capra hircus*		
a	(Windsor breed)			
b	(Saanen breed)		S	
12	Man	*Homo sapiens*		
13	Reindeer	*Rangifer tarandus*		Luick, White, Gau & Jenness (1974); White & Luick (1976);
14	Red deer	*Cervus elaphus*		Arman, Kay, Goodall & Sharman (1974)
15	Pig (domestic breeds)	*Sus scrofa*		
16	Horse	*Equus caballus*	S	
a	(pony)			
b	(draught)			
17	Ox	*Bos taurus*		
a	(beef breeds)			
b	(Jersey breed)		S	
c	(Friesian breed)		S	
d	(Guernsey breed)		S	
18	Water buffalo	*Bubalus bubalis*		
19	Bactrian camel	*Camelus bactrianus*		Dzhumagulov & Baîmukanov (1971)

* When no letter is shown in figures breed is unspecified.
† See Linzell (1972) for source unless otherwise stated.

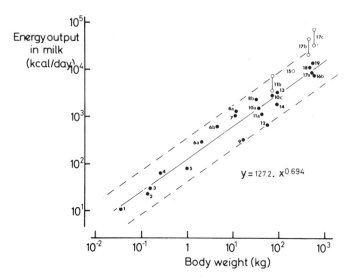

FIG. 2. Relation between energy output in milk and body weight. Plotted as in Fig. 1. See Table I for details of the species and breeds.

For the unselected animals listed in Table I, the following relationships were calculated (Figs 1–3):

$$\text{Milk yield (kg/day)} = 0.084.\ \text{body weight (kg)}^{0.77} \qquad (1)$$

$$\text{Energy output in milk (kcal/day)} = 127.2.\ \text{body weight (kg)}^{0.69} \quad (2)$$

$$\text{Mammary weight (kg)} = 0.045.\ \text{body weight (kg)}^{0.82} \qquad (3)$$

Therefore, these analyses confirm that smaller animals have higher milk yields, higher energy outputs in milk and more mammary tissue than large animals. Blaxter (1971) has commented on the similarity of these exponents of body weight and that which relates metabolism to body weight (approximately 0.75); he argued that this would be expected since the metabolic rate of the young is a simple proportion of that of the mother. A comparison of the exponents of body weight (Table II) indicates that there is a significant difference between that for the relation to mammary weight and that for the relation to energy output, thus indicating that the mammary glands of large animals tend to secrete milk with a relatively lower calorific content; this can also be seen from the data of Payne & Wheeler (1968) where birth weight of the young was compared with the calorific value of the milk. However, it is clearly unwise to take such comparisons too far from these limited data.

While the analyses show that lactation imposes greater metabolic demands on small animals than on large ones,, they do not illustrate in simple terms the magnitude. This is done in Fig. 4, which shows on log-linear co-ordinates, milk yield, energy loss and mammary weight per

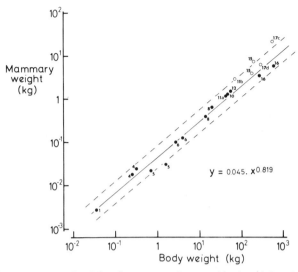

FIG. 3. Relation between total weight of mammary tissue and body weight. Plotted as in Fig. 1. See Table I for details of the species and breeds.

unit body weight, calculated from Equations 1–3, in animals of different sizes. Thus the demands are approximately 16-fold higher per unit body weight in an animal the size of a mouse than in one the size of a cow. We can also predict that the daily milk yield, as a percentage of body weight, would vary in land mammals from 1·25% in an elephant to 28% in a pygmy shrew.

TABLE II

Comparison of exponents of body weight in the dimensional analysis of milk yield, energy output in milk and mammary weight

A. Exponents of body weight v.			
milk yield	$0·77 \pm 0·022$	(22)*	
energy output in milk	$0·69 \pm 0·041$	(20)	
mammary weight	$0·82 \pm 0·026$	(14)	
B. Comparison of exponents:			
	Milk yield	Energy output	
milk yield	—	—	
energy output	$P>0·1$ n.s.	—	
mammary weight	$P>0·1$ n.s.	$P<0·05$	

* ±s.e., number of observations in parentheses.
n.s., not significant.

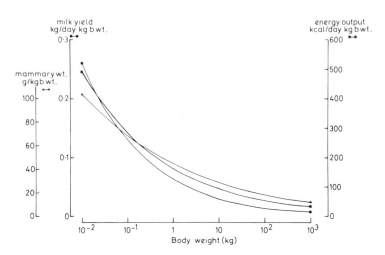

FIG. 4. Relation of milk yield, energy output in milk and mammary weight to body weight, plotted on log-linear co-ordinates and calculated from Equations 1–3 in the text.

The substrates required for milk secretion and energy metabolism can be obtained from an increased rate of food intake and from body stores. Since the daily output of energy in milk is a significant proportion of the total calorific content of a small animal but a much smaller proportion in a large animal, it would appear that an increased food intake must be the main mechanism for supporting lactation in small animals whereas large animals can draw on body reserves to a much greater extent. If this is so then at one extreme it can be calculated that a pygmy shrew (*Sorex minutus*) weighing 5 g must more than double its food consumption during lactation and eat perhaps more than four times its own weight per day (see Southern, 1964 for basic data).

Scaling studies also show that in general the duration of lactation is longer in larger animals (see Blaxter, 1971). Therefore lactation is characterized by a relatively short period of intense metabolic demand in small animals and by a longer period of lesser demand in large animals.

Effects of Domestication

It is obvious from Figs 1–3 that breeds which have been selected for milk yield fall above the calculated regression lines, and often outside the 95% confidence limits, for the unselected animals. As an illustration of the success of cattle breeders in raising milk yield, a genetic character with a heritability of only 0·2–0·3, it can be calculated that a typical Friesian cow has been effectively converted, as far as milk yield is concerned, to the metabolic size of a dog, and the world record holder to the size of a rat. Therefore it seems hardly surprising that such animals are prone to the so-called production or metabolic diseases which appear to result from the demands of lactation being in excess of the supply of substrates.

Milk Yield in Relation to Mammary Weight

When milk yield is plotted against mammary weight (Fig. 5), the following relationship is apparent:

$$\text{Milk yield (kg/day)} = 1\cdot67 \cdot \text{body weight (kg)}^{0\cdot95} \qquad (4)$$

In this analysis all data were used, not just those from unselected animals. The exponent 0·95 is not significantly different from 1·0 and therefore

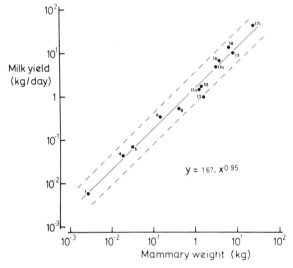

FIG. 5. Relation between milk yield and total weight of mammary tissue. The regression line and 95% confidence limits are for the species and breeds shown in Table I.

there is clearly a direct relationship between milk yield and the amount of mammary tissue. It is also evident that, on average, each gram of mammary tissue produces 1·67 g milk per day. However, there is some variation in this ratio not only at different stages of lactation but also between species. For example, in rodents and lagomorphs that have been studied, the ratio is 2·2–2·4 g per g, whereas in artiodactyls it is 1·25–1·95. It is tempting to speculate that milk production per unit weight of mammary tissue is also scaled according to body weight but until more data accumulate this would be an unwise extrapolation, especially since there is some suggestion that horse mammary tissue may produce more than twice its own weight of tissue per unit time (J. L. Linzell, unpubl. obs.).

CARDIOVASCULAR EFFECTS OF LACTATION

Basic Changes

The metabolic demands of lactation are associated with effects on the cardiovascular system of the mother. In general these changes are an increase in mammary blood flow to supply substrates to the gland, an increase in the rate of blood flow through the alimentary tract and liver, and a rise in cardiac output. The most complete studies of this aspect of lactation have been done in the rat by Hanwell & Linzell (1973a,b) and only limited data are available for other animals.

Cardiac output

In the rat, cardiac output, which is already somewhat higher in late pregnancy than in virgin animals (Linzell, 1970), increases markedly during the first few days of lactation (Chatwin, Linzell & Setchell, 1969; Hanwell & Linzell, 1973a) (Fig. 6). This increase is reflected in a higher

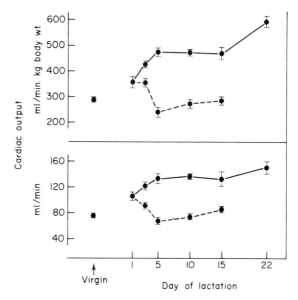

FIG. 6. Changes in cardiac output during lactation in rats. The continuous lines show data for lactating animals and the dashed lines for animals in which the young were removed shortly after parturition. Mean ± s.e.m. (From Hanwell & Linzell, 1973a, with permission.)

rate of blood flow through all organs of the body in the first few days of lactation; later the distribution is altered to favour those organs most intimately concerned with lactation (see below). Changes in cardiac output can affect mammary blood flow in addition to any local effects on mammary blood vessels, and Hanwell (1972) argued that such changes are largely responsible for restoring mammary blood flow after relatively long periods of separation from the young and also for inducing the rapid increase in mammary blood flow at the onset of lactation in the rat.

It would appear that the elevation of cardiac output is induced by hormones. In support of this view is that the increase has been found to be proportional to the number of young and, therefore, to the intensity of the suckling stimulus, and that the exponential decrease following separation from the litter to levels characteristic of non-lactating rats suggests some form of humoral control; moreover, cardiac output was found not to fall under pentobarbitone anaesthesia, whereas a fall was apparent in virgins (Hanwell, 1972). Direct evidence for the involvement of lactogenic hormones (prolactin and/or growth hormone) released from the pituitary in response to suckling has also been obtained. When young rats were returned to their mothers, in which cardiac output had decreased as a result of separation, cardiac output increased again after about four hours. If, instead of the young being returned, prolactin or growth hormone was injected, cardiac output again increased to normal lactating levels with a similar time course (Fig. 7). Similarly the fall following

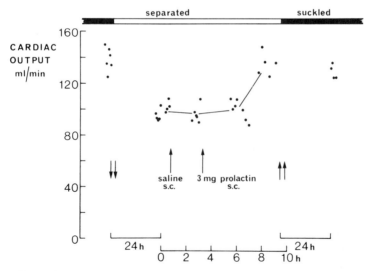

FIG. 7. Effects of prolactin on cardiac output in a lactating rat which had been separated from its young.

separation was prevented by the injection of these hormones. Furthermore, the administration of prolactin or growth hormone raised cardiac output in virgin rats (Hanwell & Linzell, 1972, 1973b).

The question remains as to whether lactogenic hormones affect the heart directly to increase stroke volume (heart rate is not increased, Hanwell, 1972) or whether metabolites or other hormones, released by prolactin or growth hormone, are responsible. If a metabolite is involved present evidence suggests that it does not originate from the mammary glands because of the effect in virgin rats where the degree of mammary development is meagre, and because it was found that cardiac output was restored to normal lactating levels within four hours of resuckling following separation, whereas at that time mammary metabolic activity was still significantly depressed (Hanwell, 1972). There is also evidence, obtained by Horrobin and his co-workers from *in vitro* studies, that prolactin can directly affect cardiac function (see Horrobin, 1974). Therefore, at present, it seems to us that a direct effect of lactogenic, anterior pituitary hormones on the heart to increase cardiac output during lactation is the most likely explanation of the mechanism.

In the Saanen goat, a dairy breed, cardiac output is also elevated during lactation, although it reaches similar levels in the last two months of pregnancy (Linzell, 1974).

Gut and liver blood flow

Up to day 5 of lactation in the rat, the blood flow through the intestinal tract and liver increased in a manner comparable to that of cardiac output and these organs received no greater proportion of the cardiac output. After day 5 these organs grew and the proportion of the cardiac output they received increased so that blood flow per unit weight of tissue then remained steady at an elevated level compared with non-lactating animals (Fig. 8) (Hanwell & Linzell, 1973a). These findings are of course to be expected in view of the increased food consumption and growth of the liver and alimentary tract in lactating animals (see, for example, Kennedy, Pearce & Parrott, 1958; Fell, Smith & Campbell, 1963; Smith & Baldwin, 1974).

Mammary blood flow

The rate of blood flow through the mammary glands increases markedly at the onset of milk secretion. In goats this is achieved by local vasodilatation (see Linzell, 1974) but in the rat there is a rise in cardiac output as well as local vasodilatation. As lactation proceeds, the mammary glands receive a larger proportion of the cardiac output, which is associated with an increase in weight of the glands (Fig. 8) (Hanwell & Linzell, 1973a).

The mechanisms involved in local vasodilatation are probably complex. The rate of mammary blood flow is related to the rate of milk secretion. Although we lack definitive evidence it seems likely, by analogy with other organs, that the active mammary gland produces metabolites which act as vasodilators. Blood flow and the supply of substrates then increases so that the cells produce more milk and more vasoactive metabolites, which in turn increase blood flow further. Therefore, the whole mechanism would act as a positive feedback system which must eventually be limited by the capabilities of the secretory cells to produce milk; this capability may be influenced by such factors as the degree of

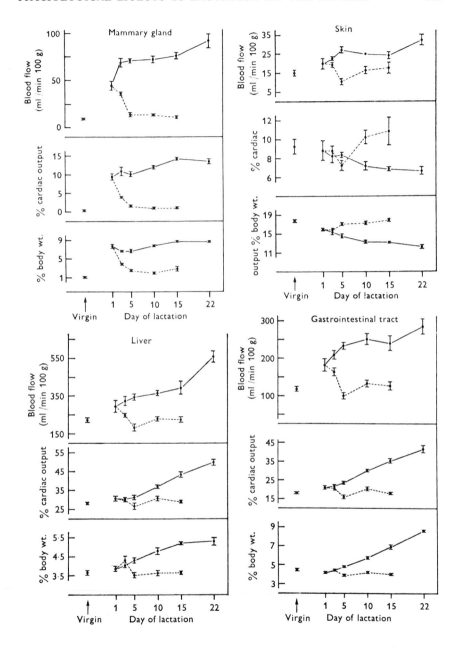

FIG. 8. Changes in blood flow, fractional flow (% cardiac output) and weight of mammary glands, skin, liver and gastro-intestinal tract during lactation. The continuous lines show data for lactating animals and the dashed lines for animals in which the young were removed shortly after parturition. Mean ± s.e.m. (From Hanwell & Linzell, 1973a, with permission.)

hormonal stimulation, the availability of some metabolic substrates and local inhibition by product accumulation.

Like cardiac output, milk secretion and the high mammary blood flow have been shown to depend greatly on the suckling stimulus in the rat. Following separation from the young, the rate of milk accumulation in the glands, cardiac output and the proportion of the cardiac output received by the glands decreased. However, when the young were not removed but the removal of milk was prevented by sealing the teats with adhesive, mammary blood flow, cardiac output and the rate of milk secretion remained high. These and similar studies led to the suggestion that the dependence of mammary blood flow and milk secretion on suckling may be a mechanism by which milk production can be rapidly adjusted to changes in the demands of the young. They also show that in the rat the reduction in mammary blood flow that occurs during separation from the young is not due to compression of the capillaries by accumulated milk but to withdrawal of the suckling stimulus (Hanwell & Linzell, 1973b). However this latter conclusion may not apply to all other animals since, in the goat, the pressure of accumulating milk appears to be involved rather than the lack of the suckling or milking stimulus (M. Peaker, unpubl. obs.). This difference between rats and goats may be related to the frequency of suckling (i.e. almost continuous in the rat but at intervals of some hours in the goat), with mammary metabolism being dependent upon the very frequent release of lactogenic hormones in the rat.

Other organs

As with the mammary glands, gastro-intestinal tract and liver, blood flow to other organs (muscle, heart and kidney) increases in early lactation in the rat as cardiac output rises. Later however, as cardiac output is redistributed to favour those organs most intimately concerned with lactation, these other organs receive a smaller proportion and blood flow per unit weight of tissue decreases. However, skin blood flow remains high (Fig. 8) and in view of the fact that the thermoregulatory mechanisms are not fully developed until approximately 18 days of age in the rat (see Hahn, 1956) it is possible that this increased skin blood flow represents an important means of increasing the transfer of heat from the mother to the young in those species in which the young are altricial. This is reminiscent of the brood-patch in birds which is highly vascularized and facilitates heat transfer to the eggs and young.

Scaling of Cardiovascular Changes

In the first part of this article, the scaling of the metabolic demands of lactation in relation to body size was discussed. Therefore the question arises of whether other processes associated with lactation are scaled in a similar manner. For example, is cardiac output increased to a lesser extent in large animals than in small animals? For such a comparison to be made a knowledge of cardiac output in lactating and in non-lactating, non-pregnant animals is required. Moreover, since cardiac output is depressed in non-lactating animals under anaesthesia, the data must be from conscious, undisturbed animals. From Hanwell's (1972) data it can be calculated that in conscious virgin rats mean cardiac output is 36 ml/min/100 g, and 49–60 ml/min/100 g from days 10 to 22 of lactation, i.e. 1·4–1·7 times that in virgins. Similarly, from Linzell's (1974) data for conscious goats, cardiac output is approximately 10 ml/min/100 g in non-lactating, non-pregnant animals and 17 ml/min/100 g during the first four months of lactation, i.e. 1·7 times that of non-lactating goats and a similar figure to that for the rat. At first sight it would appear that the proportional increase is similar in both species and therefore not scaled according to body weight. In other words, since cardiac output is related to body weight (exponent approximately 0·78) (see Günther, 1975), the regression line relating cardiac output in lactation to body weight should in a log–log plot be higher than, but parallel to, that for non-lactating animals. However, it must be borne in mind that the goat data were from a dairy breed and the possibility must be considered that cardiac output is higher in animals that have been selected for high yields. Clearly the acquisition of more data is required for meaningful comparisons of this nature to be made.

REFERENCES

Arman, P., Kay, R. N. B., Goodall, E. D. & Sharman, G. E. M. (1974). The composition and yield of milk from captive red deer. *J. Reprod. Fertil.* **37**: 67–74.

Blaxter, K. L. (1971). The comparative biology of lactation. In *Lactation*: 51–69. Falconer, I. R. (ed.). London: Butterworths.

Brody, S. (1945). *Bioenergetics and growth with special reference to the efficiency complex in domestic animals.* New York: Reinhold Publishing Corp.

Brody, S. & Nisbet, R. (1938). Growth and development with special reference to domestic animals. XLVII. A comparison of the amounts and energetic efficiencies of milk production in rat and dairy cow. *Res. Bull. Mo. Agric. exp. Sta.* No. 285.

Chatwin, A. L., Linzell, J. L. & Setchell, B. P. (1969). Cardiovascular changes during lactation in the rat. *J. Endocr.* **44**: 247–254.

Ciba (1976). *Breast-feeding and the mother.* Amsterdam: Associated Scientific Publishers.

Dzhumagulov, I. K. & Baîmukanov, A. B. (1971). [Physiological characters of lactation and milking rate in the camel.] *Vest. sel'.-khoz. Nauki, Alma-Ata* **14**: 46–49. [In Russian.]

Fell, B. F., Smith, K. A. & Campbell, R. M. (1963). Hypertrophic and hyperplastic changes in the alimentary canal of the lactating rat. *J. Path. Bact.* **85**: 179–188.

Günther, B. (1975). Dimensional analysis and theory of biological similarity. *Physiol. Rev.* **55**: 659–699.

Hahn, P. (1956). Effect of environmental temperature on the development of thermoregulatory mechanisms in infant rats. *Nature, Lond.* **178**: 96–97.

Hanwell, A. (1972). *Cardiovascular aspects of lactation in the rat.* Ph.D. thesis: University of Cambridge.

Hanwell, A. & Linzell, J. L. (1972). Elevation of the cardiac output in the rat by prolactin and growth hormone. *J. Endocr.* **53**: lvii–lviii.

Hanwell, A. & Linzell, J. L. (1973a). The time course of cardiovascular changes in lactation in the rat. *J. Physiol., Lond.* **233**: 93–109.

Hanwell, A. & Linzell, J. L. (1973b). The effects of engorgement with milk and of suckling on mammary blood flow in the rat. *J. Physiol., Lond.* **233**: 111–125.

Horrobin, D. F. (1974). *Prolactin 1974.* Lancaster: Medical and Technical Publishing.

Kennedy, G. C., Pearce, W. M. & Parrott, D. M. V. (1958). Liver growth in the lactating rat. *J. Endocr.* **17**: 158–165.

Linzell, J. L. (1970). Cardiac output and organ blood flow in late pregnancy and early lactation in rats. In *Lactogenesis*: 153–156. Reynolds, M. & Folley, S. J. (eds). Philadelphia: University of Philadelphia Press.

Linzell, J. L. (1972). Milk yield, energy loss in milk, and mammary gland weight in different species. *Dairy Sci. Abstr.* **34**: 351–360.

Linzell, J. L. (1974). Mammary blood flow and methods of identifying and measuring precursors of milk. In *Lactation* **1**: 143–225. Larson, B. L. & Smith, V. R. (eds). New York and London: Academic Press.

Luick, J. R., White, R. G., Gau, A. M. & Jenness, R. (1974). Compositional changes in the milk secreted by grazing reindeer. I. Gross composition and ash. *J. Dairy Sci.* **57**: 1325–1333.

Martin, R. D. (1968). Reproduction and ontogeny in tree-shrews (*Tupaia belangeri*), with reference to the general behaviour and taxonomic relationships. *Z. Tierpsychol.* **25**: 409–495.

Medawar, P. B. (1967). *The art of the soluble.* London: Methuen.

Payne, P. R. & Wheeler, E. F. (1968). Comparative nutrition in pregnancy and lactation. *Proc. Nutr. Soc.* **27**: 129–138.

Smith, N. E. & Baldwin, R. L. (1974). Effects of breed, pregnancy, and lactation on weight of organs and tissues in dairy cattle. *J. Dairy Sci.* **57**: 1055–1060.

Southern, H. N. (ed.) (1964). *The handbook of British mammals.* Oxford: Blackwell.

White, R. G. & Luick, J. R. (1976). Glucose metabolism in lactating reindeer. *Can. J. Zool.* **54**: 55–64.

Symp. zool. Soc. Lond. (1977) No. 41, 313–331

Maternal Behaviour in Mammals

ELIZABETH SHILLITO WALSER

Department of Applied Biology,
ARC Institute of Animal Physiology,
Babraham, Cambridge, England

SYNOPSIS

Maternal behaviour occurs in three phases. First there is the time before parturition when the female becomes physiologically and behaviourally ready to respond to the new-born young. In this phase some mammals build nests or seek a suitable shelter in which the young can be born. Other mammals may move away from members of their group just before parturition, and some make no behavioural changes at all.

The second and major phase of maternal behaviour is the time of lactation, when the young depend on their mother for nourishment. This can be seen as a series of behaviour patterns which are elicited by the different stages of the developing young and which relate to the way of life of the animal concerned. Helpless, naked infants elicit retrieving and nest building and are cleaned and cared for by the female. Older animals or those born with fur but with poor locomotion are still carried and cleaned, but when the young can move well, signals for communication and behaviour patterns which keep the young and parent together become more important.

Thirdly, there is the weaning phase of maternal behaviour when the young become independent of their mother. In some animals weaning occurs quickly and the young disperse. Some animals stay with the parents longer although lactation has ended, and they become incorporated into a larger group. Some animals are actively rejected or abandoned.

INTRODUCTION

Maternal care is not the prerogative of mammals. Many other viviparous and egg-laying animals show behaviour changes that bring about incubation and care of the young after birth or hatching. In mammals it is only the females that incubate the embryos and nourish them by lactation from mammary glands and this makes the role of the female obligatory.

Even with the best structure, function and development of the mammary glands, lactation is of no significance without the appropriate behaviour from the mother which allows suckling to take place. In a similar way, the behaviour of the young is also of importance; the two behaviour patterns have to fit together before the offspring of any mammal can be nourished. Infant mammals are born at various stages of development and the maternal side of behaviour is more important with the least developed animals, while the infant's behaviour seems to be more significant as the infant becomes more developed.

Maternal behaviour occurs in three stages.

  (i) Preparatory: behaviour during gestation such as nest building.
 (ii) Lactation: behaviour related to caring for offspring.
(iii) Weaning: behaviour which helps the young to be
                    independent.

A survey of behaviour patterns within the therian mammals will show that there is a range of behaviour varying from only slight and brief care to a complete change of life.

BEHAVIOUR DURING GESTATION

In the monotremes and some small marsupials the pouch only develops during pregnancy, and the echidna, *Tachyglossus aculeatus*, carries the egg in the pouch for 10 days before it hatches (Griffiths, McIntosh & Coles, 1969). In the marsupials which have pouches, there is an increase in pouch cleaning, which is maintained after parturition as long as the young are in the pouch. Many marsupials, such as the marsupial mouse, opossum and bandicoot, build nests in which the young are kept when they have become too large to be carried. So in these animals and in the mammals which have young that are born blind and naked, nest building is an important feature of maternal behaviour. Nests are used for protection from predators and to keep the babies warm. Maternal nests are usually larger and more dense than ordinary sleeping nests. In the tree-shrews, *Tupaia* sp., the male builds a maternal nest (Martin, 1966) but usually female mammals carry nest material and build the nest in a suitable place. Female rabbits line the breeding nest with hair plucked from their chests. The carrying of nest material may be a prelude to carrying the offspring after parturition.

Other mammals seek for a suitable shelter in which to give birth to their young. The felids find a dry, covered place. The canids excavate burrows, and many of the seals come ashore before parturition. In animals living in groups, the females may move away from others such as in the red deer and rhinoceros.

An increase in aggression is often seen before parturition as the female defends the territory surrounding the nest. Prairie dogs (*Cynomys*) guard the nest sites until the young leave the nest (King, 1955). Female mustelids become dominant to the males when pregnant (Ewer, 1973), and the canids which form pairs (such as foxes) defend definite areas.

In contrast there are a few animals which make very little change in their behaviour and give birth with other animals around them, for example, caribou (*Rangifer arcticus*) (Lent, 1966); this is seen most frequently in the domesticated animals.

LACTATION AND BEHAVIOUR RELATED TO NOURISHMENT AND CARE OF THE OFFSPRING

In all mammals the female provides the food supply for her offspring and so lactation is the one common factor in the behaviour patterns of maternal care. In some animals the time when suckling takes place is the only time of contact between the mother and young; in others the life of the mother is changed completely by the presence of her family. In marsupials and monotremes the young are born in a very premature condition and spend many days or weeks in the pouch attached to a teat. Many mammals, for example rodents and insectivores, are also born naked and blind although at a further stage of development than the marsupials. These offspring are described as "altricial" which is an adjective of the latin noun "*altrix*" meaning "a feeder or nurse". Young that are born without hair need to be kept warm if they are to survive, and the majority are kept in nests. Some rodents and many of the carnivores and primates are born with hair and in some cases their eyes may be open, but their locomotion is poor and movement uncoordinated. These offspring are described as "semi-altricial"; they may be hidden in dens or nests or they may be carried by the parents. Finally, most of the ungulates and cetaceans are born well developed and fully able to move on their own; the

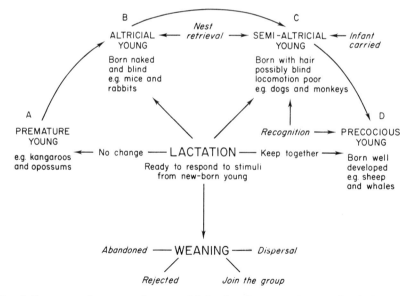

FIG. 1. Summary of patterns of maternal behaviour in mammals. Lactation is the central function. The young are born at various stages of development or pass through each stage of development and are finally weaned. Stages A to C are sheltered in a pouch or nest or den. In stages C and D behaviour patterns related to keeping mother and young together are important.

young are described as precocious. Each of these four categories demands an appropriate amount of maternal care. These are summarized in Fig. 1 and will be discussed in turn.

Monotremes and Marsupials

The premature birth of the young causes very little change in behaviour in the marsupials which have been studied. Russell & Giles (1974) observed that female Tammar wallabies (*Macropus eugenii*) showed an increase in pouch cleaning in the first 40 days of pouch life. They comment that even the presence of a new joey weighing 0·5 g causes some change in behaviour. Young marsupials crawl into the pouches of their mothers and attach themselves to a teat. Sharman (1970) found that red kangaroo joeys can be moved to another teat up to two days after birth but after that only the suckled teat secretes milk. Some kangaroos have overlapping periods of lactation as they can feed one joey in the pouch and one at foot, and the two glands secrete milk concurrently although the milk is of different composition. The female kangaroos do not seem to recognize individual young, and Merchant & Sharman (1966) transferred joeys to different pouches and also crossed red and grey kangaroos and swamp wallabies quite successfully. The females accepted the joeys and responded to their calls although they were different from normal. The young may fall out of the pouch or be ejected during pouch cleaning and the mother has to be in the right posture for the young to re-enter the pouch quickly. Young kangaroos and wallabies are carried in the pouch at first and then when older, they follow on foot, but some marsupials such as koala bears and opossums carry their babies when they are too large for the pouch. Bandicoots carry their young for 50 days and then leave them in a nest (Stodart, 1966); finally the young start to follow and forage with their mother. Marsupial mice also leave the young in a nest after 40 days when they have become too large for the pouch (Godfrey, 1969); they are fed for about a month after that. One marsupial which does not have a pouch—the opossum, *Marmosa cinerea*—depends on the strength of the babies to hold on while she carries them; they hold the teats with a strong grip and also have a grasping response with their feet when the female covers them. Beach (1939) describes an interesting retrieval experiment in which he scattered the young of a female opossum and mixed them with juvenile rats. The female gathered them up by pushing the young under her and the possum babies held on. The rats did not have the grasping response to a surface above them, and so got left behind.

After carrying the egg around in the pouch, the echidnas carry the newly hatched young until it develops spines. The monotremes do not have teats but inside the pouch are well defined milk areolae and the young suck the milk. Griffiths (1968) reports that once the young echidna is left in a burrow, the female visits it every 36 hours, and stays about half an hour. He describes how the female moves up to the young animal and nudges it with her snout until the baby lies between her forelimbs. Then she pushes it under her body and arches her back. The young echidna hangs on upside down clinging with its forepaws and sucks up the milk.

These examples of marsupials and monotremes show that a whole range of maternal care is shown as the infant develops and requires it. So there is nest building and retrieval, memory of nest sites and co-operation with young at foot, but possibly very little recognition of individual young. All these factors will be discussed in more detail in the next two sections.

Altricial Young, Born Naked and Blind

Most insectivores and the myomorph and sciuromorph rodents are born naked and blind; so are rabbits and ferrets and bears. Many of these mammals make nests to live in but at parturition nest-building activity increases and the breeding nest is well developed. It seems that the nest is recognized rather than the individuals which occupy it. It is quite easy to foster the young of most rodents by putting them into the nest belonging to the foster parents, rather as marsupial young can be fostered by changing them in the pouch, but this alters with age and becomes more difficult with older pups. Many wild animals will move their nests if they have been disturbed, for example, hedgehogs, mice and voles. The carrying or retrieving ability of lactating mothers has been well investigated in laboratory rodents. Retrieval usually stops when the eyes of the young open; this is associated with a greater mobility of the offspring and the end of ultrasonic calls.

Rowell (1960) investigated retrieval behaviour in the golden hamster, *Mesocricetus auratus*, and found that any pups between seven and 14 days old were more likely to be retrieved than attacked, but when the pups' eyes were open and they behaved in a more adult way they were attacked. To hamsters the nest is important and any young found in it are more likely to be accepted than eaten; Rowell (1961) found that some non-lactating females may tolerate pups in the nest.

Many rodents make ultrasonic calls when young and Noirot (1969) working with mice has suggested that the pups make specific ultrasonic calls when they are lost; these calls elicit retrieval from adult mice. She found that non-pregnant female mice and male mice retrieved pups of one or two days but they kept dropping them and were not so efficient as a lactating animal. Fraser & Barnett (1975) have demonstrated that lactating mice show well-ordered gathering of young mice when a nest is present, but in unfamiliar surroundings the carrying is undirected and the babies are frequently dropped. The indiscriminate retrieval of day-old mice is probably related to the tendency in these rodents to have communal nests. Sayler & Salmon (1971) studied communal nursing in house mice and found that two lactating female mice combined their nests when barriers separating the nests were removed. The young animals which were nursed communally sucked more than young raised by a single mother, whereas the females nursed less and had larger mammary glands.

The maternal behaviour of the laboratory rat is well known and well documented. The young rats also use ultrasonic calls and Allin & Banks (1972) found that 50% of lactating females tested, left their nests and located the source of calls when recorded sounds were played back. Male rats and virgin females merely oriented to the sound, but the lactating females began to search for the source of the sound. Deis (1968) showed that female lactating rats responded to auditory stimuli from other lactating rats suckling their litters. Milk ejection was induced in a female separated from her litter. When the litter was replaced after 15 min the young obtained more milk during the next 30 min than they had done during a control period; deafened mothers did not show this response.

Maintenance of the nest and retrieval of the young are two aspects of maternal care. A third aspect is the need to remember the position of the nest or shelter and the change of activity associated with lactation, for the female has to visit the nest to feed her young if she does not spend all the time there. Small mice, voles and rats spend a lot of time in the nest and suckle frequently. These animals know the geography of the area in which they live and have a well developed kinaesthetic sense. Presumably they forage near to the breeding nest and visit it regularly. Rabbits stay near the breeding area but they only visit to suckle the litter once a day. Venge (1963) found that female rabbits delay the first suckling for 16–24 h after parturition and that the baby rabbits can live for several days without food. However, the drive to feed the young is very strong, and Findlay (1969) found that when milk was emptied from the mammary glands of rabbits without active nursing, the does would subsequently nurse their litters despite lack of milk.

Another example of a mammal which feeds her babies infrequently is the tree-shrew. In these animals the male builds the maternal nest and the female gives birth to two young in it. She suckles the young until their stomachs are distended with about 6 g of milk. After that she visits the nest every two days and the visits are very short. The young must suck very rapidly for they get 10–20 g milk in less than 10 min. The tree-shrews show no carrying or retrieving behaviour and the young animals groom themselves (Martin, 1966). Presumably the rapid and infrequent visits to the nests are the most efficient way to avoid the attention of predators.

Once the young animals have grown fur and can see, hear and move around, they leave the nest and either become independent or follow their mother about, returning to the nest for shelter. In social rodents such as beavers the young stay with the family group for two years. In prairie dogs the young animals become part of the coterie; both sexes are friendly to the young and they have a gradual weaning. After leaving the burrow in which they are born, the young pups may spend the night in a burrow belonging to another female; they are accepted by all the females and fed wherever they are (King, 1955). The acceptance of any juvenile does not mean that each mother cannot recognize her own young, but that they tolerate other young of the same group; animals from a strange colony are attacked.

Most females which give birth to altricial and semi-altricial young clean them by licking and they keep the nests clean by consuming urine and faeces from the infants. Often the stimulus of maternal licking is needed before the young will excrete at all.

Semi-Altricial Young, Born with Fur and Possibly Sight, but with Poor Locomotion

Some of the mammals in this section include nest builders, so their maternal behaviour patterns include retrieval of the young, and a geographical memory of the nest site is important. Most of the carnivores make a nest or find a shelter at the time of parturition. The canids dig burrows but the felids use only natural cover. Many animals move their young to a different shelter as they grow up. Raccoons build the first maternal nest in the trees and then move to a ground nest when the young are active but not skilled enough to move in the trees (Ewer, 1973).

As female carnivores stay with their young most of the time, it is difficult for them to obtain food and the problem is solved in various ways related to the social behaviour of the animal. Some form pairs while they breed and the male brings back food for the female, as in the fox and the Canadian lynx. In the carnivores which hunt in groups, such as wolves and hunting dogs, the pack regurgitates food for the females when they return and they also give the pups food when they are old enough (van Lawick-Goodall & van Lawick-Goodall, 1970).

The social behaviour of lions also reflects on their maternal behaviour. Lions live in "prides" made up of five to nine adult females and two or three males and the cubs of the last two years. They hunt co-operatively, but the females hunt more than the males who then gain access to the food (Bertram, 1975). At parturition the females leave the open area and move into thick vegetation where the cubs are reared for four to six weeks. Then they mix with females and cubs of their own age and the cubs suck from any female for about six months. The chance of survival for cubs is increased if there are other families of the same age in the pride because then one female can stay with the cubs while the others hunt; otherwise the cubs are left alone.

The suckling behaviour of domestic cats and dogs has been more accurately observed. Puppies do not seem to have a "teat order" and suck from any teat on their mother. Kittens, however, establish a teat order in the first few days of life (Ewer, 1960). The ownership of a teat is maintained until the kittens leave the nest and feed independently at about 32 days.

The second group of animals in this category includes most of the primates which have young born with fur and vision, but they carry the young with them rather than leave them in a nest. The male marmoset monkeys of South America carry the infants and just give them to the female for nursing (McBride, 1971), but usually the females are possessive and protect their young from other interested monkeys. Jensen, Gordon & Wolfheim (1975) have studied nursing behaviour in infant pig-tailed monkeys (*Macaca nemestrina*). They found that the closest positions of mother and infant relative to each other led to continued closeness, such as nursing combined with maternal cradling. When nursing took place with the infant standing near, that is outside the mother's lap, the infant was likely to move away. J. van Lawick-Goodall (1968) describes how chimpanzee infants initiate feeding by searching for the nipple and then the mother responds by cradling the infant. Schaller (1963) found that female gorillas looked after their infants with constant watchfulness and great care. Young gorillas are very helpless and do not seem to respond to moving objects or reach for things until they are about one month old. The females hold the babies close to them and carry them with one arm holding them to their chests when they move. The infants cling to the fur of their mothers at one month and ride on their backs at three months. When they suck they use both nipples and Schaller reports attempts to suck from strange females. Juveniles are tolerated by other members of the group, but their mothers keep their own young very close to them until they are five months old, so they may only briefly contact a stranger. It seems to be common for other members of primate groups, particularly juvenile females, to be interested in new-born infants. Dunbar & Dunbar (1974) report this in gelada baboons and J. van Lawick-Goodall (1968) describes it in chimpanzees.

Precocious Young, Born Well Developed

Hystricomorph rodents

Many hystricomorph rodents give birth to well developed young that start to eat solid food within two or three days of birth. Nursing continues for several weeks but the young can often survive without it. The bond between mother and young is thus much more one of contact than need for nourishment. Newson (1966) reports that female coypu (*Myocastor coypus*) can survive if weaned at five days but in captivity they suckle for six to ten weeks. Female coypu have four pairs of teats above the mid-lateral line so the young can suckle without the female altering her normal crouching position.

The initial maternal care of cleaning and licking the new-born infant is seen in these rodents. Kleiman (1972) describes an increased restlessness and purring at parturition in the green acouchi (*Myoprocta pratti.*) The young are licked by the mother and the purring vocalization is a contact call to the young. They in turn are active and vocal which stimulates maternal behaviour. Although the young are precocious, there is a long nursing period and Kleiman suggests that the frequent suckling is not always related to milk supply in the older juveniles. When the female is willing to nurse she lies on one side to expose her nipples and she purrs. The young have a teat order and prefer the inguinal teats of the four available mammae.

In contrast to the acouchi, young guinea-pigs are more loosely attached to their mother. Fullerton, Berryman & Porter (1974) found that young guinea-pigs showed no preference for their mother and some natural fostering occurred. There is more contact and suckling and general responsiveness between mother and infant initially, than between aliens, but the young can groom themselves on the day of birth.

Aquatic mammals

Precocious young are born to all the aquatic mammals apart from beavers which have to teach their young to swim. The whales and dolphins have very long gestation periods and the calves are large at birth. The large whalebone whale calves suck for six months until the baleen is long enough to function (Matthews, 1952). Harrison (1969) reports that lactation in the toothed whales continues for up to a year in some species. All suckling occurs under water but near to the surface. Animals are fed two or three times an hour and Harrison quotes a consumption figure of 600 litres of milk a day. Observations of parturition in captive dolphins show that other dolphins are interested in the new-born animal and may help with the birth. McBride & Kritzler (1951) observed parturition in the bottlenose dolphin and found that the dam whirls round after the calf is born and this movement breaks the umbilical cord. Normal infants rise to the surface to breathe without help, but if the young is abnormal or weak the female or other animals near will lift it up to the surface. Nursing starts about four hours after birth. Suckling is brief—the female rolls on one side and glides while the infant moves with its mother. There is a lot of vocal communication in the form of whistling between the mother and infant. If the calf gets separated from its mother it circles quickly near the surface. The young follow the dorsal fins of their mothers and they can move very quickly soon after birth.

Other aquatic mammals are the manatees and dugongs although not much is known about the latter. Moore (1956) describes suckling in the manatees; these were alleged to be the mermaids as they have pectoral mammary glands and legend had it that they suckled in a vertical position. Moore reports that feeding occurs under water in a horizontal position. The calf lies at an angle to the cow's body sucking at the pectoral nipple and the cow puts her flipper laterally and forward over the calf.

The seals are also aquatic mammals but they return to the land to breed, except for the common seal, *Phoca vitulina*. These seals may give birth in the sea or on temporarily-exposed sand banks. The pups are able to swim at birth and may suck in the water, but Wilson (1974) describes how the female initiates suckling by hauling out onto land; the pup then follows. Wilson also describes the following responses of the pups to their mothers who signal with their flippers or by nodding their heads. There is a progressive decrease in the mother leading the pup as the time for weaning approaches at three weeks. Grey seals (*Halichoerus grypus*) haul themselves onto sheltered beaches in the autumn and give birth to one pup which stays on the beach and within a very small area of the birth site. The females do not feed while they are lactating but they remain just offshore from the beach except when feeding the pup. Fogden (1968, 1971) has described the maternal behaviour of grey seals. If the colony is undisturbed the females always feed their own pups and the maternal–filial bond is formed very quickly after birth because the female does not spend much time with the new-born pup. The pups sleep most of the time and wake up hungry. They call for food and the females respond by coming ashore. When the mother is near she smells the pup and presents her nipples for the pup to feed. Fogden considers that the bond depends on sound and smell and that the pups are not selective and will solicit food from any female. When there is a general disturbance pups get lost and the females get confused and will allow strange pups to feed.

The female northern elephant seals (*Mirounga angustirostrus*) will feed alien young when they have lost their own pups, but normally they keep to their own infants. Le Boeuf, Whiting & Gantt (1972) state that the mother and pup stay close together and that suckling occurs every day until weaning at 27 days. The females make a warbling vocalization to the pups as they are born and turn to touch them with their noses. As in other seals there is no licking or cleaning of the new-born infant.

Californian sea-lions (*Zalophus californianus*) are very selective and feed only their own young (Peterson & Bartholomew, 1967). When the female comes ashore she makes a characteristic "attraction" call and several pups orient and approach. She selects only her own and does not allow any other pup to suck. Recognition of individual pups is also important in the Alaska fur seal *Callorhinus ursinus* because after an initial period of five or six days the females go back to the sea and stay away five or eight days at a time (Bartholomew, 1959; Bartholomew & Hoel, 1953). Two points are common to all these different breeds of seals. One is that vocalization and scent are important in recognition of the pups and the other is that the pups do not seem to recognize, or be selective in sucking from, their own mothers.

Ungulates

Ungulates have precocious young which stand very quickly after birth and orient to their mother and start to search for the teats. Orientation is helped by a drive for the new-born animal to put its head under the nearest object, which is usually its mother. The teats are located by searching movements of the head with licking and sucking movements on smooth hairless surfaces. Stephens & Linzell (1974) observed the orientation of kids sucking from female goats which had one mammary gland transplanted to the neck. Domestic goats have low pendulous udders and kids often have to be directed to the teat; but when the glands were hanging from the neck they were in a better position. Stephens found that the transplanted gland was sucked easily and the kids developed a teat preference so when there were two kids, one sucked from the front and one from the back. Bareham (1975) found that vision was very important for teat location in lambs, so that orientation may well be visual, particularly when the animals have once found the teat.

Female ungulates learn to recognize their own young quickly and normally feed them exclusively. There are two basic types of infant behaviour in the ungulates; these are known as "hiders" and "followers". The latter group is made up of the animals which live in open grassland such as the equids, large bovids and ovids. Their young are able to move well and follow their mothers soon after birth so that mother and infant are always close together. The "hiders" are more often the animals, deer for example, which live in forested areas or where there is some cover. Their young are able to move, but they remain still and hide while the females move away to feed. Clutton-Brock & Guinness (1975) have described the behaviour of red deer (*Cervus elaphus*) at calving time. The doe remains within 50 m of the new-born calf for the first few hours but later on spends most of the time about 100 m away and returns to feed the calf two to four times a day. The calves begin to move with their mother after ten days and spend progressively more time with them. Lent (1974) has reviewed mother–infant relationships in ungulates. He suggests that the nest period which occurs in the Suidae is equivalent to the hiding phase of development seen in the deer.

The Suidae are unusual in the ungulates in that they build a nest and the young stay in and near it for some days. They are able to move well, but need the nest for warmth. Wild pigs (*Sus scrofa*) make sleeping nests as well as maternal nests. The females leave the group in the breeding season and give birth to piglets in the nest, which is defended strongly. They suckle and give birth lying on their side and although there is little maternal care, there is much nasal contact between the piglets and their mother (Gundlach, 1968). Warthogs (*Phacochoerus aethiopicus*) are similar in their behaviour and they all have a teat order and well defined sucking behaviour, as in domestic pigs (Frädrich, 1974). The teat order is established in domestic pigs after about a week. Each piglet has its own teat which it uses regardless of which side the sow is lying. Fraser (1973) has shown that stimulation of the anterior teats is important for normal lactation, and if they are unoccupied the female will get up prematurely. The way in which the teat order is established is still a matter of speculation. The position relative to the sow seems to be important, as piglets attempting to suck from an alien sow will try to get the teat which they normally occupy. Experiments with an artificial sow at Babraham showed that piglets had a very marked teat order on the dummy sow, in spite of the fact that the teats were taken out and washed every day. So taste and smell do not seem to be the most important features, although they may be functional on the real mother. The sucking pattern of piglets is well defined. The sow calls the piglets to feed by slow grunts which she continues as she lies on her side. The piglets scramble for position, then they start to nuzzle the teat and gland with their noses. The grunts of the sow increase in frequency until they are coming very quickly and sound sharp and clear. At this time the piglets hold the teats in their mouths and then milk ejection occurs in the female. The piglets become tense and rigid and their ears stand up and they suck very quickly. The grunts of the female slow down after milk ejection and when the piglets have sucked they resume their nuzzling. The time spent nuzzling is important for the amount of milk produced at the following feed. The rapid grunting of the sow is related to milk ejection and the piglets come to associate it with food and will get very excited when such grunts are played to them from a tape recorder. They can recognize their mother's voice from the voices of other sows and in the experiments with the artificial sow, it was possible to train the piglets to respond to different recorded grunts from the sow in contrast to the grunts from their own mother. Piglets reared completely on the dummy sow came running to it when the tape recorder was played. The drive to suck in piglets is very strong and sometimes the sow will lie down to let the piglets nuzzle without milk ejection occurring. In these "no milk" feeds the rapid grunting does not occur and the sow just grunts slowly all the time. The piglets reared on the artificial sow showed much excess sucking and nuzzling. When the udder was available all day, they would lie and nuzzle for a short time every hour or so regardless of whether milk was there or not. Excess sucking behaviour seems to have developed as a comfort behaviour in many animals.

Vocalization between mother and young occurs in many ungulates which are otherwise normally silent. Sowls (1974) has described the social behaviour of the collared peccary (*Dicotyles tajacu*) and reports that there is a constant communication between the female and her young. They make purring and grunting sounds to each other. The females nurse the young standing up and suckling occurs at frequent intervals. The young have a strong following response but in the wild the female runs away from danger while the piglets freeze.

In the ungulates which are "follower" types, the new-born young will follow anything. Once they are on their feet their muscular co-ordination develops quickly and soon they can run with the herd. The maternal attachment forms quickly and it is the female which maintains contact with the infant. In caribou the female signals to her calf by bobbing her head and grunting (Lent, 1966). Both domestic and feral goats form a maternal bond very quickly. Klopfer & Gamble (1966) found that if a mother remained with her young for five minutes after birth she would subsequently accept her young. This was so even if the goat was partially anosmic, although recognition did seem to be based on chemical contact. Rudge (1970) described maternal behaviour in feral goats. He found that they seem to be intermediate between "hiders" and "followers" as the kid follows its mother, but she hides it under a bush before going off to feed. The kids are left for at least one-and-a-half hours and the nannies have a very precise memory of the hiding position. The kid will respond to bleats from the adult or to pretend bleats, and comes out of hiding to feed; then the mother and infant rest together. Hiding stops after four days and the kid follows its mother as in domesticated goats.

Hafez & Lineweaver (1968) studied the behaviour of young calves. In beef cows, the calves change teats and butt more frequently as nursing ends, possibly because the milk supply is not so plentiful as in dairy cows. Calves have no teat preference and suck between four and six times daily. Kiley (1976) has studied fostering and adoption in beef cattle. Cows will sometimes allow another calf to suck but it does not develop as well as the cow's own calf. Kiley calls this "fostering" in contrast to "adoption" when the strange calf grows better and is treated more as the cow's own calf. Sight appears to be important in recognition of the calf and blindfolding of the cow is recommended in order to facilitate adoption.

WEANING BEHAVIOUR

The end of lactation is associated with a change in behaviour on both the maternal and filial side and the extent of the breaking up of the maternal–offspring bond is related to the social life of the mammal concerned.

In mammals which are solitary for most of their life, the drive to move away from others develops as the young mature and they disperse voluntarily. Many mice and voles leave the nest very soon after their eyes open. The females stop maintaining the nest so the maternal nest disintegrates with the family moving in and out, and the centre for the family aggregation disappears. Some females move away from the family and build another nest with the advent of another family; this is so in water voles and prairie dogs. Grizzly bears abandon their young by chasing them up a tree, as they do in cases of danger, and then just leave them there (McBride, 1971). Seals leave their pups on the beach.

In animals that live in groups the young become incorporated into the herd. Nevertheless the female still shows some rejection of the young when they try to suck; this may just be a movement away from the soliciting juvenile or a definite butting away. van Lawick-Goodall & van Lawick-Goodall (1970) describe weaning in hyaenas. It is a stressful time because the pups have depended on the female for milk and have not had food brought back for them as in the jackals and hunting dogs. The females nip the pups when they try to suck. Weaning is long and difficult in the animals which are predators because not only have the young to learn to hunt, but also they are not strong enough to kill for some time. So there is often a period of attachment between the parents and juveniles during which there is no suckling, but the parents are still supplying food. The felids do not regurgitate food as do the canids, and they prepare their young to get food by carrying back small animals such as mice and insects (Ewer, 1973). The young of pack hunters start hunting by running with the group, and although they do not join in the kill they are allowed to feed from the carcass.

In family groups with a matriarchal bias the young males are chased away, or leave the family group first. In red deer the hinds stay together and the males become solitary. Water buffalo have a similar social grouping and the young males are butted away before the young females. In contrast the plains and mountain zebras have close family units controlled by the stallions, and the young mares leave at two years when they are stolen by other stallions. The young stallions stay with the family for four years before joining a young stallion group (Klingel, 1974). Animals which live in groups must come to know and recognize the young within the group, and they become accepted, while young from other colonies may be attacked, prairie dogs and rabbits for example. Mykytowycz & Dudzinski (1972) have described the dispersal of young rabbits. While they are in the nest they learn the odours of their mother and when they emerge from the burrow they keep within her territory which is marked by her own secretions. The bucks and does of the warren mark the young animals with their urine and chin secretions and so as they get older and their activity increases they are accepted by more and more adults. The kittens of a subordinate doe which has littered outside the main warren may have difficulty in getting accepted because they lack the group odour.

Juvenile animals of many species survive the stress of weaning and integration into groups by adopting subordinate postures which appease aggression. Many of these are related to soliciting behaviour and juvenile play, and speculation on their origin and ritualization is interesting. The form of submissive behaviour in dogs and wolves is ritualized and symbolized cub behaviour. Schenkel (1967) suggests that begging for food from the mother is transferred to begging for food from animals in the group. The submissive lying down behaviour is a similar posture to that adopted by cubs when they are being cleaned by their mothers.

DISCUSSION

There are many different aspects to the study of maternal behaviour. The physiology of lactation is an important part and many patterns of behaviour are related to it. For example, the sucking behaviour of the infant includes nuzzling or treading the mammary gland or butting with the head, and this stimulates the release of oxytocin which then induces milk ejection. It is interesting that pigs can over-ride this chain of events and not let any milk down although the piglets are nuzzling the udder.

The structure and function of the glands vary and it is interesting to speculate how this came about. How is it that rabbits can suckle once a day, tree-shrews once every two days and seals once a week, while in some animals the glands regress if they are not sucked several times a day? What is the mechanism which brings the female back to the young to feed them? Is it the feeling of milk in the glands or is it part of a built-in activity pattern?

So many parts of maternal behaviour are related to the social life of the animal concerned. Some animals are only sociable while there is a family present. Many felids rear their kittens and then become solitary again. Mice and voles may never even recognize individuals but just know the nest and accept its occupants.

The carnivores generally keep together as a family for longer than herbivores because it is difficult for the juveniles to catch enough food to survive. However, many ungulates live in groups and so the juveniles may become part of that group and even associate with their mothers after lactation has ended. It is interesting that communal nursing occurs more frequently in the carnivores than in the ungulates. This is probably related to the method of obtaining food, but it may also be the result of the rapid bonding which occurs in precocious animals.

Why is it advantageous for the mother and young to recognize each other? The new-born animals will follow any adult and try to suck from them but they learn that only their own mother will feed them. Rhinoceroses and deer detach from the group and give birth to their young away from others (Owen-Smith, 1974; Clutton-Brock & Guinness, 1975). Zebra mares spend much time chasing away other adults, so the young become attached only to their own mother. This must be a satisfactory way of providing nutrition for the offspring, yet if all the lactating females accepted any lamb or fawn or kid there would be no problem of orphans in the group. The primary advantage of the maternal–filial bond seems to be for protection and care of the young. The females look for their offspring, stay near and will turn and fight for them. Would this happen if the juveniles were communally fed?

The methods of communication which have developed in connection with epimeletic behaviour are interesting. Vocalization is particularly important in animals with precocious young. Many of the ungulates vocalize with loud calls to attract attention from a distance and make soft rumbling or purring sounds when mother and young are close. The rumbling sound made by ewes to newborn lambs is a different call from other vocalizations and seems to help in the orientation of the lamb to the ewe. Selman, McEwan & Fisher (1970) describe three calls made by cows to new-born calves. Many animals can recognize individuals by voice. Espmark (1971) found a difference between reindeer calves' voices. Lambs can recognize their mother's voice when the ewes are out of sight (Shillito, 1975) although there is much more evidence that sheep use vision in recognizing their lambs at a distance greater than one metre (Shillito & Alexander, 1975; Alexander, 1977). Many deer and bovids use visual signals to get their young to approach or follow. Bontebok, roe deer and caribou signal by moving their heads up and down (David, 1975; Lent, 1974).

The well defined pattern of vocalization during lactation in domestic pigs may be the way in which the behaviour of many precocious young is synchronized. The other animals which give birth to precocious young only have one or two infants and so efficient feeding behaviour occurs more easily. Ewes with twin lambs will feed them only together after the first week of life, and the lambs learn to look for their sibling before running to the ewe for food (Ewbank, 1964). Seals recognize their young by voice and use vocalization to find them after coming ashore.

Animals which keep their young near all the time, primates for example, do not need vocalizations to bring mother and young together. In these animals maternal behaviour has a great influence on socialization. Hansen (1966) working with rhesus monkeys found that infants reared on a surrogate were socially retarded; Evans (1967) found that when pig-tailed macaque infants were reared in an incubator the young were lacking in all categories of social interaction. This contrasts with guinea-pigs which showed no defects in their social behaviour after being reared in isolation (Harper, 1968).

The state of the development of the young influences which behaviour patterns will be useful. The altricial young respond to warmth and probably the smell of milk. Animals born with their eyes closed may just respond to scent and sounds; when the young can see they are able to orient to objects as well as to their mother. The interesting point about maternal behaviour is that it is a way of life which appears at a particular time. It is well adapted to ensure that young animals are fed, cleaned and protected and brought up until they have developed enough to look after themselves, and then the behaviour wanes and disappears until another reproductive cycle starts again.

REFERENCES

Alexander, G. (1977). Role of auditory and visual clues in mutual recognition between ewes and lambs in Merino sheep. *Appl. Anim. Ethol.* **3**: 65–81.

Allin, J. T. & Banks, E. M. (1972). Functional aspects of ultrasound production by infant albino rats. *Anim. Behav.* **20**: 175–185.

Bareham, J. (1975). The effect of lack of vision on suckling behaviour of lambs. *Appl. Anim. Ethol.* **1**: 245–250.

Bartholomew, G. A. (1959). Mother-young relations and the maturation of pup behaviour in the Alaska fur seal. *Anim. Behav.* **7**: 163–171.

Bartholomew, G. A. & Hoel, P. G. (1953). Reproductive behaviour of the Alaska fur seal, *Callorhinus ursinus. J. Mammal.* **34**: 417–436.

Beach, F. A. (1939). Maternal behaviour of the pouchless marsupial *Marmosa cinerea. J. Mammal.* **20**: 315–322.

Bertram, B. C. R. (1975). Social factors influencing reproduction in wild lions. *J. Zool., Lond.* **177**: 463–482.

Clutton-Brock, T. H. & Guinness, F. E. (1975). Behaviour of Red deer (*Cervus elaphus* L.) at calving time. *Behaviour* **55**: 287–300.

David, J. H. M. (1975). Observations on mating behaviour, parturition, suckling and the mother-young bond in the Bontebok (*Damaliscus dorcas dorcas*). *J. Zool., Lond.* **177**: 203–223.
Deis, R. P. (1968). The effect of an exteroceptive stimulus on milk ejection in lactating rats. *J. Physiol., Lond.* **197**: 37–46.
Dunbar, R. I. M. & Dunbar, P. (1974). Behaviour related to birth in wild gelada baboons (*Theropithecus gelada*). *Behaviour* **50**: 185–191.
Espmark, Y. (1971). Individual recognition by voice in reindeer mother-young relationship. Field observations and playback experiments. *Behaviour* **40**: 295–301.
Evans, C. S. (1967). Methods of rearing and social interactions in *Macaca nemestrina*. *Anim. Behav.* **15**: 263–266.
Ewbank, R. (1964). Observations on the suckling habits of twin lambs. *Anim. Behav.* **12**: 34–37.
Ewer, R. F. (1960). Suckling behaviour in kittens. *Behaviour* **15**: 146–162.
Ewer, R. F. (1973). *The carnivores*. Ithaca: Cornell University Press.
Findlay, A. L. R. (1969). Nursing behaviour and the condition of the mammary gland in the rabbit. *J. comp. Physiol. Psychol.* **69**: 115–118.
Fogden, S. C. L. (1968). Suckling behaviour in the Grey seal (*Halichoerus grypus*) and Northern elephant seal (*Mirounga angustirostris*). *J. Zool., Lond.* **154**: 415–420.
Fogden, S. C. L. (1971). Mother-young behaviour at Grey seal breeding beaches. *J. Zool., Lond.* **164**: 61–92.
Frädrich, H. (1974). A comparison of behaviour in the Suidae. In *The behaviour of ungulates and its relation to management*. (Papers of Int. Symp. at Calgary, Canada, 1971). Geist, V. & Walther, F. (eds). *IUCN Publs* **24** (1): 133–143.
Fraser, D. (1973). The nursing and suckling behaviour of pigs. 1. The importance of stimulation on the anterior teats. *Br. vet. J.* **129**: 324–334.
Fraser, D. G. & Barnett, S. A. (1975). Pup-carrying by laboratory mice in an unfamiliar environment. *Behav. Biol.* **14**: 353–360.
Fullerton, C., Berryman, J. C. & Porter, R. H. (1974). On the nature of mother-infant interactions in the guinea-pig (*Cavia porcellus*). *Behaviour* **48**: 145–156.
Godfrey, G. (1969). Reproduction in a laboratory colony of the marsupial mouse *Sminthopsis larapinta* (Marsupialia: Dasyuridea). *Aust. J. Zool.* **17**: 637–654.
Griffiths, M. (1968). *Echidnas*. Oxford: Pergamon Press.
Griffiths, M., McIntosh, D. L. & Coles, R. E. A. (1969). The mammary gland of the echidna *Tachyglossus aculeatus* with observations on the incubation of the egg and on the newly hatched young. *J. Zool., Lond.* **158**: 371–386.
Gundlach, H. (1968). Brukfürsorge, Brutpflege, Verhaltensontogenese und Tagesperiodik beim Europäischen Wildschwein (*Sus scrofa* L.). *Z. Tierpsychol.* **25**: 955–995.
Hafez, E. S. E. & Lineweaver, J. A. (1968). Suckling behaviour in natural and artificially fed neonate calves. *Z. Tierpsychol.* **25**: 187–198.
Hansen, E. W. (1966). The development of maternal and infant behaviour in the rhesus monkey. *Behaviour* **27**: 107–149.
Harper, L. V. (1968). The effects of isolation from birth on the social behaviour of guinea pigs in adulthood. *Anim. Behav.* **16**: 58–64.
Harrison, R. J. (1969). Reproduction and reproductive organs. In *The biology of marine mammals*: 253–348. Andersen, H. T. (ed.). New York & London: Academic Press.
Jensen, G. D., Gordon, B. N. & Wolfheim, J. (1975). Nursing behaviour in infant monkeys: a sequence analysis. *Behaviour* **55**: 115–127.

Kiley, M. (1976). Fostering and adoption in beef cattle. *Br. Cattle Breed. Cl. Digest* **31**: 42–55.

King, J. A. (1955). Social behaviour, social organisation and population dynamics in a Black-tailed prairie-dog town in the Black Hills of South Dakota. *Contr. Lab. Vertebr. Biol. Univ. Mich.* No. 67: 1–123.

Kleiman, D. (1972). Maternal behaviour of the Green acouchi *Myoprocta pratti* Pocock). A South American caviomorph rodent. *Behaviour* **43**: 48–84.

Klingel, H. (1974). A comparison of the social behaviour of the Equidae. In *The behaviour of ungulates and its relation to management*. (Papers of Int. Symp. at Calgary, Canada, 1971). Geist, V. & Walther, F. (eds). *IUCN Publs* **24** (1): 124–132.

Klopfer, P. H. & Gamble, J. (1966). Maternal "imprinting" in goats. The role of chemical senses. *Z. Tierpsychol.* **23**: 588–592.

Lawick-Goodall, J. van (1968). The behaviour of free-living chimpanzees in the Gombe stream reserve. *Anim. Behav. Monogr.* **1** (3): 161–311.

Lawick-Goodall, H. van & Lawick-Goodall, J. van (1970). *Innocent killers*. London: Collins.

Le Boeuf, B. J., Whiting, R. J. & Gantt, R. F. (1972). Perinatal behaviour of Northern elephant seal females and their young. *Behaviour* **43**: 121–156.

Lent, P. C. (1966). Calving and related social behaviour in the barren ground caribou. *Z. Tierpsychol.* **23**: 701–756.

Lent, P. C. (1974). Mother-infant relationships in ungulates. In *The behaviour of ungulates and its relation to management*. (Papers of Int. Symp. at Calgary, Canada, 1971). Geist, V. & Walther, F. (eds). *IUCN Publs* **24** (1): 14–55.

Martin, R. D. (1966). Tree shrews: Unique reproductive mechanism of systematic importance. *Science, N.Y.* **152**: 1402–1404.

Matthews, L. H. (1952). *British mammals*. London: Collins.

McBride, G. (1971). *Animal families*. U.S.A.: The Readers Digest Association Inc.

McBride, A. F. & Kritzler, H. (1951). Observations on pregnancy, parturition and post-natal behaviour in the bottle nose dolphin. *J. Mammal* **32**: 251–266.

Merchant, J. C. & Sharman, G. B. (1966). Observations on the attachment of marsupial pouch young to the teats and on the rearing of pouch young by foster-mothers of the same or different species. *Aust. J. Zool.* **14**: 593–609.

Moore, J. C. (1956). Observations of manatees in aggregations. *Am. Mus. Novit.* No. 1811: 1–24.

Mykytowycz, R. & Dudzinski, M. L. (1972). Aggressive and protective behaviour of adult rabbits *Oryctolagus cuniculus* (L.) towards juveniles. *Behaviour* **43**: 97–120.

Newson, R. M. (1966). Reproduction in the feral coypu (*Myocastor coypus*). *Symp. zool. Soc. Lond.* No. 15: 323–334.

Noirot, E. (1969). Serial order of maternal responses in mice. *Anim. Behav.* **17**: 547–550.

Owen-Smith, R. N. (1974). The social system of the white rhinoceros. In *The behaviour of ungulates and its relation to management*. (Papers of Int. Symp. at Calgary, Canada 1971). Geist, V. & Walther, F. (eds). *IUCN Publs* **24** (1): 341–351.

Peterson, R. & Bartholomew, G. (1967). The natural history and behaviour of the California sea lion. *Spec. Publs Am. Soc. Mammal.* **1**: 1–79.

Rowell, T. E. (1960). On the retrieving of young and other behaviour in lactating Golden hamsters. *Proc. zool. Soc. Lond.* **135**: 265–282.

Rowell, T. E. (1961). Maternal behaviour in non-maternal Golden hamsters (*Mesocricetus auratus*). *Anim. Behav.* **9**: 11–15.

Rudge, M. R. (1970). Mother and kid behaviour in feral goats (*Capra hircus* L.). *Z. Tierpsychol.* **27**: 687–692.

Russell, E. &. Giles, D. (1974). The effects of young in the pouch on pouch cleaning in the Tammar wallaby *Macropus eugenii* Desmarest (Marsupialia). *Behaviour* **51**: 1–37.

Sayler, A. & Salmon, M. (1971). An ethological analysis of communal nursing by the house mouse (*Mus musculus*). *Behaviour* **40**: 62–85.

Schaller, G. B. (1963). *The mountain gorilla. Ecology and behaviour.* Chicago: University of Chicago Press.

Schenkel, R. (1967). Submission: Its features and function in the wolf and dog. *Am. Zool.* **7**: 319–329.

Selman, I. E., McEwan, A. D. & Fisher, E. W. (1970). Studies on natural suckling in cattle during the first eight hours post partum. 1. Behavioural studies (Dams). *Anim. Behav.* **18**: 276–283.

Sharman, G. B. (1970). Reproductive physiology of marsupials. *Science, N.Y.* **167**: 1221–1228.

Shillito, E. E. & Alexander, G. (1975). Mutual recognition in ewes and lambs (*Ovis aries*). *Appl. Anim. Ethol.* **1**: 151–165.

Shillito, E. E. (1975). A comparison of the role of vision and hearing in lambs finding their own dams. *Appl. Anim. Ethol.* **1**: 369–377.

Sowls, L. (1974). Social behaviour of the collared peccary *Dicotyles tajacu* (L.). In *The behaviour of ungulates and its relation to management.* (Papers of Int. Symp. at Calgary, Canada, 1971). Geist, V. & Walther, F. (eds). *IUCN Publs* **24** (1): 144–165.

Stephens, D. B. & Linzell, J. L. (1974). The development of sucking behaviour in the new born goat. *Anim. Behav.* **22**: 628–633.

Stodart, E. (1966). Management and behaviour of breeding groups of the marsupial *Perameles nasuta* Geoffroy in captivity. *Aust. J. Zool.* **14**: 611–623.

Venge, O. (1963). The influence of nursing behaviour and milk production on early growth in rabbits. *Anim. Behav.* **11**: 500–506.

Wilson, S. (1974). Mother-young interactions in the Common seal *Phoca vitulina vitulina*. *Behaviour* **48**: 23–36.

Symp. zool. Soc. Lond. (1977) No. 41, 333–339

The Management of Young Mammals

M. R. BRAMBELL and D. M. JONES

The Zoological Society of London, Regent's Park, London, England

SYNOPSIS

In this paper we review the impact that lactation has upon the way mammals are managed in the zoo. Bearing in mind that good wild animal management is aimed towards providing the animals with the means to solve their own problems rather than providing the intense supervision to which laboratory-managed animals might be subjected, we discuss the nutritional, behavioural, social and immunological provisions which need to be considered in order to maintain healthy and self-reproducing groups of mammals.

We also discuss artificial rearing, a field in which the good zoo has little experience, but which does demand on occasion a considerable degree of expertise.

INTRODUCTION

Between birth and the time the young animal can survive on the adult diet, it is dependent on its mother's milk for its nutrition. This period can be divided into three phases: a neonatal phase at the onset of milk production, the main phase of total dependence on milk and the phase of partial dependence or weaning.

The nutritional demands of the young animal on its mother during lactation depend on the length of time that lactation lasts, on the growth of the young animal during that period and on the reserves carried by the young animal at the start of the period. The demand on the mother will also depend on the number in the litter, the relative size of the young animals and on how active the young need to be to survive the lactation period.

Under good management zoo animals should be able to solve their nutritional, behavioural and social problems without day-to-day decisions having to be made on their behalf by their keepers. The ideal is to set up the enclosure, populate it with an integrated social group, provide a diet adequate for all the animals present, and arrange for the minimum of supervision and servicing needed to ensure that all is well. The aim is that every young mammal should be reared by its mother and, with the great majority of young born in the Zoological Society's collection, this is fortunately what happens. It follows that if the management of the social group is good, fewer problems will be encountered with lactation.

Depending on the variation in these factors for each species, we can expect lactation to range from a short period in which, nutritionally, a comparatively incomplete milk may be secreted, to a long period in which, for practical purposes, we must assume that all the body tissues of the young animal have to be derived from milk. The maternal demands will range from a period involving a slight increase in her nutritional load to one where the extra demand of lactation totally dominates her nutritional economy.

THE INITIAL PHASE

Although a mother may carry her litter to full term and give birth normally, she does not necessarily have sufficient capacity to provide the young with milk. Amongst the large cats for instance, first litters often do not thrive as well as subsequent litters; it seems that the mother may not be physiologically and behaviourally adapted to lactation until she has had (and lost) a first litter. Of 15 lion cubs born in first litters at Regent's Park since 1960, 13 showed some signs of inadequate nutrition; but of 21 lion cubs born in second and subsequent litters only seven showed symptoms which may have had a nutritional origin. The female cheetah which began the breeding of that species at Whipsnade had three cubs in her first litter. All three suffered developmental abnormalities which appeared to be nutritional in origin, yet the cubs of her second and subsequent litters showed no abnormalities. It is not clear if this physiological and behavioural "apprenticeship" is common in other groups, but our impression is that it is widespread.

Even if the mother is physiologically able to support her litter, there is a risk that milk production and suckling will be inhibited if she cannot nest or deposit her young in a secure place. She may be too restless to allow the young to reach her teats or she may hold her young too tightly to allow it to explore for the nipples. Any animal under stress may fail to let down her milk to her infant, and if the manager becomes too anxious about this situation his attention may be counterproductive. The phenomenon is well known in the farming world where mares may reject foals and where whole herds of pigs can show an "epidemic" agalactia, curable by sending the pigman on holiday. There is no doubt that the best management advice is to leave well alone initially and let the mother settle into lactation by herself. It is notable that zoos where great efforts are made to observe maternal behaviour after birth handrear a higher proportion of their young than we have to do in the Society's zoos.

The length of time a mother may take to come fully into milk may be much longer than is generally realized. Humans have taken as long as five days without any apparent adverse effect on the baby, and many gorillas do not show signs of being in milk for 24 hours or more. The braver managers have allowed gorilla infants to go without milk for almost 48 hours before interference was attempted. One American zoo believes in leaving them for 96 hours after birth, and so far they have not had to remove a baby for handrearing (L. Fisher, pers. comm.). Our general policy is to allow larger mammals between two and three days before interfering.

Perhaps the most spectacular example of an apparently delayed onset of lactation is that of the tree-shrews. These animals give birth to their young in a separate nest to that occupied by both adults, and the female only returns to suckle once every 24 to 48 hours. Until these two facts were recognized it was always assumed that the young had been abandoned after birth and that, to survive, they had to be reared artificially (Martin, 1966).

The manager can encounter some unexpected problems in the early stages of lactation. There is a real danger of ungulate calves imprinting themselves on the wrong animal, or even on a human attendant if its mother has had to be immobilized for surgical assistance at the time of birth. In addition, the recovering mother will not have consciously experienced the birth and may therefore take no interest in the calf, not realizing it to be her own. For different reasons duikers do not seem to distinguish their own young from any other, nor do young distinguish their own mothers (Ralls, 1970). In the wild no disadvantage arises from this, but if a number of females and young are kept together in captivity this lack of discrimination can lead to social problems in the group.

The immunoglobins in the colostrum of many animals are absorbed whole through the alimentary tract of the neonate and form an important contribution to the defence of the young animal against potential pathogens of which the mother has had recent experience. This passive transfer of immunity through milk takes place in marsupials, insectivores, some rodents, carnivores, and perissodactyl and artiodactyl ungulates. In some of these groups there has also been passive transfer of immunity prior to birth, reducing the importance of the post-natal transfer (Brambell, 1970). The capacity to absorb immunoglobulin unchanged is usually lost between 24 and 48 hours after birth, but in the case of the hedgehog it can last for 24 days after birth (Morris, 1963).

When handreared animals are removed from their place of birth they tend to be kept in clean environments where they are usually challenged by a different range of organisms from those experienced by the mother. For these animals the lack of colostrum seems to be less important in practical terms than is generally assumed, provided that first-rate hygienic standards are maintained during the rearing period. By weaning they are immunologically competent and can cope with the return to "field" conditions. Nevertheless when it is practicable we collect serum from the mother to give to the young animal, either in its food or by intravenous infusion.

THE MAIN PHASE

The period in which the young is totally dependent on the mother for all its energy requirements may place a considerable burden on the energy metabolism of the mother. To maintain the mother through this period it is necessary to understand the added stresses which could arise and to offset these by providing more food, more security and less competition. The more we understand of the natural history of the species, the better able we are to meet its requirements in captivity.

Apart from marsupials and many of the primates and bats which carry their unweaned young more or less continuously, most of the smaller mammal species keep their young in nests or at least in protectable retreats until they are mobile. Most of the larger species with relatively well developed independent young nurse them in an area where predators can be easily spotted. There is a difference in behaviour between large species living in a habitat with dense tree and shrub cover where there is a tendency for the young to lie concealed between feeds, and those living in open country where the young remain with the adults.

As we have already commented, there is a range in the importance of lactation in this phase. At one end we have the young guinea-pig which has a 50:50 chance of survival even if it gets no milk at all, so good are its reserves at birth and so short the time before it can use solid foods (J. L. Linzell, pers. comm.). At the other extreme a baby kangaroo weighs less than a gram at birth and is dependent solely on its mother's milk for at least 190 days by when it will have grown to 5 kg. In contrast, the kangaroo's eutherian equivalent, the bovid, may produce milk with relatively low concentrations of iron, copper and fat-soluble vitamins; the normal new-born calf has adequate reserves of these in the liver to maintain it over the pre-weaning period. The kangaroo joey cannot possibly have such reserves and all of its requirements must reach it through milk.

Other contributors to the symposium have commented on the kangaroo's adaptation to support simultaneously two young at different stages of development and on different milks. Lactation lasts for very much longer in the marsupials and is adversely affected by lowering the nutritional plane of the mother (Tyndale-Biscoe, 1973). In some species we know that stress can cause premature rejection of the young from the pouch before they are thermally independent.

Looking from the mother's point of view we have at one extreme the elephants and rhinoceroses in which a mother weighing two tonnes may give birth to a baby weighing 70 kg. This will grow to 500 kg in the next eight months almost solely on its mother's milk. The mother's basal metabolism is in the region of $0\cdot009$ kcals/g/day. The energy required by the baby to maintain its basal metabolism and for the growth of new tissue will add about $0\cdot003$ kcal/g/day, i.e. 33%, to the mother's metabolic load. At the other extreme a female common shrew weighing 11 g may have a litter of eight, each weighing 0·5 g and each growing to 7 g in the next three weeks, i.e. to a total weight of five times the weight of the mother. This means that the mother's basal metabolism of about 0·2 kcal/g/day has an added load of 0·9 kcal/g/day imposed on it by the requirements of her litter, making a total of over 1 kcal/g/day, an increase of 500%. If the shrew is being kept under marginal nutritional conditions it is probable that she would be unable to maintain the required energy output to the end of normal lactation.

THE THIRD PHASE: WEANING

Weaning is the period of changeover from total dependence on milk to total dependence on the adult diet. It is usually a gradual process but it may take place abruptly. In tree-shrews and in many seals the mother ceases to suckle the young before they have started to feed on the adult diet. It is interesting that in both groups, whose ways of life are very different, the mother only suckles after long intervals, and her milk contains much more fat than that of most other mammals. When weaning occurs these young have only those reserves of energy already in their bodies on which to draw until they are able to fend for themselves.

In most mammals weaning is a more gradual process and for those species with simple stomachs the overlapping of the two types of diet does not lead to any serious problems. It is in those groups which have complicated forestomachs adapted for the microbial digestion of cellulose or chitin that there is a conflict between the two diets. Milk requires very little enzymic action to render it absorbable. Cellulose requires an enormous amount of enzymic action before the mammal can draw any benefit from having eaten it. If the milk is given the same treatment as the cellulose, either the milk is effectively destroyed by too much enzymic action, or the cellulose is not digested. In plant-eating animals with a simple stomach the cellulose is not subjected to microbial fermentation until it has passed the small intestine where the milk is absorbed, and has reached the large intestine. However in the kangaroos and wallabies, the leaf-eating langurs and colobus monkeys, the sloths, the whalebone whales (which feed on chitin-bound krill), the sirenians and the ruminating artiodactyls, microbial fermentation takes place anterior to the small intestine. In all these animals there is a similar mechanism to separate the milk from the plant foods. Connecting the oesophageal opening of the stomach to the glandular part of the stomach are two muscular ridges. When these contract they effectively form a tube along the stomach wall. This oesophageal groove contracts on the stimulus of sucking and drinking and does not relax until the session is over. Thus milk is diverted from the oesophagus directly to the absorptive part of the gut.

Without this adaptation the fermentation of cellulose in the anterior stomach would be impossible because the growing animal would not have time to establish its adult digestive mechanism before it starved. From time to time individual young animals fail to react properly to the stimulus of sucking, as when they have been frightened during attempts to persuade them to accept bottle feeding. Milk may pass into the fermentation chamber and the animal may die from digestive malfunction, with foetid and flocculent milk filling the developing rumen.

Weaning can present other problems. The young animal has yet to establish itself within the social group. It may be inhibited from feeding properly because of its low ranking. It may not be able to eat fast enough. If the food is wrongly presented for its size, it may have difficulty in handling the food. It is at this time that many animals which have hitherto done well because of their mothers' performances start to lose condition and to starve. The most spectacular example we have experienced is that of weaning sea-lions. Unlike seals these animals remain suckling from their mothers until they are almost one year old; they learn to feed on fish by playing with them long before they need to swallow them. In the zoo this presents a very difficult problem. Even if the young sea-lion is seen to be playing with fish, this is no guarantee that it is ready to eat them. Since the age of weaning can vary from eight months to two years, great patience is needed before a youngster can be left to fend for itself among the adults.

HANDREARING

We will not try to review this field but will make one or two points which we think are important. Experience bears out that the maintenance of good hygiene and a standard consistency of milk are more important than substituting a correct composition. Where milks of the right composition are available we unhesitatingly recommend them, but to struggle to alter a composition may lead to lack of consistency and to a lowering of hygiene. Sometimes it is impossible to feed a milk whose composition approaches that of the species. For example the fat content of sea-lion milk is about 35% (Ben Shaul, 1962) but reconstituted milk with this level of fat has a cheese-like consistency and cannot flow.

Even wallabies whose milks are very different from eutherian milks thrive well on powdered human milk substitutes (for example Ostermilk 2) with vitamin E supplement, although it should be pointed out that the wallabies were well advanced in lactation before they first needed to be handreared. Standard bottled domestic cows' milk has provided excellent results with antelopes, horses, camels and giraffes. As the lactation develops the young become more tolerant of changes in consistency and the milk can be thickened with human baby cereal feed (for example Farex). Only with musk ox amongst the bovids have we had trouble with feeding milk alone, but with the addition of lactobacillus to the milk the young musk oxen thrive.

Though several monkeys and great apes have been reared on powdered baby milks or on diluted condensed milk (Carnation), we have had the least trouble with the baby gorilla born in July 1976 and handreared since she was nearly a month old. On the advice of Dr Beryl Corner we used Wyeth's SMA Gold Label, a pre-sterilized ready-to-use pack in which day-to-day variations in consistency and hygiene are virtually impossible, and which has been modified to bring its salts, fats and protein into line with human requirements. Powdered milks are available for domestic dogs and cats and they can also be used for wild carnivores.

Wherever possible handrearing is to be avoided for even if the youngster survives the lactation period it probably has to mature outside a social group and thus becomes progressively more difficult to reintegrate. There is no doubt that mother-reared babies are best.

REFERENCES

Ben Shaul, D. M. (1962). The composition of the milk of wild animals. *Int. Zoo Yb.* **4**: 333–342.
Brambell, F. W. R. (1970). *The transmission of passive immunity from mother to young.* Amsterdam: North-Holland Publishing Co.
Martin, R. D. (1966). Tree shrews: unique reproductive mechanism of systematic importance. *Science, N.Y.* **152**: 1402–1404.
Morris, B. (1963). The selection of antibodies by the gut of the young hedgehog. *Proc. R. Soc.* (B) **158**: 253–360.
Ralls, K. (1970). Duikers from African forests to 'African Plains'. *Anim. Kingd.* **73**(5): 18–23.
Tyndale-Biscoe, H. (1973). *Life of marsupials.* London: Edward Arnold.

Symp. zool. Soc. Lond. (1977) No. 41, 341–358

Comparison of Milk-Like Secretions Found in Non-Mammals

A. CHADWICK

Department of Pure and Applied Zoology,
University of Leeds, Leeds, England

SYNOPSIS

The transference of nutritive secretions from one individual to another is a widespread though sporadic occurrence in the animal kingdom. In insects the ant–aphid relationship, trophallaxis and royal jelly are described. In vertebrates the nutritive secretions of the ovary and oviduct, which are important in fishes, are described and the special case of secretion of a nutritive ectodermal mucus is considered.

Although in birds most nutritive material transferred from parent to young consists of partly-digested food, the pigeon is an exception, and the origin, nature and control of secretion of "pigeon milk" is considered in detail. Avian prolactin has been purified and results obtained using a radioimmunoassay are included. Aspects of the nature of milk and its possible secretory origins are discussed.

INTRODUCTION

Taking the term *milk* in a narrow sense would eliminate from consideration virtually all the cases in invertebrates where there occurs a secretion having some affinity with that produced in lactation in mammals, even if it were possible to include one or two of the milk-like secretions in non-mammalian vertebrates. However, by taking as a definition of "non-mammalian milk" any nutritive product, not a primary or unmodified food material, which is passed between animals of the same or of different species, it is possible to include a number of interesting examples in the invertebrates which do, in fact, closely reflect some of man's own concerns with milk. Since any milk-like secretion must consist largely of water it is not surprising that examples are encountered almost exclusively in terrestrial animals, both invertebrate and vertebrate, and this itself is a highly significant factor in any discussion of the primary function of milk-like products. However, our definition is wide enough to include the many examples in aquatic vertebrates of nutritive materials passed from mother to offspring quite apart from the suckling seen in cetacean species.

Birds of course are neither viviparous nor do they lactate, despite classical allusions to the pelican. Quite apart, however, from the special position which the pigeons and doves hold in the story of research into lactation, the close parallel between some aspects of avian and mammalian physiology, in particular with respect to homothermy and to water regulation in an arid environment, warrants a careful investigation of the role of prolactin in this class.

INSECTS

In insects there are many examples of secretions manufactured and released under specific circumstances which are ingested by other members of the same species or by members of different species. Some of these secretions are clearly nutritive and serve to nourish the young larvae during a phase of rapid growth. Other secretions serve a more conjectural nutritive function but are nevertheless important in the life of the colony in certain social insects. In yet other cases the secretions and even excretions produced by one species are utilized by another species for their own purpose.

Ants and Aphids

The relationship between the ant (*Formica* sp.) and the aphid (*Lachnus* sp.) shows a ready parallel with that between man and the dairy cow. This relationship was first recorded by Huber (1810) who accurately described how an ant solicits honeydew using its antennae and how the aphid voids the fluid from its anus, upon which the fluid is immediately imbibed by the ant. Aphids, which include the common greenfly and blackfly, are homopteran plant parasites—sucking insects which eat phloem sap from plant cells. The sap is normally under positive hydrostatic pressure in the plant and will flow into the insect's alimentary canal without suction through the piercing stylet once the cell has been located, though this latter process may take several hours. Once feeding, an aphid can ingest well over its own weight in food every hour, so most of this fluid must pass straight through the alimentary canal to be voided periodically through the anus.

Because of the ready food supply in summer the population growth rate of aphids can be very high, as all gardeners know. In warm weather aphids reproduce parthenogenetically; all are female and are born as fully-formed miniature adults with the embryos of the next generation already developed in their ovaries. They take only two weeks to reach maturity and theoretically a single aphid could yield a population of millions of descendants in the course of a single summer.

The sap flowing through the insect is modified to some extent during its passage. About 50% of the nitrogen-containing compounds, which are present in only very small concentrations, are removed by the insect for protein synthesis and growth (Mittler, 1958). The original sugar content of the phloem sap is between 10% and 25% sucrose; this sugar is modified in the alimentary canal to yield oligosaccharides including glucosucrose.

The honeydew, which is secreted by several other types of phytophagous insects as well as by many species of aphid, is normally shaken off or ejected some distance onto adjacent areas of the plant, where it accumulates. A secretion of this nature from a coccid bug *Trabutina mannipara* is almost certainly the biblical manna, and it is still collected as food by Arabs in some parts of the Middle East (Wilson, 1971); Australian aborigines are also said to collect a similar substance. Many species of insect, mainly ants, collect honeydew and use it as food. Whereas an ant might normally fall upon and devour a small insect like an aphid, they are seen instead to stroke the flanks with their antennae. The aphids respond by raising their abdomen and expelling the honeydew, which is then removed by the mandibles of the ant. Many types of ant husband their aphids, herding them to selected areas and protecting them from predators, and even, in the case of root-sucking aphids, preparing shelter underground for them and letting them out occasionally for an airing. Aphids appear to benefit from this protection, growing and proliferating more rapidly under the care of the ants (Way, 1963). In certain species of ant small castes "milk" the aphids, transferring their product to larger and faster intermediaries which take it back to the ant colony. Some species of ant living in arid environments even have repletes—ants which act as storage reservoirs and which become turgid with a store of honeydew in their abdomens. Stumper (1961) showed that these honeypot ants store honeydew in cool damp weather and release it back into the colony in hot dry weather, inferring a water-storage function.

Thus there is quite extensive analogy between the mutualism of the ant and the aphid and the organization of the dairy industry despite the discrepancy in the nature of the raw material employed!

Bees and Wasps

Some insects do produce nutritive secretions specifically for the purpose of providing for the rapid growth and development of their young, and this is seen particularly well in certain other hymenopterans. Royal jelly is perhaps the most well known, albeit a rather specialized example, of these secretions. It is produced by the hypopharyngeal and mandibulary glands, and perhaps by other glands as well, in the female worker bee and if fed to a female larva it will produce a queen. It appears only to be produced when the amount of queen substance (9 keto-*trans*-2-decenoic acid) released by the active queen falls to a level insufficient to satisfy the workers' daily requirement. The workers first of all build queen cells and they then start to feed the larvae developed from eggs laid in them on royal jelly, a watery fluid containing sugar, proteins, B vitamins, ADP, ATP, RNA, DNA and many other substances including royal jelly acid (10-hydroxy-*trans*-2-decenoic acid) (Barbier, 1968). This food brings about the very rapid growth of the much larger queen bee (only 16 days to reach maturity compared with 21–24 days for a worker), which has much reduced mouthparts, eyes, antennae and brain, but large mandibulary glands for the production of queen substance and a well developed reproductive apparatus capable of laying 1000 eggs per day.

In other cases the secretions of the salivary glands are transferred between individual members of social insect colonies, a process called, rather imprecisely, trophallaxis. Even members of different species of insect sometimes engage in this. In certain wasps the secretions of the salivary glands of the larvae appear to be yielded up as a reward to the adult which solicits this fluid from the larvae but only receives it after it has fed the larva (Wheeler, 1928). Thus mutual benefit and social cohesion result. *Vespa orientalis* shows a remarkable extension of the importance of this process in the life of the colony which almost reverses the normal dependence for food of the offspring upon the adult (Jekan, Bergmann, Ishay & Gitter, 1968). The adult wasps appear to lack chymotrypsin and carboxypeptidases A and B and they cannot carry out gluconeogenesis; the larvae, on the other hand, possess these enzymes and the ability to synthesize sugars, so a biochemical division of labour exists. The adults, which can utilize a sugar-rich diet but do not need much protein, feed the larvae, which then use the protein in the diet for growth. In return the larvae supply, by trophallaxis, a saliva rich in glucose and trehalose with a sugar concentration of 9%; this is three times that of larval haemolymph. The saliva contains only one-fifth of the haemolymph concentration of amino acids and proteins.

These few examples of nutritive secretions in insects must suffice to show that there is a fairly widespread occurrence and complex use of such secretions in certain invertebrates. Whether any underlying theme regarding the usefulness of such secretions in the relatively arid terrestrial environment can be detected is a matter of conjecture. What must be remembered is that evidence as to their direct control by hormones is so far totally lacking.

VERTEBRATES

The subphylum Vertebrata represents only a tiny fraction of the variety of animal life—there are more beetle species than all other animal species including the host of other insects. However in this Symposium vertebrates are of special interest as the parent group of animals which lactate; furthermore they include more than one class of animals which have conquered the terrestrial environment, a significant point in the context of milk secretion.

Is it possible to recognize precursors of the milk secretion of mammals in descendants of their lower vertebrate ancestors? There are a few cases in the lower vertebrates in which secretions of comparable nutritive value are encountered. These cases are not widespread and the secretions are not, of course, homologous with milk since the mammary glands and their secretion are diagnostic mammalian characters. They are nevertheless worthy of consideration and they may throw light on the way in which milk secretion might have arisen in primitive, but now extinct, mammals. So far as I am aware there are no cases of milk-like secretions, fitting my definitions as to nutritive value, to be found either in living amphibians or in living reptiles, although in ovoviviparous forms the transfer of water at least must be envisaged. Why this should be will be referred to later but it is worth bearing in mind that both these classes consist today of relics of groups which were formerly important but have been largely replaced by mammals and birds, leaving only a few representatives, which are nevertheless successful in their specialized modes of life. What formerly existed in the way of nutritive secretions in these groups we cannot say at present but it seems not unlikely, for reasons that will become clear, that such adaptations have occurred in the past.

Fishes

Although, to judge from the succcession of different types throughout the geological past, the selective pressure on fishes has been just as great as on the terrestrial vertebrates, nevertheless the basic fish organization has remained the predominant vertebrate form in the aquatic environment, despite small incursions by reptiles, birds and mammals. Thus it appears that representation of all the different patterns of parent–offspring relationship has survived to the present time. Parental care in fishes ranges from a total lack of concern for the egg and the larva, relying on external fertilization in the surface layers of the sea, to true viviparity and some concern for the interests of the offspring. In many cases, ranging from the Siamese fighting fish (*Betta splendens*) through the common stickleback (*Gasterosteus aculeatus*) to the African mouth-breeder (*Tilapia mossambica*), considerable parental care is exercised, even though the eggs are fertilized and hatch outside the mother's body. Whilst in most of these cases there is no semblance of any milk-like secretion, it is worth noting that there is a good deal of evidence that the administration of prolactin produces significant changes in behaviour in fishes (see, for example, Blüm & Fiedler, 1965). On the other hand there are many cases in which viviparity or ovoviviparity is accompanied by total unconcern for the offspring after parturition. Do milk-like secretions play any part in this diverse picture? If we take the nutritive secretion derived from whatever organ as our model then there certainly are a number of examples occurring particularly in viviparous and ovoviviparous cartilaginous and bony fishes. Several species are fully adapted to viviparity, with the yolk sac of the embryo being much reduced and pseudoplacentae supplying all the exchange and absorption needed by the developing embryo, for example, the teleost *Heterandria formosa* and the elasmobranch *Cetorhinus* sp. (Matthews & Marshall, 1956). However, other species appear to be less well adapted and growth of the embryo depends on the ingestion by mouth of secretory products formed by the ovary in which it grows, for example, in the *Jenynsiidae* and the *Goodeidae*.

In the viviparous blenny, *Zoarces*, fertilization is internal and the eggs hatch after about three weeks. The young remain, however, in a cavity in the ovary for at least a further three months and at parturition they are relatively large, about one-sixth the length of the parent. The fluid surrounding them is secreted by the ovary and it nourishes them and provides for their growth. In some elasmobranchs secretions from villi on the inner wall of the oviduct serve to nourish the developing young, and in one Indian ray, at least, villi are observed to pass into the spiracles of the embryo and pour the so-called "uterine milk" down the throat of the foetus (Jenkins, 1925). In other cases the villus appears to grow continuously and to be digested at its distal end by the young. What part the pituitary glands play in regulating these secretions is as yet unknown, but a close similarity can be observed between the hypertrophied ovarian follicular cells after ovulation in some of these fishes and the *corpora lutea* in mammals. Whilst mammalian milk can be in no way homologous with this uterine milk, as the latter is not integumentary in origin, it may be remarked in passing that in man a uterine origin for milk was conjectured in ancient times and that Leonardo da Vinci may have held this belief since he illustrated a duct joining the uterus to the nipple in his sketch of the internal organs of a human couple in copulation.

There is one well documented case in teleost fishes where a nutritive secretion of ectodermal origin is produced and where, moreover, its endocrine control parallels that in mammals. This is the case of the cichlid fish *Symphysodon discus*, a beautiful and popular tropical aquarium fish from South America. It is not alone in producing ectodermal mucus in response to prolactin but the curious history of attempts to breed this fish under artificial conditions is worth recalling. Normally the adults are removed after spawning in this type of fish to prevent the eggs and fry from being eaten by them. In the case of the discus this results in total failure to rear the fry, but if the parents are allowed to remain the fry can be reared successfully. After hatching the fry appear to pick at the side of one of the parents and they do in fact eat the copious mucous secretion which appears at this time on both parents. They live off this secretion for about five weeks, feeding first on one parent, then the other, the brood being very neatly transferred as the adults swim along side by side (Hildemann, 1959). The nature and composition of the mucus has not been studied but the considerable synthetic ability of ectodermal mucous cells in fishes is well known and their proliferation and activity has been shown in some cases, including that of the discus fish, to be induced by treatment with mammalian prolactin (see review by Nicoll & Bern, 1971). Thus there is a striking analogy not only between the general character of the secretion and its cellular origin in fishes and in mammals but also between the endocrine factor which controls its production.

Birds

Very few birds show any secretory activity even remotely comparable with lactation. Nest building and parental care are highly developed in most birds, only a very few abandoning their eggs to be incubated by heat of purely environmental origin. Although not viviparous, birds are warm-blooded, and in one of the two major divisions of birds a great deal of parental attention as regards warmth and feeding is demanded by the newly hatched offspring. Usually, food carried to the nest by the parents is regurgitated in a more or less digested state in response to signals from the young. Water will also be obtained in the same way but the requirements are relatively low. In some birds, for example the parrot family, the food has been so far digested before regurgitation that it resembles milk and is sometimes referred to as such, but nevertheless it appears merely to be the product of digestive processes in the fore-gut.

Pigeon

It is only in the columbiform birds, the doves and pigeons, that a true nutritive secretion, specifically produced for feeding to the nestlings, is encountered; this is the well known "pigeon milk". The importance of this phenomenon in the pigeon to the history of research into prolactin and milk secretion in mammals is one of the striking contributions of comparative endocrinology. Even the term prolactin was coined in connection with studies on crop gland secretory activity in the pigeon (Riddle, Bates & Dykshorn, 1932). Most of the early work on the physiology of prolactin was intimately bound up with the pigeon crop-sac response, as this provided a ready way of detecting and measuring the lactogenic hormone (Lyons & Page, 1935; Folley, Dyer & Coward, 1940).

While incubating the eggs, the walls of the crop sac in parents of both sexes start to thicken. By the time of hatching the lateral areas of the crop—the so-called crop glands— are very greatly thickened and their mucosal lining is thrown into folds. The secretion is formed over the entire area of thickened tissue and takes the form of a semi-solid cream-coloured mass composed of irregularly-shaped particles roughly 8 mm × 3 mm × 2 mm. This form probably results from the folded nature of the surface and the continuous sloughing of the entire surface layer of mucosal cells. Thus pigeon milk is a holocrine secretion formed by the wholesale degeneration or conversion of the cell structure of the outer layers of mucosal epithelium. These layers are constantly replaced by cell division in the proliferative epithelium next to the *lamina propria* (Beams & Meyer, 1931). During the first day or two after hatching the parents appear not to feed; no grain is found in their crops, but only crop milk, which they regurgitate and thrust down the throats of the nestlings. The crops of the young soon become extremely distended with this milk, which is presumably passed into the alimentary canal for digestion. Later, the young are fed on a mixture of grain and milk and by three weeks the parents' crop glands have regressed and milk has disappeared from the diet.

This sequence of changes in the crop sac of adult birds can be imitated by administering prolactin to the birds by subcutaneous injection (see, for example, Dumont, 1965; Chadwick, 1966). In addition the local injection of prolactin produces a small localized crop-sac response which is useful for detecting small amounts of the hormone (Lyons & Page, 1935; Forsyth, Folley & Chadwick, 1965). There are two main cytological characteristics of the crop-sac response: the vigorous proliferation of the squamous cells, which constantly displace previously formed cells towards the lumen of the crop; and the appearance of lipid droplets or granules within the cells as they are displaced. In the outermost layers all cell structure breaks down. The lipid is visible as refractile bodies in fresh-frozen sections and it can be stained with lipophilic dyes. The rapid cell proliferation results in the folding of the epithelium, as there appears to be no room for the cells on the basement membrane without this. The degree of thickening and folding and the amount of visible lipid are proportional to the amount of prolactin injected, as indeed is the total weight of the crop (Folley, Dyer & Coward, 1940; Nicoll, 1967; Chadwick & Hall, 1975). The prolactin referred to is mainly of mammalian origin, usually ovine, but recently it has been shown that purified avian prolactin stimulates the pigeon crop in an identical manner (Chadwick & Jordan, 1971; Scanes, Bolton & Chadwick, 1975). On the other hand it has often been shown that prolactin from the pituitaries of poikilothermic vertebrates is deficient in the ability to stimulate fully the pigeon crop (see, for example, Le Blond & Noble, 1937; Nicoll & Bern, 1968; Chadwick, 1970). In view of the relative absence of stainable lipid from the crop-sac mucosal response to lower vertebrate "prolactin" preparations, it has been suggested that the hormone molecules might possess in differing degrees the ability to initiate a process of conversion in the basal nutritive layer which leads to the appearance of sudanophilic droplets (Chadwick & Jordan, 1971).

Crop milk. The chemical composition of "pigeon milk" has not received a great deal of attention and whilst carbohydrate and protein components are undoubtedly present only the nature and composition of the lipid fraction have been extensively studied (Dumont, 1965; Chadwick & Jordan, 1971). Crop milk is composed of about 25% solids and 75% water, not differing significantly from the crop-sac tissue itself, either in the stimulated or in the unstimulated condition (Table I). However, lipid forms a higher percentage (51%) of the solids in the milk than of the solids from stimulated crop tissue (24%); unstimulated crop-sac tissue contains an even lower fat percentage (14%). A progressive rise in the fat content of the crop appears to be a feature of the crop-sac response. This, allied to the progressive disappearance of all cell structure in the formation of the crop milk recalls a condition of fatty degeneration. The resulting food is

TABLE I

Comparison of the water content and of the fat content of crop glands, crop "milk" and depot fat from injected and non-injected pigeons

Material	Treatment (intradermal injection)	No. of pigeons	Crop-sac response*	Dry wt: wet wt (%)	Lipid: dry wt (%)
Crop epithelium	Water	10	0	29·7 ± 1·4	14·2 ± 1·0(7)
	33 μg ovine prolactin	18	3–4	27·4 ± 0·6	23·6 ± 1·2(13)
	Non-mammalian pituitary preparations	10	2–3	25·9 ± 1·6	26·5 ± 1·5(10)
Crop "milk"	—	7	—	25·6 ± 1·1	50·7 ± 5·8(4)
Depot fat	—	7	—	88·3 ± 0·2	94·4 ± 0·4(2)

* Response on the scale 0 to 4 (Forsyth *et al.*, 1965);　† = not significant;　‡ = $P < 0.001$; numbers in parentheses = number of estimates. (From Chadwick & Jordan, 1971, with permission.)

clearly a rich energy source and in addition contains much metabolic or oxidative, as well as free, water. Table I shows that there was no difference between the lipid content of crop tissue proliferating in response to treatment with mammalian prolactin and that responding to non-mammalian pituitary preparations. It has already been pointed out that in the former case visible lipid droplets were prominent whereas in the latter visible lipid was absent. Dumont (1965) showed that the stainable lipid consisted of a neutral unsaturated triglyceride. The fatty acid composition of the triglycerides was investigated in an attempt to discover a reason for this apparent discrepancy. Whilst in the ealier publication (Chadwick & Jordan, 1971) it was remarked that an insufficient number of individual fatty acid determinations had been made to enable an assessment of the significance of the small differences observed, nevertheless, by pooling the observations on a rational basis some significant differences have emerged (Table II). It can be seen that by pooling, firstly, untreated controls with saline-injected control data, secondly, all poikilothermic animal pituitary preparations (frog, teleost and cartilaginous fish), thirdly, all homeothermic animal pituitary responses (ovine prolactin, pigeon prolactin, parent crop), and fourthly, milk from both parent and nestling, that reasonably large numbers of comparable examples were obtained. Depot fat was included for comparison. Oleic acid formed 40–50% of the total fatty acid content in every case. In depot fat it was 49% and it was present at a similar concentration in unstimulated crop epithelium from control pigeons. The proportion of oleic acid was also unchanged in hypertrophied crops which showed no visible lipid, i.e. crop

TABLE II

Fatty acid composition of triglycerides extracted from pigeon crop epithelium, crop "milk" and depot fat

Source of lipid extract	No. of pigeons	Crop-sac response	C_{10}–C_{14}	C_{16}	Fatty acids present: % means ± s.e. $C_{16:1}$	C_{18}	$C_{18:1}$	$C_{18:2}$	$C_{18:3}$–C_{22}
Depot fat	2	—	2·5	18·5	9·1	8·8	48·8	11·2	5·3
Control crop	4	0	3·1	20·3	8·4	7·6	48·6	9·6	3·9
			±1·0	±0·8	±0·4	±0·3	±0·2	±1·1	±0·6
Stim. crop*	8	3‡	2·6	18·2	7·4	7·1	49·8	10·1	3·1
			±0·6	±0·5	±0·2	±0·4	±1·5	±0·8	±1·0
Stim. crop†	8	4§	1·5	20·8	9·2	8·1	44·0	11·9	5·6
			±0·7	±0·5	±0·7	±0·5	±0·6	±0·8	±1·0
"Milk"	4	—	1·0	19·8	12·0	8·8	44·2	9·8	5·0
			±0·1	±0·1	±0·1	±0·4	±0·5	±0·1	±0·2

* Crop epithelium from pigeons injected with prolactin preparations from poikilothermic animals (see text).
† Crop epithelium from pigeons injected with prolactin preparations from homeothermic animals (see text).
‡ and §: Crop sac responses not equivalent to each other (see text).
‖ $P < 0.03$.
¶ $P < 0.002$.

tissue stimulated by fish or amphibian pituitary preparations. However in crop milk itself and in the crop mucosal epithelium from pigeons injected with avian or mammalian prolactins the proportion of oleic acid was significantly reduced (44%). Thus there appears to be some shift in the proportions of this unsaturated fatty acid, and also possibly in other fatty acids, in association with the occurrence of visible lipid droplets in the cells and with the natural production of crop milk.

Effects of prolactin. There is no reason to doubt that the pigeon crop-sac response is specific for prolactin, particularly in view of the recent report that highly-purified pituitary prolactin from the domestic fowl is capable of promoting a crop response and that the potency is comparable to that of highly-purified mammalian prolactin (Scanes, Bolton & Chadwick, 1975). Reports of crop stimulation by human growth hormone and human placental lactogen (Chadwick, Folley & Gemzell, 1961; Josimovich & Maclaren, 1962; Forsyth et al., 1965) may be explained by the prolactin-like activity of these hormones (Lewis, Singh & Seavey, 1972). The ability of thyrotrophin-releasing hormones (TRF) to stimulate crop-sac proliferation in pigeons (Chadwick & Hall, 1975) is probably a consequence of the prolactin-releasing potential of TRF, and these observations also support the concept that the mechanism of prolactin release in birds is through a releasing factor rather than an inhibiting factor, as is seen in mammals (Hall, Chadwick, Bolton & Scanes, 1975). Despite this difference between mammals and birds the situation in the pigeon must be seen as being remarkably parallel to that in the mammal; secretory glands develop under hormonal influence in both cases and nutritive products, on which the juveniles feed for at least a short initial period during their independent existence, are released. Moreover, it is equally striking that in both cases the phenomenon is only important for a very limited period in the life history of the individual despite its overriding importance for the survival of the species. Therefore, it may well be asked whether prolactin has other functions which demand its presence in both male and female at all stages of the life of the individual. A wealth of information has slowly been accumulating regarding the multitude of roles of prolactin in mammals in addition to that of promoting milk secretion (see reviews by Horrobin, 1973, 1974), and it is appropriate here to consider what has emerged recently regarding prolactin physiology in birds.

Control of prolactin secretion in birds

The establishment of a radioimmunoassay for prolactin, capable of measuring the concentration of prolactin in body fluids as well as in pituitary glands and in incubation media (Scanes, Chadwick & Bolton, 1976), has permitted the investigation of hormone levels in different physiological states in the domestic fowl. Prolactin has also been measured in other galliforms, including the grouse *Lagopus*, the turkey *Meleagris* and the quail *Coturnix*, but not, unfortunately, in the pigeon, as the assay appears to be too specific to allow this. Lack of cross-reaction in pigeon and duck preparations indicates species differences at least in antigen sites on the hormone molecules; in addition no cross-reaction is obtained in ovine prolactin radioimmunoassay systems. We have reported that prolactin levels are, in general, higher in young chickens than in adults and that whilst they are higher in males than females up to 12 weeks of age, on reaching maturity the laying hen has a much higher prolactin level than the adult cockerel (Scanes, Chadwick *et al.*, 1976; Harvey, Scanes, Falconer, Bolton & Chadwick, 1977). When hens were subjected to a salt load, either orally or by intravenous injection, the plasma prolactin concentration increased by about 50%; intravenous water loads had no effect. When an acid extract of chicken hypothalamic tissue was administered intravenously and serial blood samples taken from the contralateral brachial vein, the time course of the dramatic rise and subsequent fall in plasma prolactin could be followed. Fifteen minutes after treatment, the plasma prolactin concentration attained its maximum—to reach a level 18-fold that obtaining previously (Bolton, Chadwick, Hall & Scanes, 1976). More recently even higher levels, representing up to 40-fold increases, have been recorded within 10 min of treatment with hypothalamic extracts, a striking confirmation of the predominance of releasing substances in hypothalamic extracts of birds, a situation which is in complete contrast to that in mammals.

The probable effectiveness of TRF as a releasing factor for prolactin in the pigeon has already been referred to. In the chicken experiments *in vivo* so far have been equivocal, with prolactin levels increasing in some cases (for example, Bolton, Chadwick & Scanes, 1976) but not in others. Occasionally the increase appeared after a delay of over an hour. However, TRF does promote prolactin release from chicken pituitaries when added to the incubation medium (Hall *et al.*, 1975). The possible link between prolactin secretion and plasma electrolyte levels, indicated from observations on animals as diverse as teleost fish (Ball & Ensor, 1967) and man (Horrobin, 1973), has been investigated in the chicken (Bolton, Chadwick & Scanes, 1976). Results from many different experiments have shown that whenever the plasma prolactin concentration is raised there is a significantly correlated rise in plasma potassium concentrations and, overall in the pooled results, in the plasma sodium concentration as well. The possibility that plasma levels of prolactin in the chicken are affected by stress has been investigated, in view of the known effects of stressful stimuli on milk secretion in mammals (Nicoll, Talwalker & Meites, 1960; Chadwick, 1971), and work done in collaboration with Mr S. Harvey, Dr C. G. Scanes and Miss G. Border (Table III) has shown that fasting decreases prolactin levels in plasma whereas exposure to ether vapour, to heat or to cold, results in an increased plasma prolactin concentration. The low concentration in fasted chickens was also associated with depressed plasma sodium and potassium concentrations (Harvey, Scanes, Chadwick & Bolton, in prep.). Recently, we have sub-

Table III

Effect of various stresses on plasma prolactin in the domestic fowl

Stressful agent	Period of exposure	No. of animals	Plasma prolactin (ng/ml): control value	experimental value or % control value
Ether vapour	0	5	57·2 ± 18.6	
	5 min			85·8 ± 11·2*
	40 min			76·6 ± 8·6*
Heat	0	6	101·0 ± 11·2(100%)	
	15 min			122% ± 39%
	30 min			138% ± 16%
	45 min			154% ± 55%
	60 min			155% ± 15%*
	75 min			154% ± 34%
Handling	0	6	159·0 ± 14·6	
	30 min			164·9 ± 9·9
	60 min			157·3 ± 9·2
Cold	0	6	79·4 ± 18·6	
	2 min			313·2 ± 64·3‡
Fasting	0	5	65·4 ± 10·3	
	12 hr			21·1 ± 3·4†
	24 hr			19·9 ± 3·7‡
	48 hr			21·0 ± 0·9‡

* $P < 0·05$.
† $P < 0·01$.
‡ $P < 0·001$.

jected chickens to water deprivation for periods of 6 and 12 h and have observed spectacular rises in plasma prolactin levels as a result of this treatment (Table IV). This result has wide implications for prolactin physiology in birds, and an aspect of water metabolism associated with the reproductive cycle which is at present receiving our attention is the period

Table IV

Effect of water deprivation on plasma prolactin concentration in 5-week old cockerels

Period of water deprivation (h)	Plasma prolactin (ng/ml) mean ± s.e. (n) control	dehydrated
6	204·5 ± 6·2(8)	310·2 ± 25·2(8)*
12	208·6 ± 6·7(8)	439·8 ± 40·9(8)†

* $P < 0·01$.
† $P < 0·001$.

in the egg laying cycle during which water is taken up by the albumen of the egg in the oviduct.

In collaboration with Dr P. J. Sharp of the ARC Poultry Research Centre, Edinburgh, and with my colleague Dr Scanes, a preliminary study has been carried out on normal bantam hens throughout a single natural cycle of laying a clutch of eggs, brooding them, hatching them and rearing the young, in which prolactin levels were determined in blood samples taken every four days. Table V shows that there was a significant fall in the plasma prolactin concentration at the end of the period of broodiness when the chicks were hatched. This is the first time that evidence has been presented concerning plasma prolactin levels in the broody fowl, and they support the concept that prolactin may be involved in broodiness in birds (Riddle, 1963).

TABLE V

Plasma prolactin concentrations in three hens during a single breeding cycle of 12 weeks duration

Phase of cycle	No. of observations	Level of prolactin (% of overall mean level for each bird ±s.e.)
Non-laying	33	$103 \cdot 3 \pm 8 \cdot 8$
Laying	21	$103 \cdot 5 \pm 9 \cdot 1$
Incubating clutch of eggs	26	$111 \cdot 6 \pm 7 \cdot 9$ $\Big\}*$
Rearing chicks	24	$81 \cdot 8 \pm 4 \cdot 9$

Blood samples taken approximately every 4 days.
* $P < 0 \cdot 01$.

CONCLUSIONS

What conclusions can be drawn regarding the origin, nature, and usefulness of milk-like secretions? The first must be that the existence of functional relationship between parent and offspring in the form of a nutritive secretion may be diagnostic of mammals but it is not confined solely to that class. Throughout the vertebrates the dependence of the young, after hatching or parturition, on parental secretions has arisen independently more than once and even invertebrates are not devoid of examples. The occurrence of the phenomenon in aquatic vertebrates may have a rather different origin from that in terrestrial animals. In ovoviviparous fishes it is clearly associated with the provision of a sheltered environment for the eggs after hatching, to ensure their growth and development free from predation. In these cases the number of eggs fertilized is greatly reduced compared with the oviparous species, which ensure their survival in a different way. In the very unusual case of the discus fish, what appears to be the elaboration of the not uncommon ability of epidermal glands to secrete copious mucus, has been adopted for nutritive purposes. This is of special interest in relation to the fact that prolactin undoubtedly plays an important role in freshwater survival and salt regulation in some teleost fishes, and that it has proven ability to influence skin mucus secretion in several fish species. Mucus itself may play some part in forming a barrier to free diffusion of salts and thus may directly assist in osmoregulation. This leads one to suggest that a hormonally-controlled ectodermal gland secretion, normally involved in salt and water regulation, has been adapted to serve the purpose of nutrition of the young, a situation not without parallel in the terrestrial vertebrates.

Mammary-gland secretion must first have developed in mammal-like reptiles and may have developed independently more than once. These mammal-like reptiles were mostly rather small inconspicuous animals and must have encountered to a marked extent the major problem of terrestrial animals—that of dehydration—particularly as they were presumably warm-blooded and probably used evaporative cooling as one of their temperature-regulating mechanisms, as is the case today in both birds and mammals. The risk of dehydration is perhaps greatest during the first few days of an animal's life, either after hatching from an egg or after a live birth. If so, the adaptation of a secretion from skin glands to provide the young with water might well have proved of selective advantage. The existence of such glands, in the form of the sweat gland, used for temperature regulation, provides a not unlikely candidate for such an adaptation, particularly since embryological evidence from living mammals persuades some workers that mammary glands have such an origin at least in the monotremes (Raynaud, 1961). We should therefore look to fossils of this group of reptiles to see if there is any supporting evidence, such as skin impressions showing glandular structures. A consequence of this view is the rather surprising conclusion that the nutritive role of milk-like secretions is of secondary importance. This is, however, supported by the relatively small proportion of the secretion that the nutritive components represent, even though in mammals these tend to receive the greatest attention.

The production of milk and milk-like secretions must be a significant factor in the salt and water metabolism of the parent and it is perhaps not surprising that the hormone prolactin has become associated with the process. The hormone has clearly been implicated in salt and water metabolism in all groups of vertebrates through to birds and mammals, though in the latter cases it plays a role secondary to antidiuretic hormone and the mineralocorticoids. This may account for prolactin being associated with milk-like secretions wherever they have arisen, the hormone being both already present and adapted to fill such a role. This hypothesis at least explains why amphibians and reptiles possess prolactin even though they do not possess well defined milk-like secretions. It is perhaps unwise to select individual attributes from the many roles for the hormone in these groups suggested by workers (see review by Nicoll & Bern, 1971), but it is tempting to remark that the water-drive effect in amphibians has connotations of both water balance and reproduction (C. S. Chadwick, 1941). Viewed in this light the provision of milk-like secretions in amphibians and reptiles would seem to offer some advantages. The fact that it does not occur today is no proof that it did not occur in the geological past, particularly as more primitive vertebrates do show this phenomenon. Positive evidence would indeed be hard to obtain in the absence of fossilized soft parts, though it is not impossible that impressions of the body would show up any morphological adaptations if they could be properly interpreted. The evidence recently put forward for homothermy in dinosaurs (Bakker, 1971) might indicate that this group did possess appropriate secretory glands.

In the light of these contentions, how can it be explained that birds do not resort to lactation-like phenomena to any significant extent? The lactogenic hormone is clearly present and intimately involved in salt and water metabolism. Birds, of course, entirely lack viviparous or ovoviviparous representatives, perhaps owing to the exigencies of flight (Chadwick, 1969). However, the adoption of egg laying by a fully-terrestrial vertebrate presupposes an adequate system for the detoxification and disposal of nitrogenous waste material in the closed environment of the egg during incubation, a problem which does not arise during an intra-uterine existence. Thus the acquisition of uricotelic nitrogenous excretion on the one hand permits the adoption of oviparity on land and, on the other hand, greatly reduces the demand for water after hatching. This then might be suggested as the reason for the absence of lactation in birds; furthermore it might also be inferred that the full acquisition of uricotelic excretion is delayed until a few days after hatching in columbiforms, a contention open to experimental verification.

REFERENCES

Bakker, R. T. (1971). Dinosaur physiology and the origin of mammals. *Evolution* **27**: 636–658.
Ball, J. N. & Ensor, D. M. (1967). Specific action of prolactin on plasma sodium levels in hypophysectomised *Poecilia latipinna* (Teleostei). *Gen. comp. Endocr.* **8**: 432–440.
Barbier, M. (1968). Biochimie de l'abeille. In *Traité de biologie de l'abeille* **1**: 378–409. Chavin. R. (ed.). Paris: Manon et Cie.
Beams, H. W. & Meyer, R. K. (1931). The formation of pigeon 'milk'. *Physiol. Zoöl.* **4**: 486–500.
Blüm, V. & Fiedler, K. (1965). Hormonal control of reproductive behaviour in some cichlid fish. *Gen. comp. Endocr.* **5**: 186–196.
Bolton, N. J., Chadwick, A., Hall, T. R. & Scanes, C. G. (1976). Effect of chicken and rat hypothalamic extracts on prolactin secretion in the chicken. *I.R.C.S. Med. Sci.* **4**: 495.

Bolton, N. J., Chadwick, A. & Scanes, C. G. (1976). Plasma electrolytes and prolactin in the domestic fowl. *I.R.C.S. Med. Sci.* **4**: 516.

Chadwick, A. (1966). Prolactin-like activity in the pituitary gland of the frog. *J. Endocr.* **34**: 247–255.

Chadwick, A. (1969). Effects of prolactin in homoiothermic vertebrates. *Gen. comp. Endocr.* Suppl. **2**: 63–68.

Chadwick, A. (1970). Pigeon crop sac-stimulating activity in the pituitary of the flounder (*Pleuronectes flesus*). *J. Endocr.* **47**: 463–469.

Chadwick, A. (1971). Lactogenesis in pseudo-pregnant rabbits treated with adrenocorticotrophin and adrenal corticosteroids. *J. Endocr.* **49**: 1–8.

Chadwick, A., Folley, S. J. & Gemzell, C. A. (1961). Lactogenic activity of human pituitary growth hormone. *Lancet* **1961** (ii): 241–243.

Chadwick, A. & Hall, T. R. (1975). Thyrotrophin releasing factor and prolactin release in the pigeon *in vivo*. *J. Endocr.* **74**: 26–27P.

Chadwick, A. & Jordan, B. J. (1971). The lipids of the crop epithelium of pigeons after injection with prolactin from the pituitaries of different vertebrate classes. *J. Endocr.* **49**: 51–58.

Chadwick, C. S. (1941). Further observations on the waterdrive in *Triturus viridescens*. II. Induction of the water drive with the lactogenic hormone. *J. exp. Zool.* **86**: 175–187.

Dumont, J. N. (1965). Prolactin-induced cytologic changes in the mucosa of the pigeon crop during crop 'milk' formation. *Z. Zellforsch. mikrosk. Anat.* **68**: 755–782.

Folley, S. J., Dyer, F. J. & Coward, K. H. (1940). The assay of prolactin by means of the pigeon crop-gland response. *J. Endocr.* **2**: 179–193.

Forsyth, I. A., Folley, S. J. & Chadwick, A. (1965). Lactogenic and pigeon crop-stimulating activities of human pituitary growth hormone preparations. *J. Endocr.* **31**: 115–126.

Hall, T. R., Chadwick, A., Bolton, N. J. & Scanes, C. G. (1975). Prolactin release *in vitro* and *in vivo* in the pigeon and in the domestic fowl following administration of synthetic thyrotrophin releasing factor (TRF). *Gen. comp. Endocr.* **25**: 298–306.

Harvey, S., Scanes, C. G., Chadwick, A. & Bolton, N. J. (in prep.). *Influence of fasting, glucose and insulin on plasma levels of growth hormone and prolactin in the domestic fowl.*

Harvey, S., Scanes, C. G., Falconer, J., Bolton, N. J. & Chadwick, A. (1977). Variations in circulating levels of growth hormone, prolactin and somatamedin during growth in the domestic fowl. *J. Endocr.* **73**: 10P.

Hildemann, W. H. (1959). A cichlid fish *Symphysodon discus*, with unique nurture habits. *Am. Nat.* **93**: 27–34.

Horrobin, D. F. (1973). *Prolactin: physiology and clinical significance.* Lancaster: Medical and Technical Publishing.

Horrobin, D. F. (1974). *Prolactin 1974.* Lancaster: Medical and Technical Publishing.

Huber, P. (1810). *Recherchés sur les moeurs des fourmis indigènes.* Paris: J. J. Paschard.

Jekan, R., Bergmann, E. D., Ishay, J. & Gitter, S. (1968). Proteolytic enzyme activity in the various colony members of the Oriental hornet, *Vespa orientalis*, F. *Life Sci.* **7**:929–934.

Jenkins, J. T. (1925). *The fishes of the British Isles.* London: Warne.

Josimovich, J. B. & Maclaren, J. A. (1962). Presence in the human placenta and term serum of a highly lactogenic substance immunologically related to pituitary growth hormone. *Endocrinology* **71**: 209–220.

Le Blond, C. P. & Noble, G. K. (1937). Prolactin-like reaction produced by hypophyses of various vertebrates. *Proc. Soc. exp. Biol. Med.* **36**: 517–518.

Lewis, U. J., Singh, R. N. P. & Seavey, B. K. (1972). Problems in purification of human prolactin. In *Prolactin and carcinogenesis*: 4–12. Boyns, A. R. & Griffiths, K. (eds). Cardiff: Alpha Omega Alpha.

Lyons, W. R. & Page, E. (1935). Detection of mammotrophin in the urine of lactating women. *Proc. Soc. exp. Biol. Med.* **32**: 1049–1050.

Matthews, L. H. & Marshall, F. H. A. (1956). Cyclical changes in the reproductive organs of the lower vertebrates. In *Marshall's physiology of reproduction* **1**: 156–225. Parkes, A. S. (ed.). London: Longmans Green.

Mittler, T. E. (1958). Studies on the feeding and nutrition of *Tuberolachnus salignus* (Gmelin) (Homoptera, Aphididae) III. The nitrogen economy. *J. exp. Biol.* **35**: 626–638.

Nicoll, C. S. (1967). Bioassay of prolactin: analysis of the pigeon crop-sac response to local prolactin injection by an objective and quantitative method. *Endocrinology* **80**: 641–655.

Nicoll, C. S. & Bern, H. A. (1968). Further analysis of the occurrence of pigeon crop sac-stimulating activity (prolactin) in the vertebrate adenohypophysis. *Gen. comp. Endocr.* **11**: 5–20.

Nicoll, C. S. & Bern, H. A. (1971). On the actions of prolactin among the vertebrates: is there a common denominator? In *Lactogenic hormones*: 299–374. Wolstenholme, G. E. W. & Knight, J. (eds). Edinburgh: Churchill Livingstone.

Nicoll, C. S., Talwalker, P. K. & Meites, J. (1960). Initiation of lactation in rats by nonspecific stresses. *Am. J. Physiol.* **198**: 1103–1106.

Raynaud, A. (1961). Morphogenesis of the mammary gland. In *Milk: the mammary gland and its secretion* **1**: 3–46. Kon, S. J. & Cowie, A. T. (eds). New York and London: Academic Press.

Riddle, O. (1963). Prolactin in vertebrate function and organisation. *J. natn. Cancer Inst.* **31**: 1039–1110.

Riddle, O., Bates, R. E. & Dykshorn, S. W. (1932). A new hormone of the anterior pituitary. *Proc. Soc. exp. Biol. Med.* **29**: 1211–1212.

Scanes, C. G., Bolton, N. J. & Chadwick, A. (1975). Purification and properties of an avian prolactin. *Gen. comp. Endocr.* **27**: 371–379.

Scanes, C. G., Chadwick, A. & Bolton, N. J. (1976). Radioimmunoassay of prolactin in the plasma of the domestic fowl. *Gen. comp. Endocr.* **30**: 12–20.

Stumper, R. (1961). Radiobiologische Untersuchungen über den Sozialen Nährungsaushalt der Honigameise *Proformica nasuta* (Nyl). *Naturwissenschaften* **48**: 735–736.

Way, M. J. (1963). Mutalism between ants and honeydew-producing Homoptera. *A. Rev. Ent.* **8**: 307–344.

Wheeler, W. M. (1928). *The social insects: their origin and evolution*. London: Kegan Paul.

Wilson, E. O. (1971). *The insect societies*. Cambridge, Mass.: Harvard University Press.

Author Index

Numbers in italics refer to pages in the References at the end of each article.